Birkhäuser

Trends in Mathematics

Trends in Mathematics is a series devoted to the publication of volumes arising from conferences and lecture series focusing on a particular topic from any area of mathematics. Its aim is to make current developments available to the community as rapidly as possible without compromise to quality and to archive these for reference.

Proposals for volumes can be submitted using the Online Book Project Submission Form at our website www.birkhauser-science.com.

Material submitted for publication must be screened and prepared as follows:

All contributions should undergo a reviewing process similar to that carried out by journals and be checked for correct use of language which, as a rule, is English. Articles without proofs, or which do not contain any significantly new results, should be rejected. High quality survey papers, however, are welcome.

We expect the organizers to deliver manuscripts in a form that is essentially ready for direct reproduction. Any version of $\TeX$ is acceptable, but the entire collection of files must be in one particular dialect of $\TeX$ and unified according to simple instructions available from Birkhäuser.

Furthermore, in order to guarantee the timely appearance of the proceedings it is essential that the final version of the entire material be submitted no later than one year after the conference.

Advances in Applied Analysis

Sergei V. Rogosin
Anna A. Koroleva
Editors

Birkhäuser

Editors
Sergei V. Rogosin
Department of Economics
Belarusian State University
Minsk, Belarus

Anna A. Koroleva
Department of Economics
Belarusian State University
Minsk, Belarus

ISBN 978-3-0348-0416-5 ISBN 978-3-0348-0417-2 (eBook)
DOI 10.1007/978-3-0348-0417-2
Springer Basel Heidelberg New York Dordrecht London

Library of Congress Control Number: 2012946916

Mathematics Subject Classification (2010): Primary: 30-02, 35-02, 11M06, 26A33, 30E25, 30G25, 76D27; secondary: 22E46, 30D05, 30K05, 31A30, 35A10, 35R11

Printed on acid-free paper

Springer is part of Springer Science+Business Media (www.springer.com)

Contents

Preface

In January 2010, the Council of Young Scientists of the Belarusian State University organized the 3rd International Winter School "Modern Problems of Mathematics and Mechanics". Young researchers, graduate, master and post-graduate students from Belarus, Lithuania, Poland and Ukraine participated in this school. They attended lectures of well-known experts in Analysis and its Applications. Six cycles of 3–4 lectures each were presented by Dr. V. Kisil (Leeds University, UK), by Prof. A. Laurinčikas (Vilnius University, Lithuania), Prof. Yu. Luchko (Beuth Technical University of Applied Sciences, Berlin, Germany), Prof. V.Mityushev (Krakow Pedagogical Academy, Poland), Prof. S. Plaksa (Institute of Mathematics, National Academy of Sciences, Ukraine) and Dr. S. Rogosin (Belarusian State University, Minsk, Belarus).

The book is made up of extended texts of the lectures presented at the School. These lectures are devoted to different problems of modern analysis and its applications. Below we briefly outline the main ideas of the lectures. Since they have an advanced character, the authors tried to make them self-contained.

A cycle of lectures by Dr. V. Kisil "Erlangen Program at Large: An Overview" describes a bridge between modern analysis and algebra. The author introduces objects and properties that are invariant under a group action. He begins with conformal geometry and develops a special functional calculus. He uses, as a characteristic example, a construction of wavelets based on certain algebraic techniques.

Prof. A. Laurinčikas deals with the notion of universality of functions. His cycle is called "The Riemann zeta-function: approximation of analytic functions". He shows that one of the best examples of universality is the classical Riemann zeta-function. So this lecture can be considered as describing the connection between Analysis and Number Theory.

A cycle of lectures by Prof. Yu. Luchko "Anomalous diffusion: models, their analysis, and interpretation" presents a model of anomalous diffusion. This model is given in terms of differential equations of a fractional order. The obtained equations and their generalizations are analyzed with the help of both the Laplace-Fourier transforms (the Cauchy problems) and the spectral method (initial-boundary-value problems).

Prof. V. Mityushev presents in his cycle "R-linear and Riemann–Hilbert problems for multiply connected domains" elements of constructive analysis related to the solution of boundary value problems for analytic functions. He pays particular attention to further application of the obtained results in the theory of 2D composite materials and porous media.

Another type of applications are presented in the cycle of lectures by Prof. S. Plaksa "Commutative algebras associated with classic equations of mathematical physics". In his work he develops a technique for application of the theory of monogenic functions in modern problems of mathematical physics. In particular, he studies axial-symmetric problems of the mechanics of continuous media.

Dr. S. Rogosin describes some modern ideas that can be applied to the study of certain free boundary problems ("2D free boundary value problems"). In particular, he develops an illustrative example dealing with so-called Hele-Shaw boundary value problem. This problem is reduced to a couple of problems, namely, an abstract Cauchy–Kovalevsky problem and a Riemann–Hilbert–Poincaré problem for analytic functions.

The book is addressed to young researchers in Mathematics and Mechanics. It can also be used as the base for a course of lectures for master-students.

Advances in Applied Analysis
Trends in Mathematics, 1–94

Erlangen Program at Large: An Overview

Vladimir V. Kisil

Dedicated to Prof. Hans G. Feichtinger on the occasion of his 60th birthday

Abstract. This is an overview of the *Erlangen Program at Large*. Study of objects and properties, which are invariant under a group action, is very fruitful far beyond traditional geometry. In this paper we demonstrate this on the example of the group $\mathrm{SL}_2(\mathbb{R})$. Starting from the conformal geometry we develop analytic functions and apply these to functional calculus. Finally we link this to quantum mechanics and conclude by a list of open problems.

Mathematics Subject Classification (2010). Primary 30G35; Secondary 22E46, 30F45, 32F45, 43A85, 30G30, 42C40, 46H30, 47A13, 81R30, 81R60.

Keywords. Special linear group, Hardy space, Clifford algebra, elliptic, parabolic, hyperbolic, complex numbers, dual numbers, double numbers, split-complex numbers, Cauchy-Riemann-Dirac operator, Möbius transformations, functional calculus, spectrum, quantum mechanics, non-commutative geometry.

> A mathematical idea should not be petrified in a formalised axiomatic setting, but should be considered instead as flowing as a river.
>
> *Sylvester* (1878)

1. Introduction

The simplest objects with non-commutative (but still associative) multiplication may be 2×2 matrices with real entries. The subset of matrices *of determinant one* has the following properties:

- form a closed set under multiplication since $\det(AB) = \det A \cdot \det B$;
- the identity matrix is the set; and
- any such matrix has an inverse (since $\det A \neq 0$).

On leave from the Odessa University.

In other words those matrices form a *group*, the $\mathrm{SL}_2(\mathbb{R})$ group [97] – one of the two most important Lie groups in analysis. The other group is the Heisenberg group [42]. By contrast the $ax+b$ group, which is often used to build wavelets, is only a subgroup of $\mathrm{SL}_2(\mathbb{R})$, see the numerator in (1.1).

The simplest non-linear transforms of the real line – *linear-fractional* or *Möbius maps* – may also be associated with 2×2 matrices [8, Ch. 13]:

$$g : x \mapsto g\cdot x = \frac{ax+b}{cx+d}, \qquad \text{where } g = \begin{pmatrix} a & b \\ c & d \end{pmatrix}, \ x \in \mathbb{R}. \tag{1.1}$$

An enjoyable calculation shows that the composition of two transforms (1.1) with different matrices g_1 and g_2 is again a Möbius transform with matrix the product $g_1 g_2$. In other words (1.1) it is a (left) action of $\mathrm{SL}_2(\mathbb{R})$.

According to F. Klein's *Erlangen program* (which was influenced by S. Lie) any geometry is dealing with invariant properties under a certain transitive group action. For example, we may ask: *What kinds of geometry are related to the* $\mathrm{SL}_2(\mathbb{R})$ *action* (1.1)?

The Erlangen program has probably the highest rate of $\frac{\text{praised}}{\text{actually used}}$ among mathematical theories, due not only to the big numerator but also due to the undeservingly small denominator. As we shall see below, Klein's approach provides some surprising conclusions even for such over-studied objects as circles.

1.1. Make a guess in three attempts

It is easy to see that the $\mathrm{SL}_2(\mathbb{R})$ action (1.1) makes sense also as a map of complex numbers $z = x + \mathrm{i}y$, $\mathrm{i}^2 = -1$ assuming the denominator is non-zero. Moreover, if $y > 0$ then $g\cdot z$ has a positive imaginary part as well, i.e., (1.1) defines a map from the upper half-plane to itself. Those transformations are isometries of the Lobachevsky half-plane.

However there is no need to be restricted to the traditional route of complex numbers only. Moreover in Subsection 2.1 we will naturally come to a necessity to work with all three kinds of hypercomplex numbers. Less-known *double* and *dual* numbers, see [128, Suppl. C], have also the form $z = x + \iota y$ but different assumptions on the hypercomplex unit ι: $\iota^2 = 0$ or $\iota^2 = 1$ correspondingly. We will write ε and j instead of ι within dual and double numbers respectively. Although the arithmetic of dual and double numbers is different from the complex ones, e.g., they have divisors of zero, we are still able to define their transforms by (1.1) in most cases.

Three possible values -1, 0 and 1 of $\sigma := \iota^2$ will be referred to here as *elliptic*, *parabolic* and *hyperbolic* cases respectively. We repeatedly meet such a division of various mathematical objects into three classes. They are named by the historically first example – the classification of conic sections – however the pattern persistently reproduces itself in many different areas: equations, quadratic forms, metrics, manifolds, operators, etc. We will abbreviate this separation as *EPH-classification*. The common origin of this fundamental division of any family with one parameter can be seen from the simple picture of a coordinate line split by zero into negative and positive half-axes:

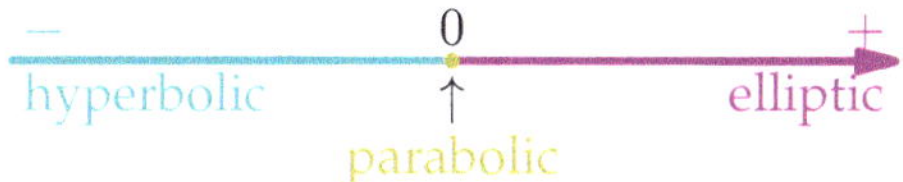

(1.2)

Connections between different objects admitting EPH-classification are not limited to this common source. There are many deep results linking, for example, the ellipticity of quadratic forms, metrics and operators, e.g., the Atiyah-Singer index theorem. On the other hand there are still a lot of white spots, empty cells, obscure gaps and missing connections between some subjects as well.

To understand the action (1.1) in all EPH cases we use the Iwasawa decomposition [97, § III.1] of $\mathrm{SL}_2(\mathbb{R}) = ANK$ into three one-dimensional subgroups A, N and K:

$$\begin{pmatrix} a & b \\ c & d \end{pmatrix} = \begin{pmatrix} \alpha & 0 \\ 0 & \alpha^{-1} \end{pmatrix} \begin{pmatrix} 1 & \nu \\ 0 & 1 \end{pmatrix} \begin{pmatrix} \cos\phi & -\sin\phi \\ \sin\phi & \cos\phi \end{pmatrix}. \tag{1.3}$$

Subgroups A and N act in (1.1) irrespectively to value of σ: A makes a dilation by α^2, i.e., $z \mapsto \alpha^2 z$, and N shifts points to left by ν, i.e., $z \mapsto z + \nu$.

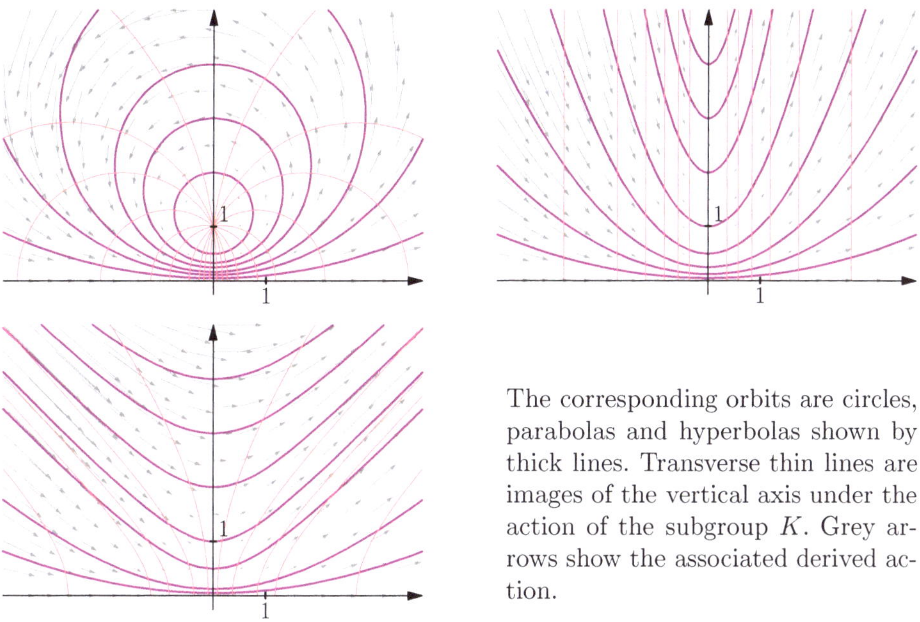

The corresponding orbits are circles, parabolas and hyperbolas shown by thick lines. Transverse thin lines are images of the vertical axis under the action of the subgroup K. Grey arrows show the associated derived action.

FIGURE 1. Action of the subgroup K.

By contrast, the action of the third matrix from the subgroup K sharply depends on σ, see Figure 1. In elliptic, parabolic and hyperbolic cases K-orbits are circles, parabolas and (equilateral) hyperbolas correspondingly. Thin traversal lines in Figure 1 join points of orbits for the same values of ϕ and grey arrows represent "local velocities" – vector fields of derived representations. We will describe some highlights of this geometry in Section 2.

1.2. Erlangen Program at Large

As we already mentioned the division of mathematics into areas is only apparent. Therefore it is unnatural to limit the Erlangen program only to "geometry". We may continue to look for $SL_2(\mathbb{R})$ invariant objects in other related fields. For example, transform (1.1) generates unitary representations on certain L_2 spaces, cf. (1.1) and Section 3:

$$g : f(x) \mapsto \frac{1}{(cx+d)^m} f\left(\frac{ax+b}{cx+d}\right). \tag{1.4}$$

For $m = 1, 2, \ldots$ the invariant subspaces of L_2 are Hardy and (weighted) Bergman spaces of complex analytic functions. All main objects of *complex analysis* (Cauchy and Bergman integrals, Cauchy-Riemann and Laplace equations, Taylor series etc.) may be obtained in terms of invariants of the *discrete series* representations of $SL_2(\mathbb{R})$ [69, § 3]. Moreover two other series (*principal* and *complimentary* [97]) play similar rôles for hyperbolic and parabolic cases [69, 82]. This will be discussed in Sections 4 and 5.

Moving further we may observe that transform (1.1) is defined also for an element x in any algebra $\mathfrak{A}$ with a unit $\mathbf{1}$ as soon as $(cx + d\mathbf{1}) \in \mathfrak{A}$ has an inverse. If $\mathfrak{A}$ is equipped with a topology, e.g., is a Banach algebra, then we may study a *functional calculus* for element x [75] in this way. It is defined as an *intertwining operator* between the representation (1.4) in a space of analytic functions and a similar representation in a left $\mathfrak{A}$-module. We will consider this in Section 6.

In the spirit of the Erlangen program, such a functional calculus is still a geometry, since it deals with invariant properties under a group action. However even for a simplest non-normal operator, e.g., a Jordan block of the length k, the obtained space is not like a space of points but is rather a space of kth *jets* [75]. Such non-point behaviour is oftenly attributed to *non-commutative geometry* and the Erlangen program provides an important input on this fashionable topic [69].

It is noteworthy that ideas of F. Klein and S. Lie are spread more in physics than in mathematics: it is a common viewpoint that laws of nature shall be invariant under certain transformations. Yet systematic use of the Erlangen approach can bring new results even in this domain as we demonstrate in Section 7. There are still many directions to extend the present work, thus we will conclude by a list of some open problems in Section 8.

Of course, there is no reason to limit the Erlangen program to the $SL_2(\mathbb{R})$ group only; other groups may be more suitable in different situations. However $SL_2(\mathbb{R})$ still possesses a big unexplored potential and is a good object to start with.

2. Geometry

We start from the natural domain of the Erlangen Program – geometry. Systematic use of this ideology allows us to obtain new results even for very classical objects like circles.

2.1. Hypercomplex numbers

Firstly we wish to demonstrate that hypercomplex numbers appear very naturally from a study of $\mathrm{SL}_2(\mathbb{R})$ action on the homogeneous spaces [85]. We begin from the standard definitions.

Let H be a subgroup of a group G. Let $X = G/H$ be the corresponding homogeneous space and $s : X \to G$ be a smooth section [55, § 13.2], which is a left inverse to the natural projection $p : G \to X$. The choice of s is inessential in the sense that, by a smooth map $X \to X$, we can always reduce one to another. We define a map $r : G \to H$ associated to p and s from the identities

$$r(g) = (s(x))^{-1} g, \qquad \text{where } x = p(g) \in X. \tag{2.1}$$

Note that X is a left homogeneous space with the G-action defined in terms of p and s as follows:

$$g : x \mapsto g \cdot x = p(g * s(x)). \tag{2.2}$$

Example 2.1 **([85]).** For $G = \mathrm{SL}_2(\mathbb{R})$, as well as for other semisimple groups, it is common to consider only the case of H being the maximal compact subgroup K. However in this paper we admit H to be any one-dimensional subgroup. Then X is a two-dimensional manifold and for any choice of H we define [64, Ex. 3.7(a)]:

$$s : (u, v) \mapsto \frac{1}{\sqrt{v}} \begin{pmatrix} v & u \\ 0 & 1 \end{pmatrix}, \qquad (u, v) \in \mathbb{R}^2,\ v > 0. \tag{2.3}$$

Any continuous one-dimensional subgroup $H \in \mathrm{SL}_2(\mathbb{R})$ is conjugated to one of the following:

$$K = \left\{ \begin{pmatrix} \cos t & \sin t \\ -\sin t & \cos t \end{pmatrix} = \exp \begin{pmatrix} 0 & t \\ -t & 0 \end{pmatrix},\ t \in (-\pi, \pi] \right\}, \tag{2.4}$$

$$N' = \left\{ \begin{pmatrix} 1 & 0 \\ t & 1 \end{pmatrix} = \exp \begin{pmatrix} 0 & 0 \\ t & 0 \end{pmatrix},\ t \in \mathbb{R} \right\}, \tag{2.5}$$

$$A' = \left\{ \begin{pmatrix} \cosh t & \sinh t \\ \sinh t & \cosh t \end{pmatrix} = \exp \begin{pmatrix} 0 & t \\ t & 0 \end{pmatrix},\ t \in \mathbb{R} \right\}. \tag{2.6}$$

Then [85] the action (2.2) of $\mathrm{SL}_2(\mathbb{R})$ on $X = \mathrm{SL}_2(\mathbb{R})/H$ coincides with Möbius transformations (1.1) on complex, dual and double numbers respectively.

2.2. Cycles as invariant families

We wish to consider all three hypercomplex systems at the same time, the following definition is very helpful for this.

Definition 2.2. The common name *cycle* [128] is used to denote circles, parabolas and hyperbolas (as well as straight lines as their limits) in the respective EPH case.

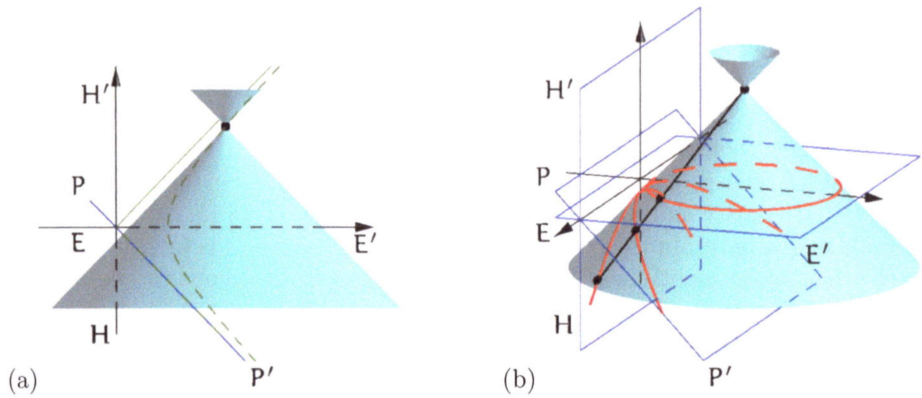

FIGURE 2. K-orbits as conic sections: circles are sections by the plane EE'; parabolas are sections by PP'; hyperbolas are sections by HH'. Points on the same generator of the cone correspond to the same value of ϕ.

It is well known that any cycle is a *conic section* and an interesting observation is that corresponding K-orbits are in fact sections of the same two-sided right-angle cone, see Figure 2. Moreover, each straight line generating the cone, see Figure 2(b), crosses corresponding EPH K-orbits at points with the same value of parameter ϕ from (1.3). In other words, all three types of orbits are generated by the rotations of this generator along the cone.

K-orbits are K-invariant in a trivial way. Moreover since actions of both A and N for any σ are extremely "shape-preserving" we find natural invariant objects of the Möbius map:

Theorem 2.3. *The family of all cycles from Definition* 2.2 *is invariant under the action* (1.1).

According to Erlangen ideology we should now study invariant properties of cycles.

Figure 2 suggests that we may get a unified treatment of cycles in all EPH cases by consideration of higher dimension spaces. The standard mathematical method is to declare objects under investigation (cycles in our case, functions in functional analysis, etc.) to be simply points of some bigger space. This space should be equipped with an appropriate structure to hold externally information which was previously considered as inner properties of our objects.

A generic cycle is the set of points $(u,v) \in \mathbb{R}^2$ defined for all values of σ by the equation

$$k(u^2 - \sigma v^2) - 2lu - 2nv + m = 0. \tag{2.7}$$

This equation (and the corresponding cycle) is defined by a point (k,l,n,m) from a *projective space* $\mathbb{P}^3$, since for a scaling factor $\lambda \neq 0$ the point $(\lambda k, \lambda l, \lambda n, \lambda m)$ defines an equation equivalent to (2.7). We call $\mathbb{P}^3$ the *cycle space* and refer to the initial $\mathbb{R}^2$ as the *point space*.

In order to get a connection with Möbius action (1.1) we arrange numbers (k,l,n,m) into the matrix

$$C^s_{\breve{\sigma}} = \begin{pmatrix} l + \breve{\imath}sn & -m \\ k & -l + \breve{\imath}sn \end{pmatrix}, \tag{2.8}$$

with a new hypercomplex unit $\breve{\imath}$ and an additional parameter s usually equal to ± 1. The values of $\breve{\sigma} := \breve{\imath}^2$ are -1, 0 or 1 independently from the value of σ. The matrix (2.8) is the cornerstone of an extended *Fillmore–Springer–Cnops construction* (FSCc) [19].

The significance of FSCc in the Erlangen framework is provided by the following result:

Theorem 2.4. *The image $\tilde{C}^s_{\breve{\sigma}}$ of a cycle $C^s_{\breve{\sigma}}$ under transformation* (1.1) *with $g \in \mathrm{SL}_2(\mathbb{R})$ is given by similarity of the matrix* (2.8)*:*

$$\tilde{C}^s_{\breve{\sigma}} = g C^s_{\breve{\sigma}} g^{-1}. \tag{2.9}$$

In other words FSCc (2.8) intertwines *Möbius action* (1.1) *on cycles with linear map* (2.9).

There are several ways to prove (2.9): either by a brute force calculation (fortunately performed by a CAS) [82] or through the related orthogonality of cycles [19], see the end of the next Subsection 2.3.

The important observation here is that our extended version of FSCc (2.8) uses a hypercomplex unit $\breve{\imath}$, which is not related to ι defining the appearance of cycles on a plane. In other words any EPH type of geometry in the cycle space $\mathbb{P}^3$ admits drawing of cycles in the point space $\mathbb{R}^2$ as circles, parabolas or hyperbolas. We may think of points of $\mathbb{P}^3$ as ideal cycles while their depictions on $\mathbb{R}^2$ are only their shadows on the wall of Plato's cave.

Figure 3(a) shows the same cycles drawn in different EPH styles. We note the first-order contact between the circle, parabola and hyperbola in their intersection points with the real line. Informally, we can say that EPH realisations of a cycle look the same in a vicinity of the real line. It is not surprising since cycles are invariants of the hypercomplex Möbius transformations, which are extensions of $\mathrm{SL}_2(\mathbb{R})$-action (1.1) on the real line.

Points $c_{e,p,h} = (\frac{l}{k}, -\breve{\sigma}\frac{n}{k})$ are the respective e/p/h-centres of drawn cycles. They are related to each other through several identities:

$$c_e = \bar{c}_h, \quad c_p = \frac{1}{2}(c_e + c_h). \tag{2.10}$$

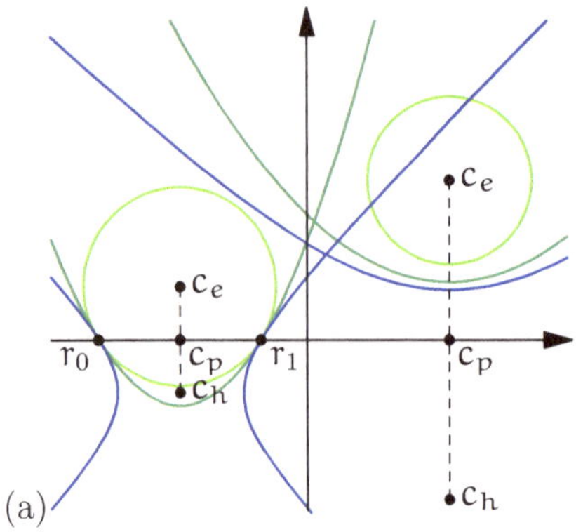

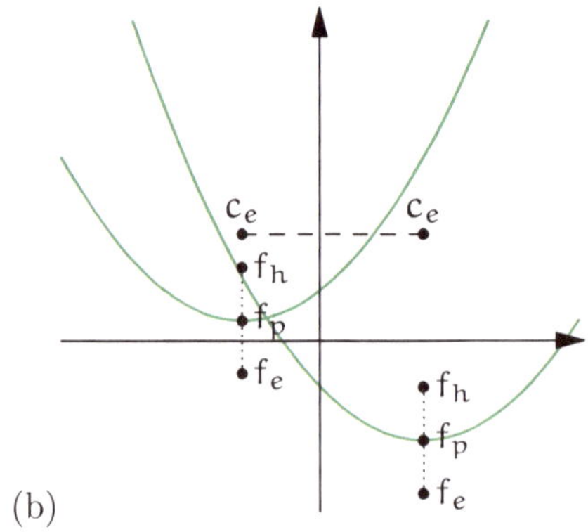

FIGURE 3. (a) Different EPH implementations of the same cycles defined by quadruples of numbers.
(b) Centres and foci of two parabolas with the same focal length.

Figure 3(b) presents two cycles drawn as parabolas; they have the same focal length $\frac{n}{2k}$ and thus their e-centres are on the same level. In other words *concentric* parabolas are obtained by a vertical shift, not scaling as an analogy with circles or hyperbolas may suggest.

Figure 3(b) also presents points, called e/p/h-foci:

$$f_{e,p,h} = \left(\frac{l}{k}, -\frac{\det C^s_{\breve{\sigma}}}{2nk} \right), \tag{2.11}$$

which are independent of the sign of s. If a cycle is depicted as a parabola then h-focus, p-focus, e-focus are correspondingly geometrical focus of the parabola, its vertex, and the point on the directrix nearest to the vertex.

As we will see, cf. Theorems 2.6 and 2.8, all three centres and three foci are useful attributes of a cycle even if it is drawn as a circle.

2.3. Invariants: algebraic and geometric

We use known algebraic invariants of matrices to build appropriate geometric invariants of cycles. It is yet another demonstration that any division of mathematics into subjects is only illusive.

For 2×2 matrices (and thus cycles) there are only two essentially different invariants under similarity (2.9) (and thus under Möbius action (1.1)): the *trace* and the *determinant*. The latter was already used in (2.11) to define a cycle's foci. However due to the projective nature of the cycle space $\mathbb{P}^3$ the absolute values of trace or determinant are irrelevant, unless they are zero.

Alternatively we may have a special arrangement for normalisation of quadruples (k, l, n, m). For example, if $k \neq 0$ we may normalise the quadruple to $(1, \frac{l}{k}, \frac{n}{k}, \frac{m}{k})$ with a highlighted cycle's centre. Moreover in this case $\det C^s_{\breve{\sigma}}$ is equal to the square of the cycle's radius, cf. Section 2.6. Another normalisation

$\det C^s_{\breve{\sigma}} = 1$ is used in [58] to get a nice condition for touching circles. Moreover, the Kirillov normalisation is preserved by the conjugation (2.9).

We still get important characterisation even with non-normalised cycles, e.g., invariant classes (for different $\breve{\sigma}$) of cycles are defined by the condition $\det C^s_{\breve{\sigma}} = 0$. Such a class is parametrised only by two real numbers and as such is easily attached a to certain point of $\mathbb{R}^2$. For example, the cycle $C^s_{\breve{\sigma}}$ with $\det C^s_{\breve{\sigma}} = 0$, $\breve{\sigma} = -1$ drawn elliptically represents just a point $(\frac{l}{k}, \frac{n}{k})$, i.e., (elliptic) zero-radius circle. The same condition with $\breve{\sigma} = 1$ in hyperbolic drawing produces a null-cone originated at point $(\frac{l}{k}, \frac{n}{k})$:

$$(u - \tfrac{l}{k})^2 - (v - \tfrac{n}{k})^2 = 0,$$

i.e., a zero-radius cycle in hyperbolic metric.

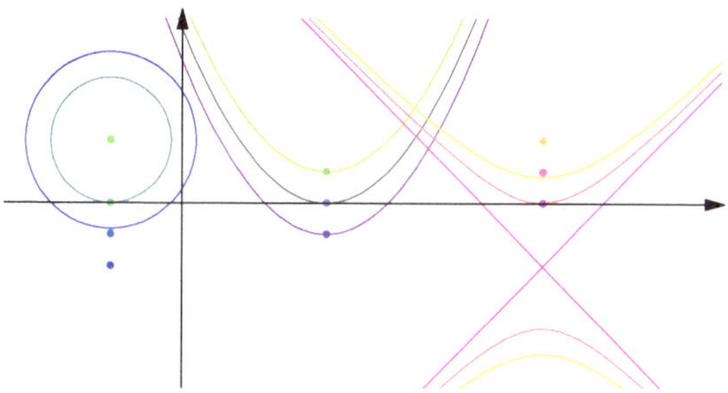

FIGURE 4. Different σ-implementations of the same $\breve{\sigma}$-zero-radius cycles and corresponding foci.

In general for every notion there are (at least) nine possibilities: three EPH cases in the cycle space times three EPH realisations in the point space. Such nine cases for "zero radius" cycles is shown in Figure 4. For example, p-zero-radius cycles in any implementation touch the real axis.

This "touching" property is a manifestation of the *boundary effect* in the upper half-plane geometry. The famous question on hearing a drum's shape has a sister:

Can we see/feel the boundary from inside a domain?

Both orthogonality relations described below are "boundary aware" as well. It is not surprising after all since $SL_2(\mathbb{R})$ action on the upper half-plane was obtained as an extension of its action (1.1) on the boundary.

According to the categorical viewpoint internal properties of objects are of minor importance in comparison to their relations with other objects from the same class. As an illustration we may cite the proof of Theorem 2.4 sketched at the end of the next section. Thus from now on we will look for invariant relations between two or more cycles.

2.4. Joint invariants: orthogonality

The most expected relation between cycles is based on the following Möbius invariant "inner product" built from a trace of product of two cycles as matrices:

$$\left\langle C^s_{\breve{\sigma}}, \tilde{C}^s_{\breve{\sigma}} \right\rangle = \operatorname{tr}(C^s_{\breve{\sigma}} \tilde{C}^s_{\breve{\sigma}}). \tag{2.12}$$

By the way, an inner product of this type is used, for example, in GNS construction to make a Hilbert space out of a C^*-algebra. The next standard move is given by the following definition.

Definition 2.5. Two cycles are called *$\breve{\sigma}$-orthogonal* if $\left\langle C^s_{\breve{\sigma}}, \tilde{C}^s_{\breve{\sigma}} \right\rangle = 0$.

The orthogonality relation is preserved under the Möbius transformations, thus this is an example of a *joint invariant* of two cycles. For the case of $\breve{\sigma}\sigma = 1$, i.e., when geometries of the cycle and point spaces are both either elliptic or hyperbolic, such an orthogonality is the standard one, defined in terms of angles between tangent lines in the intersection points of two cycles. However in the remaining seven ($= 9 - 2$) cases the innocent-looking Definition 2.5 brings unexpected relations.

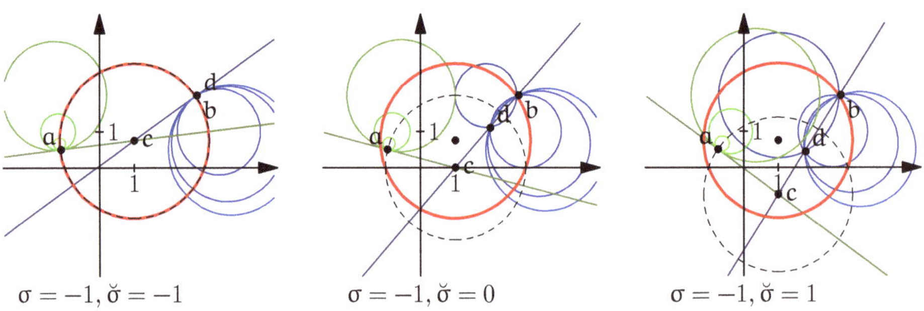

FIGURE 5. Orthogonality of the first kind in the elliptic point space. Each picture presents two groups (green and blue) of cycles which are orthogonal to the red cycle $C^s_{\breve{\sigma}}$. Point b belongs to $C^s_{\breve{\sigma}}$ and the family of blue cycles passing through b is orthogonal to $C^s_{\breve{\sigma}}$. They all also intersect in the point d which is the inverse of b in $C^s_{\breve{\sigma}}$. Any orthogonality is reduced to the usual orthogonality with a new ("ghost") cycle (shown by the dashed line), which may or may not coincide with $C^s_{\breve{\sigma}}$. For any point a on the "ghost" cycle the orthogonality is reduced to the local notion in the terms of tangent lines at the intersection point. Consequently such a point a is always the inverse of itself.

Elliptic (in the point space) realisations of Definition 2.5, i.e., $\sigma = -1$ is shown in Figure 5. The left picture corresponds to the elliptic cycle space, e.g., $\breve{\sigma} = -1$. The orthogonality between the red circle and any circle from the blue or green families is given in the usual Euclidean sense. The central (parabolic in the cycle space) and the right (hyperbolic) pictures show the non-local nature of the orthogonality. There are analogous pictures in parabolic and hyperbolic point spaces as well, see [82, 90].

This orthogonality may still be expressed in the traditional sense if we will associate to the red circle the corresponding "ghost" circle, which is shown by the dashed line in Figure 5. To describe a ghost cycle we need the *Heaviside function* $\chi(\sigma)$:

$$\chi(t) = \begin{cases} 1, & t \geq 0; \\ -1, & t < 0. \end{cases} \tag{2.13}$$

Theorem 2.6. *A cycle is $\breve{\sigma}$-orthogonal to cycle $C^s_{\breve{\sigma}}$ if it is orthogonal in the usual sense to the σ-realisation of "ghost" cycle $\hat{C}^s_{\breve{\sigma}}$, which is defined by the following two conditions:*

i. *$\chi(\sigma)$-centre of $\hat{C}^s_{\breve{\sigma}}$ coincides with $\breve{\sigma}$-centre of $C^s_{\breve{\sigma}}$.*
ii. *Cycles $\hat{C}^s_{\breve{\sigma}}$ and $C^s_{\breve{\sigma}}$ have the same roots, moreover* $\det \hat{C}^1_\sigma = \det C^{\chi(\breve{\sigma})}_\sigma$.

The above connection between various centres of cycles illustrates their meaningfulness within our approach.

One can easily check the following orthogonality properties of the zero-radius cycles defined in the previous section:

i. Due to the identity $\langle C^s_{\breve{\sigma}}, C^s_{\breve{\sigma}} \rangle = \det C^s_{\breve{\sigma}}$, zero-radius cycles are self-orthogonal (isotropic) ones.
ii. A cycle $C^s_{\breve{\sigma}}$ is σ-orthogonal to a zero-radius cycle $Z^s_{\breve{\sigma}}$ if and only if $C^s_{\breve{\sigma}}$ passes through the σ-centre of $Z^s_{\breve{\sigma}}$.

As we will see, in the parabolic case there is a more suitable notion of an infinitesimal cycle which can be used instead of zero-radius ones.

2.5. Higher-order joint invariants: f-orthogonality

With appetite already whetted one may wish to build more joint invariants. Indeed for any polynomial $p(x_1, x_2, \ldots, x_n)$ of several non-commuting variables one may define an invariant joint disposition of n cycles ${}^jC^s_{\breve{\sigma}}$ by the condition

$$\operatorname{tr} p({}^1C^s_{\breve{\sigma}}, {}^2C^s_{\breve{\sigma}}, \ldots, {}^nC^s_{\breve{\sigma}}) = 0.$$

However it is preferable to keep some geometrical meaning of constructed notions.

An interesting observation is that in the matrix similarity of cycles (2.9) one may replace element $g \in \mathrm{SL}_2(\mathbb{R})$ by an arbitrary matrix corresponding to another cycle. More precisely the product $C^s_{\breve{\sigma}} \tilde{C}^s_{\breve{\sigma}} C^s_{\breve{\sigma}}$ is again the matrix of the form (2.8) and thus may be associated to a cycle. This cycle may be considered as the reflection of $\tilde{C}^s_{\breve{\sigma}}$ in $C^s_{\breve{\sigma}}$.

Definition 2.7. A cycle $C^s_{\breve{\sigma}}$ is f-orthogonal (focal orthogonal) *to* a cycle $\tilde{C}^s_{\breve{\sigma}}$ if the reflection of $\tilde{C}^s_{\breve{\sigma}}$ in $C^s_{\breve{\sigma}}$ is orthogonal (in the sense of Definition 2.5) to the real line. Analytically this is defined by:

$$\operatorname{tr}(C^s_{\breve{\sigma}} \tilde{C}^s_{\breve{\sigma}} C^s_{\breve{\sigma}} R^s_{\breve{\sigma}}) = 0. \tag{2.14}$$

Due to invariance of all components in the above definition f-orthogonality is a Möbius invariant condition. Clearly this is not a symmetric relation: if $C^s_{\breve{\sigma}}$ is f-orthogonal to $\tilde{C}^s_{\breve{\sigma}}$ then $\tilde{C}^s_{\breve{\sigma}}$ is not necessarily f-orthogonal to $C^s_{\breve{\sigma}}$.

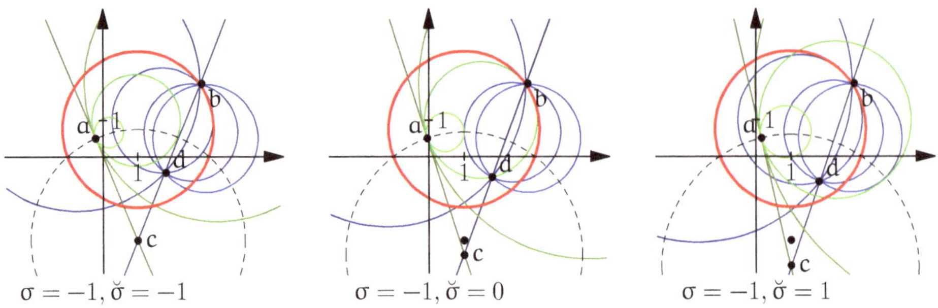

FIGURE 6. Focal orthogonality for circles. To highlight both similarities and distinctions with the ordinary orthogonality we use the same notation as that in Figure 5.

Figure 6 illustrates f-orthogonality in the elliptic point space. By contrast with Figure 5 it is not a local notion at the intersection points of cycles for all $\breve{\sigma}$. However it may be again clarified in terms of the appropriate s-ghost cycle, cf. Theorem 2.6.

Theorem 2.8. *A cycle is f-orthogonal to a cycle $C^s_{\breve{\sigma}}$ if it is orthogonal in the traditional sense to its* f-ghost cycle $\tilde{C}^{\breve{\sigma}}_{\breve{\sigma}} = C^{\chi(\sigma)}_{\breve{\sigma}} \mathbb{R}^{\breve{\sigma}}_{\breve{\sigma}} C^{\chi(\sigma)}_{\breve{\sigma}}$, *which is the reflection of the real line in $C^{\chi(\sigma)}_{\breve{\sigma}}$ and χ is the* Heaviside function (2.13). *Moreover*

i. *$\chi(\sigma)$-centre of $\tilde{C}^{\breve{\sigma}}_{\breve{\sigma}}$ coincides with the $\breve{\sigma}$-focus of $C^s_{\breve{\sigma}}$, consequently all lines f-orthogonal to $C^s_{\breve{\sigma}}$ are passing the respective focus.*
ii. *Cycles $C^s_{\breve{\sigma}}$ and $\tilde{C}^{\breve{\sigma}}_{\breve{\sigma}}$ have the same roots.*

Note the above intriguing interplay between cycle's centres and foci. Although f-orthogonality may look exotic it will naturally reappear at the end of the next section.

Of course, it is possible to define another interesting higher-order joint invariants of two or even more cycles.

2.6. Distance, length and perpendicularity

Geo*metry* in the plain meaning of this word deals with *distances* and *lengths*. Can we obtain them from cycles?

We mentioned already that for circles normalised by the condition $k = 1$ the value $\det C^s_{\breve{\sigma}} = \langle C^s_{\breve{\sigma}}, C^s_{\breve{\sigma}} \rangle$ produces the square of the traditional circle radius. Thus we may keep it as the definition of the $\breve{\sigma}$-*radius* for any cycle. But then we need to accept that in the parabolic case the radius is the (Euclidean) distance between (real) roots of the parabola, see Figure 7(a).

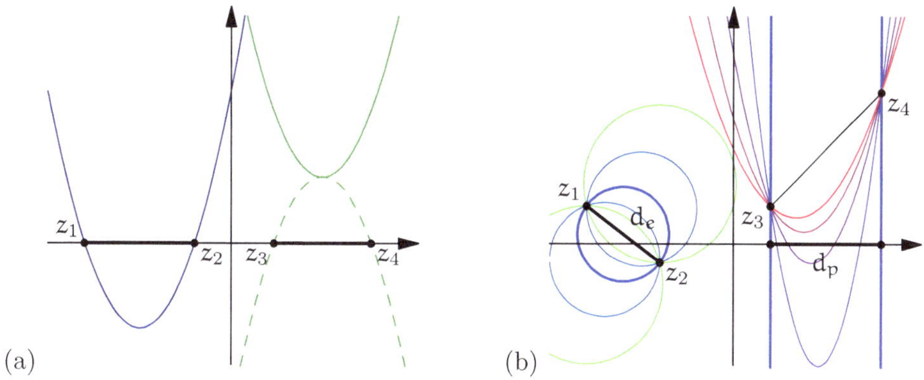

FIGURE 7. (a) The square of the parabolic diameter is the square of the distance between roots if they are real (z_1 and z_2), otherwise the negative square of the distance between the adjoint roots (z_3 and z_4). (b) Distance as extremum of diameters in elliptic (z_1 and z_2) and parabolic (z_3 and z_4) cases.

Having radii of circles already defined we may use them for other measurements in several different ways. For example, the following variational definition may be used:

Definition 2.9. The *distance* between two points is the extremum of diameters of all cycles passing through both points, see Figure 7(b).

If $\breve{\sigma} = \sigma$ this definition gives in all EPH cases the following expression for a distance $d_{e,p,h}(u, v)$ between endpoints of any vector $w = u + \mathrm{i}v$:

$$d_{e,p,h}(u, v)^2 = (u + \mathrm{i}v)(u - \mathrm{i}v) = u^2 - \sigma v^2. \tag{2.15}$$

The parabolic distance $d_p^2 = u^2$, see Figure 7(b), algebraically sits between d_e and d_h according to the general principle (1.2) and is widely accepted [128]. However one may be unsatisfied by its degeneracy.

An alternative measurement is motivated by the fact that a circle is the set of equidistant points from its centre. However the choice of "centre" is now rich: it may be either point from three centres (2.10) or three foci (2.11).

Definition 2.10. The *length* of a directed interval $\overrightarrow{AB}$ is the radius of the cycle with its *centre* (denoted by $l_c(\overrightarrow{AB})$) or *focus* (denoted by $l_f(\overrightarrow{AB})$) at the point A which passes through B.

This definition is less common and has some unusual properties like non-symmetry: $l_f(\overrightarrow{AB}) \neq l_f(\overrightarrow{BA})$. However it comfortably fits the Erlangen program due to its $\mathrm{SL}_2(\mathbb{R})$-*conformal invariance*:

Theorem 2.11 ([82]). *Let l denote either the EPH distances* (2.15) *or any length from Definition* 2.10. *Then for fixed y, $y' \in \mathbb{R}^\sigma$ the limit*

$$\lim_{t\to 0} \frac{l(g\cdot y, g\cdot(y+ty'))}{l(y,y+ty')}, \qquad \text{where } g \in \mathrm{SL}_2(\mathbb{R}),$$

exists and its value depends only on y and g and is independent of y'.

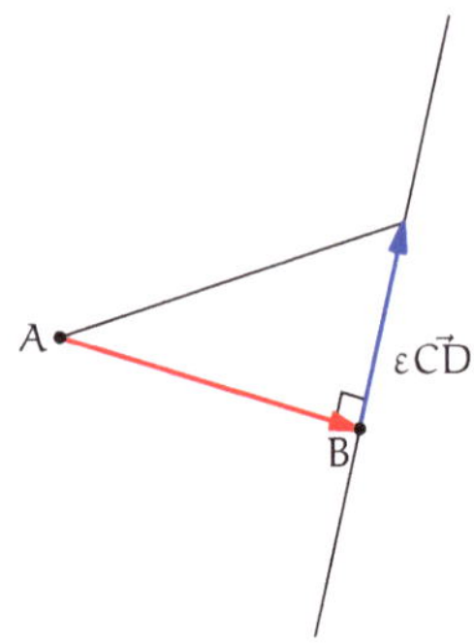

FIGURE 8. Perpendicular as the shortest route to a line.

We may return from distances to angles recalling that in the Euclidean space a perpendicular provides the shortest root from a point to a line, see Figure 8.

Definition 2.12. Let l be a length or distance. We say that a vector $\overrightarrow{AB}$ is l-*perpendicular* to a vector $\overrightarrow{CD}$ if function $l(\overrightarrow{AB}+\varepsilon\overrightarrow{CD})$ of a variable ε has a local extremum at $\varepsilon = 0$.

A pleasant surprise is that l_f-perpendicularity obtained thought the length from focus (Definition 2.10), already defined in Section 2.5, coincides with f-orthogonality as follows from Theorem 2.8(i). It is also possible [59] to make $\mathrm{SL}_2(\mathbb{R})$ action isometric in all three cases.

Further details of the refreshing geometry of Möbius transformation can be found in the paper [82]and the book [90].

All these studies are waiting to be generalised to high dimensions; quaternions and Clifford algebras provide a suitable language for this [82, 109].

3. Linear representations

For consideration of symmetries in analysis, it is natural to start from *linear representations*. The previous geometrical actions (1.1) can be naturally extended to such representations by induction [55, § 13.2; 64, § 3.1] from a representation of a subgroup H. If H is one-dimensional then its irreducible representation is a character, which is always supposed to be complex valued. However hypercomplex numbers naturally appeared in the $\mathrm{SL}_2(\mathbb{R})$ action (1.1), see Subsection 2.1 and [85]. Why then should we admit only $\mathrm{i}^2 = -1$ to deliver a character?

3.1. Hypercomplex characters

As we already mentioned, the typical discussion of induced representations of $SL_2(\mathbb{R})$ is centred around the case $H = K$ and a complex-valued character of K. A linear transformation defined by a matrix (2.4) in K is a rotation of $\mathbb{R}^2$ by the angle t. After identification $\mathbb{R}^2 = \mathbb{C}$ this action is given by the multiplication $e^{\mathrm{i}t}$, with $\mathrm{i}^2 = -1$. The rotation preserves the (elliptic) metric given by

$$x^2 + y^2 = (x + \mathrm{i}y)(x - \mathrm{i}y). \tag{3.1}$$

Therefore the orbits of rotations are circles; any line passing the origin (a "spoke") is rotated by the angle t, see Figure 9.

Dual and double numbers produce the most straightforward adaptation of this result.

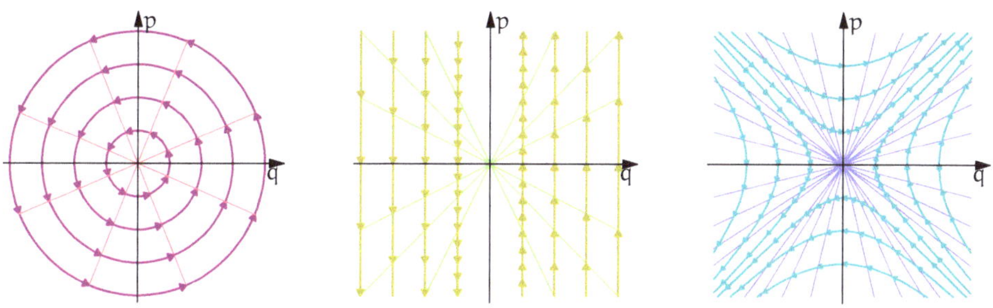

FIGURE 9. Rotations of algebraic wheels, i.e., the multiplication by $e^{\iota t}$: elliptic (E), trivial parabolic (P_0) and hyperbolic (H). All blue orbits are defined by the identity $x^2 - \iota^2 y^2 = r^2$. Thin "spokes" (straight lines from the origin to a point on the orbit) are "rotated" from the real axis. This is symplectic linear transformation of the classical phase space as well.

Proposition 3.1. *The following table shows correspondences between three types of algebraic characters:*

Elliptic	*Parabolic*	*Hyperbolic*
$\mathrm{i}^2 = -1$	$\varepsilon^2 = 0$	$\mathrm{j}^2 = 1$
$w = x + \mathrm{i}y$	$w = x + \varepsilon y$	$w = x + \mathrm{j}y$
$\bar{w} = x - \mathrm{i}y$	$\bar{w} = x - \varepsilon y$	$\bar{w} = x - \mathrm{j}y$
$e^{\mathrm{i}t} = \cos t + \mathrm{i}\sin t$	$e^{\varepsilon t} = 1 + \varepsilon t$	$e^{\mathrm{j}t} = \cosh t + \mathrm{j}\sinh t$
$\lvert w\rvert_e^2 = w\bar{w} = x^2 + y^2$	$\lvert w\rvert_p^2 = w\bar{w} = x^2$	$\lvert w\rvert_h^2 = w\bar{w} = x^2 - y^2$
$\arg w = \tan^{-1}\frac{y}{x}$	$\arg w = \frac{y}{x}$	$\arg w = \tanh^{-1}\frac{y}{x}$
unit circle $\lvert w\rvert_e^2 = 1$	*"unit" strip* $x = \pm 1$	*unit hyperbola* $\lvert w\rvert_h^2 = 1$

Geometrical action of multiplication by $e^{\iota t}$ *is drawn in Figure 9 for all three cases.*

Explicitly parabolic rotations associated with $e^{\varepsilon t}$ act on dual numbers as follows:

$$e^{\varepsilon x} : a + \varepsilon b \mapsto a + \varepsilon(ax + b). \tag{3.2}$$

This links the parabolic case with the Galilean group [128] of symmetries of classical mechanics, with the absolute time disconnected from space.

The obvious algebraic similarity and the connection to classical kinematic is a wide spread justification for the following viewpoint on the parabolic case, cf. [40, 128]:

- the parabolic trigonometric functions are trivial:

$$\operatorname{cosp} t = \pm 1, \qquad \operatorname{sinp} t = t; \tag{3.3}$$

- the parabolic distance is independent of y if $x \neq 0$:

$$x^2 = (x + \varepsilon y)(x - \varepsilon y); \tag{3.4}$$

- the polar decomposition of a dual number is defined by [128, App. C(30')]:

$$u + \varepsilon v = u(1 + \varepsilon \frac{v}{u}), \quad \text{thus} \quad |u + \varepsilon v| = u, \quad \arg(u + \varepsilon v) = \frac{v}{u}; \tag{3.5}$$

- the parabolic wheel looks rectangular, see Figure 9.

Those algebraic analogies are quite explicit and widely accepted as an ultimate source for parabolic trigonometry [40, 98, 128]. Moreover, those three rotations are all non-isomorphic symplectic linear transformations of the phase space, which makes them useful in the context of classical and quantum mechanics [87, 88], see Section 7. There exist also alternative characters [79] based on Möbius transformations with geometric motivation and connections to equations of mathematical physics.

3.2. Induced representations

Let G be a group, H be its closed subgroup with the corresponding homogeneous space $X = G/H$ with an invariant measure. We are using notations and definitions of maps $p : G \to X$, $s : X \to G$ and $r : G \to H$ from Subsection 2.1. Let χ be an irreducible representation of H in a vector space V, then it induces a representation of G in the sense of Mackey [55, § 13.2]. This representation has the realisation ρ_χ in the space $L_2(X)$ of V-valued functions by the formula [55, § 13.2.(7)–(9)]:

$$[\rho_\chi(g)f](x) = \chi(r(g^{-1} * s(x)))f(g^{-1} \cdot x), . \tag{3.6}$$

where $g \in G$, $x \in X$, $h \in H$ and $r : G \to H$, $s : X \to G$ are maps defined above; $*$ denotes multiplication on G and $\cdot$ denotes the action (2.2) of G on X.

Consider this scheme for representations of $\mathrm{SL}_2(\mathbb{R})$ induced from characters of its one-dimensional subgroups. We can notice that only the subgroup K requires a complex-valued character due to the fact of its compactness. For subgroups N' and A' we can consider characters of all three types – elliptic, parabolic and hyperbolic. Therefore we have seven essentially different induced representations. We will write explicitly only three of them here.

Example 3.2. Consider the subgroup $H = K$; due to its compactness we are limited to complex-valued characters of K only. All of them are of the form χ_k:

$$\chi_k \begin{pmatrix} \cos t & \sin t \\ -\sin t & \cos t \end{pmatrix} = e^{-ikt}, \qquad \text{where } k \in \mathbb{Z}. \tag{3.7}$$

Using the explicit form (2.3) of the map s we find the map r given in (2.1) as follows:

$$r\begin{pmatrix} a & b \\ c & d \end{pmatrix} = \frac{1}{\sqrt{c^2+d^2}} \begin{pmatrix} d & -c \\ c & d \end{pmatrix} \in K.$$

Therefore

$$r(g^{-1} * s(u,v)) = \frac{1}{\sqrt{(cu+d)^2+(cv)^2}} \begin{pmatrix} cu+d & -cv \\ cv & cu+d \end{pmatrix},$$

where $g^{-1} = \begin{pmatrix} a & b \\ c & d \end{pmatrix} \in \mathrm{SL}_2(\mathbb{R})$. Substituting this into (3.7) and combining with the Möbius transformation of the domain (1.1) we get the explicit realisation ρ_k of the induced representation (3.6):

$$\rho_k(g)f(w) = \frac{|cw+d|^k}{(cw+d)^k} f\left(\frac{aw+b}{cw+d}\right), \quad \text{where } g^{-1} = \begin{pmatrix} a & b \\ c & d \end{pmatrix}, \ w = u + \mathrm{i}v. \tag{3.8}$$

This representation acts on complex-valued functions in the upper half-plane $\mathbb{R}^2_+ = \mathrm{SL}_2(\mathbb{R})/K$ and belongs to the discrete series [97, § IX.2]. It is common to get rid of the factor $|cw+d|^k$ from that expression in order to keep analyticity and we will follow this practise for a convenience as well.

Example 3.3. In the case of the subgroup N there is a wider choice of possible characters.

i. Traditionally only complex-valued characters of the subgroup N are considered, they are

$$\chi_\tau^{\mathbb{C}} \begin{pmatrix} 1 & 0 \\ t & 1 \end{pmatrix} = e^{\mathrm{i}\tau t}, \qquad \text{where } \tau \in \mathbb{R}. \tag{3.9}$$

A direct calculation shows that

$$r\begin{pmatrix} a & b \\ c & d \end{pmatrix} = \begin{pmatrix} 1 & 0 \\ \frac{c}{d} & 1 \end{pmatrix} \in N'.$$

Thus

$$r(g^{-1} * s(u,v)) = \begin{pmatrix} 1 & 0 \\ \frac{cv}{d+cu} & 1 \end{pmatrix}, \quad \text{where } g^{-1} = \begin{pmatrix} a & b \\ c & d \end{pmatrix}. \tag{3.10}$$

A substitution of this value into the character (3.9) together with the Möbius transformation (1.1) gives us the next realisation of (3.6):

$$\rho_\tau^{\mathbb{C}}(g)f(w) = \exp\left(\mathrm{i}\frac{\tau cv}{cu+d}\right) f\left(\frac{aw+b}{cw+d}\right),$$

where $w = u + \varepsilon v$ and $g^{-1} = \begin{pmatrix} a & b \\ c & d \end{pmatrix} \in \mathrm{SL}_2(\mathbb{R})$. The representation acts on the space of *complex*-valued functions on the upper half-plane $\mathbb{R}^2_+$, which is a subset of *dual* numbers as a homogeneous space $\mathrm{SL}_2(\mathbb{R})/N'$. The mixture of complex and dual numbers in the same expression is confusing.

ii. The parabolic character χ_τ with an algebraic flavour is provided by multiplication (3.2) with the dual number

$$\chi_\tau \begin{pmatrix} 1 & 0 \\ t & 1 \end{pmatrix} = e^{\varepsilon\tau t} = 1 + \varepsilon\tau t, \qquad \text{where } \tau \in \mathbb{R}.$$

If we substitute the value (3.10) into this character, then we receive the representation

$$\rho_\tau(g)f(w) = \left(1 + \varepsilon\frac{\tau cv}{cu+d}\right) f\left(\frac{aw+b}{cw+d}\right),$$

where w, τ and g are as above. The representation is defined on the space of dual-numbers-valued functions on the upper half-plane of dual numbers. This expression contains only dual numbers with their usual algebraic operations. Thus it is linear with respect to them.

All characters in the previous example are unitary. Then the general scheme of induced representations [55, § 13.2] implies their unitarity in proper senses.

Theorem 3.4 ([85]). *Both representations of* $\mathrm{SL}_2(\mathbb{R})$ *from Example* 3.3 *are unitary on the space of function on the upper half-plane* $\mathbb{R}^2_+$ *of dual numbers with the inner product*

$$\langle f_1, f_2\rangle = \int_{\mathbb{R}^2_+} f_1(w)\bar{f}_2(w)\,\frac{du\,dv}{v^2}, \qquad \textit{where } w = u + \varepsilon v, \tag{3.11}$$

and we use the conjugation and multiplication of function values in algebras of complex and dual numbers for representations $\overset{\mathbb{C}}{\rho_\tau}$ *and* ρ_τ *respectively.*

The inner product (3.11) is positive defined for the representation $\overset{\mathbb{C}}{\rho_\tau}$ but is not for the other. The respective spaces are parabolic cousins of the *Krein spaces* [5], which are hyperbolic in our sense.

3.3. Similarity and correspondence: ladder operators

From the above observation we can deduce the following empirical principle, which has a heuristic value.

Principle 3.5 (Similarity and correspondence). i. *Subgroups K, N' and A' play a similar rôle in the structure of the group* $\mathrm{SL}_2(\mathbb{R})$ *and its representations.*

ii. *The subgroups shall be swapped simultaneously with the respective replacement of hypercomplex unit* ι.

The first part of the Principle (similarity) does not appear to be sound alone. It is enough to mention that the subgroup K is compact (and thus its spectrum is discrete) while two other subgroups are not. However in a conjunction with the second part (correspondence) the Principle has received the following confirmations so far, see [85] for details:

- The action of $\mathrm{SL}_2(\mathbb{R})$ on the homogeneous space $\mathrm{SL}_2(\mathbb{R})/H$ for $H = K$, N' or A' is given by linear-fractional transformations of complex, dual or double numbers respectively.
- Subgroups K, N' or A' are isomorphic to the groups of unitary rotations of respective unit cycles in complex, dual or double numbers.
- Representations induced from subgroups K, N' or A' are unitary if the inner product spaces of functions with values are complex, dual or double numbers.

Remark 3.6. The Principle of similarity and correspondence resembles *supersymmetry* between bosons and fermions in particle physics, but we have similarity between three different types of entities in our case.

Let us give another illustration of the Principle. Consider the Lie algebra $\mathfrak{sl}_2$ of the group $\mathrm{SL}_2(\mathbb{R})$. Pick up the following basis in $\mathfrak{sl}_2$ [120, § 8.1]:

$$A = \frac{1}{2}\begin{pmatrix} -1 & 0 \\ 0 & 1 \end{pmatrix}, \quad B = \frac{1}{2}\begin{pmatrix} 0 & 1 \\ 1 & 0 \end{pmatrix}, \quad Z = \begin{pmatrix} 0 & 1 \\ -1 & 0 \end{pmatrix}. \tag{3.12}$$

The commutation relations between the elements are

$$[Z, A] = 2B, \qquad [Z, B] = -2A, \qquad [A, B] = -\frac{1}{2}Z. \tag{3.13}$$

Let ρ be a representation of the group $\mathrm{SL}_2(\mathbb{R})$ in a space V. Consider the derived representation $d\rho$ of the Lie algebra $\mathfrak{sl}_2$ [97, § VI.1] and let $\tilde{X} = d\rho(X)$ for $X \in \mathfrak{sl}_2$. To see the structure of the representation ρ we can decompose the space V into eigenspaces of the operator $\tilde{X}$ for some $X \in \mathfrak{sl}_2$, cf. the Taylor series in Section 5.4.

3.3.1. Elliptic ladder operators. It would not be surprising that we are going to consider three cases: Let $X = Z$ be a generator of the subgroup K (2.4). Since this is a compact subgroup the corresponding eigenspaces $\tilde{Z}v_k = \mathrm{i}kv_k$ are parametrised by an integer $k \in \mathbb{Z}$. The *raising/lowering* or *ladder operators* $L^{\pm}$ [97, § VI.2; 120, § 8.2] are defined by the following commutation relations:

$$[\tilde{Z}, L^{\pm}] = \lambda_{\pm} L^{\pm}. \tag{3.14}$$

In other words $L^{\pm}$ are eigenvectors for operators $\operatorname{ad} Z$ of adjoint representation of $\mathfrak{sl}_2$ [97, § VI.2].

Remark 3.7. The existence of such ladder operators follows from the general properties of Lie algebras if the element $X \in \mathfrak{sl}_2$ belongs to a *Cartan subalgebra*. This is the case for vectors Z and B, which are the only two non-isomorphic types of Cartan subalgebras in $\mathfrak{sl}_2$. However the third case considered in this paper, the parabolic vector $B + Z/2$, does not belong to a Cartan subalgebra, yet a sort of ladder operator is still possible with dual number coefficients. Moreover, for the hyperbolic vector B, besides the standard ladder operators an additional pair with double number coefficients will also be described.

From the commutators (3.14) we deduce that L^+v_k are eigenvectors of $\tilde{Z}$ as well:

$$\begin{aligned}\tilde{Z}(L^+v_k) &= (L^+\tilde{Z} + \lambda_+ L^+)v_k = L^+(\tilde{Z}v_k) + \lambda_+ L^+ v_k = \mathrm{i}kL^+v_k + \lambda_+ L^+ v_k \\ &= (\mathrm{i}k + \lambda_+)L^+ v_k. \end{aligned} \tag{3.15}$$

Thus action of ladder operators on respective eigenspaces can be visualised by the diagram

$$\cdots \underset{L^-}{\overset{L^+}{\rightleftarrows}} V_{\mathrm{i}k-\lambda} \underset{L^-}{\overset{L^+}{\rightleftarrows}} V_{\mathrm{i}k} \underset{L^-}{\overset{L^+}{\rightleftarrows}} V_{\mathrm{i}k+\lambda} \underset{L^-}{\overset{L^+}{\rightleftarrows}} \cdots. \tag{3.16}$$

Assuming $L^+ = a\tilde{A} + b\tilde{B} + c\tilde{Z}$ from the relations (3.13) and defining condition (3.14) we obtain linear equations with unknowns a, b and c:

$$c = 0, \qquad 2a = \lambda_+ b, \qquad -2b = \lambda_+ a.$$

The equations have a solution if and only if $\lambda_+^2 + 4 = 0$, and the raising/lowering operators are $L^\pm = \pm\mathrm{i}\tilde{A} + \tilde{B}$.

3.3.2. Hyperbolic ladder operators. Consider the case $X = 2B$ of a generator of the subgroup A' (2.6). The subgroup is not compact and eigenvalues of the operator $\tilde{B}$ can be arbitrary, however raising/lowering operators are still important [44, § II.1; 102, § 1.1]. We again seek a solution in the form $L_h^+ = a\tilde{A} + b\tilde{B} + c\tilde{Z}$ for the commutator $[2\tilde{B}, L_h^+] = \lambda L_h^+$. We will get the system

$$4c = \lambda a, \qquad b = 0, \qquad a = \lambda c.$$

A solution exists if and only if $\lambda^2 = 4$. There are obvious values $\lambda = \pm 2$ with the ladder operators $L_h^\pm = \pm 2\tilde{A} + \tilde{Z}$, see [44, § II.1; 102, § 1.1]. Each indecomposable $\mathfrak{sl}_2$-module is formed by a one-dimensional chain of eigenvalues with a transitive action of ladder operators.

Admitting double numbers we have an extra possibility to satisfy $\lambda^2 = 4$ with values $\lambda = \pm 2\mathrm{j}$. Then there is an additional pair of hyperbolic ladder operators $L_\mathrm{j}^\pm = \pm 2\mathrm{j}\tilde{A} + \tilde{Z}$, which shift eigenvectors in the "orthogonal" direction to the standard operators $L_h^\pm$. Therefore an indecomposable $\mathfrak{sl}_2$-module can be parametrised by a two-dimensional lattice of eigenvalues on the double number plane, see Figure 10.

3.3.3. Parabolic ladder operators. Finally, consider the case of a generator $X = -B + Z/2$ of the subgroup N' (2.5). According to the above procedure we get the equations

$$b + 2c = \lambda a, \qquad -a = \lambda b, \qquad \frac{a}{2} = \lambda c,$$

which can be resolved if and only if $\lambda^2 = 0$. If we restrict ourselves with the only real (complex) root $\lambda = 0$, then the corresponding operators $L_p^\pm = -\tilde{B} + \tilde{Z}/2$ will not affect eigenvalues and thus are useless in the above context. However the dual number roots $\lambda = \pm\varepsilon t$, $t \in \mathbb{R}$ lead to the operators $L_\varepsilon^\pm = \pm\varepsilon t\tilde{A} - \tilde{B} + \tilde{Z}/2$.

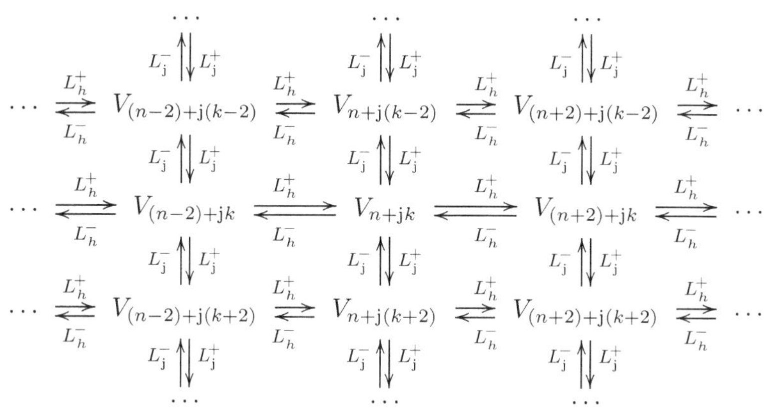

FIGURE 10. The action of hyperbolic ladder operators on a 2D lattice of eigenspaces. Operators $L_h^{\pm}$ move the eigenvalues by 2, making shifts in the horizontal direction. Operators $L_{\mathrm{j}}^{\pm}$ change the eigenvalues by $2\mathrm{j}$, shown as vertical shifts.

These operators are suitable to build an $\mathfrak{sl}_2$-module with a one-dimensional chain of eigenvalues.

Remark 3.8. The following rôles of hypercomplex numbers are noteworthy:

- the introduction of complex numbers is a necessity for the *existence* of ladder operators in the elliptic case;
- in the parabolic case we need dual numbers to make ladder operators *useful*;
- in the hyperbolic case double numbers are not required either for the existence or for the usability of ladder operators, but they do provide an enhancement.

We summarise the above consideration with a focus on the Principle of similarity and correspondence:

Proposition 3.9. *Let a vector $X \in \mathfrak{sl}_2$ generate the subgroup K, N' or A', that is $X = Z$, $B - Z/2$, or B respectively. Let ι be the respective hypercomplex unit.*

Then raising/lowering operators $L^{\pm}$ satisfying to the commutation relation $[X, L^{\pm}] = \pm\iota L^{\pm}$, $[L^-, L^+] = 2\iota X$ are

$$L^{\pm} = \pm\iota\tilde{A} + \tilde{Y}.$$

Here $Y \in \mathfrak{sl}_2$ is a linear combination of B and Z with the properties:

- $Y = [A, X]$.
- $X = [A, Y]$.
- *Killings form $K(X, Y)$ [55, § 6.2] vanishes.*

Any of the above properties defines the vector $Y \in \operatorname{span}\{B, Z\}$ up to a real constant factor.

The usability of the Principle of similarity and correspondence will be illustrated by more examples below.

4. Covariant transform

A general group-theoretical construction [2, 18, 29, 32, 66, 93, 111] of *wavelets* (or *coherent state*) starts from an irreducible square integrable representation – in the proper sense or modulo a subgroup. Then a mother wavelet is chosen to be *admissible*. This leads to a wavelet transform which is an isometry to L_2 space with respect to the Haar measure on the group or (quasi)invariant measure on a homogeneous space.

The importance of the above situation shall not be diminished, however an exclusive restriction to such a setup is not necessary, in fact. Here is a classical example from complex analysis: the Hardy space $H_2(\mathbb{T})$ on the unit circle and Bergman spaces $B_2^n(\mathbb{D})$, $n \geq 2$ in the unit disk produce wavelets associated with representations ρ_1 and ρ_n of the group $\mathrm{SL}_2(\mathbb{R})$ respectively [64]. While representations ρ_n, $n \geq 2$ are from square integrable discrete series, the mock discrete series representation ρ_1 is not square integrable [97, § VI.5; 120, § 8.4]. However it would be natural to treat the Hardy space in the same framework as Bergman spaces. Some more examples will be presented below.

4.1. Extending wavelet transform

To make a sharp but still natural generalisation of wavelets we give the following definition.

Definition 4.1. [83] Let ρ be a representation of a group G in a space V and F be an operator from V to a space U. We define a *covariant transform* $\mathcal{W}$ from V to the space $L(G, U)$ of U-valued functions on G by the formula

$$\mathcal{W} : v \mapsto \hat{v}(g) = F(\rho(g^{-1})v), \qquad v \in V,\ g \in G. \tag{4.1}$$

Operator F will be called a *fiducial operator* in this context.

We borrow the name for operator F from fiducial vectors of Klauder and Skagerstam [93].

Remark 4.2. We do not require that fiducial operator F shall be linear. Sometimes the positive homogeneity, i.e., $F(tv) = tF(v)$ for $t > 0$, alone can be already sufficient, see Example 4.14.

Remark 4.3. Usefulness of the covariant transform is in reverse proportion to the dimensionality of the space U. The covariant transform encodes properties of v in a function $\mathcal{W}v$ on G. For a low-dimensional U this function can be ultimately investigated by means of harmonic analysis. Thus $\dim U = 1$ (scalar-valued functions) is the ideal case, however, it is unattainable sometimes, see Example 4.11 below. We may have to use higher dimensions of U if the given group G is not rich enough.

Moreover, the relation between the dimensionality of U and usefulness of the covariant transform should not be taken dogmatically. Paper [91] gives an important example of a covariant transform which provides a simplification even in the case of $\dim U = \dim V$.

As we will see below the covariant transform is a close relative of the wavelet transform. The name is chosen due to the following common property of both transformations.

Theorem 4.4. *The covariant transform* (4.1) *intertwines* ρ *and the left regular representation* Λ *on* $L(G,U)$*:*

$$\mathcal{W}\rho(g) = \Lambda(g)\mathcal{W}.$$

Here Λ *is defined as usual by*

$$\Lambda(g) : f(h) \mapsto f(g^{-1}h). \tag{4.2}$$

Proof. We have a calculation similar to wavelet transform [66, Prop. 2.6]. Take $u = \rho(g)v$ and calculate its covariant transform:

$$\begin{aligned}
[\mathcal{W}(\rho(g)v)](h) &= [\mathcal{W}(\rho(g)v)](h) = F(\rho(h^{-1})\rho(g)v) \\
&= F(\rho((g^{-1}h)^{-1})v) \\
&= [\mathcal{W}v](g^{-1}h) \\
&= \Lambda(g)[\mathcal{W}v](h).
\end{aligned}$$

□

The next result follows immediately:

Corollary 4.5. *The image space* $\mathcal{W}(V)$ *is invariant under the left shifts on* G.

Remark 4.6. A further generalisation of the covariant transform can be obtained if we relax the group structure. Consider, for example, a *cancellative semigroup* $\mathbb{Z}_+$ of non-negative integers. It has a linear presentation on the space of polynomials in a variable t defined by the action $m : t^n \mapsto t^{m+n}$ on the monomials. Application of a linear functional l, e.g., defined by an integration over a measure on the real line, produces *umbral calculus* $l(t^n) = c_n$, which has a magic efficiency in many areas, notably in combinatorics [67, 96]. In this direction we also find it fruitful to expand the notion of an intertwining operator to a *token* [71].

4.2. Examples of covariant transform

In this subsection we will provide several examples of covariant transforms. Some of them will be expanded in subsequent sections, however a detailed study of all aspects will not fit into the present work. We start from the classical example of the group-theoretical wavelet transform:

Example 4.7. Let V be a Hilbert space with an inner product $\langle \cdot, \cdot \rangle$ and ρ be a unitary representation of a group G in the space V. Let $F : V \to \mathbb{C}$ be a functional $v \mapsto \langle v, v_0 \rangle$ defined by a vector $v_0 \in V$. The vector v_0 is often called the *mother wavelet* in areas related to signal processing or the *vacuum state* in a quantum framework.

Then the transformation (4.1) is the well-known expression for a *wavelet transform* [2, (7.48)] (or *representation coefficients*):

$$\mathcal{W} : v \mapsto \hat{v}(g) = \left\langle \rho(g^{-1})v, v_0 \right\rangle = \langle v, \rho(g)v_0 \rangle, \qquad v \in V,\ g \in G. \tag{4.3}$$

The family of vectors $v_g = \rho(g)v_0$ is called *wavelets* or *coherent states*. In this case we obtain scalar-valued functions on G, thus the fundamental rôle of this example is explained in Remark 4.3.

This scheme is typically carried out for a square integrable representation ρ and v_0 being an admissible vector [2, 18, 29, 32, 111]. In this case the wavelet (covariant) transform is a map into the square integrable functions [26] with respect to the left Haar measure. The map becomes an isometry if v_0 is properly scaled.

However square integrable representations and admissible vectors do not cover all interesting cases.

Example 4.8. Let $G = \text{Aff}$ be the "$ax + b$" (or *affine*) group [2, § 8.2]: the set of points (a, b), $a \in \mathbb{R}_+$, $b \in \mathbb{R}$ in the upper half-plane with the group law

$$(a, b) * (a', b') = (aa', ab' + b) \tag{4.4}$$

and left invariant measure $a^{-2}\,da\,db$. Its isometric representation on $V = L_p(\mathbb{R})$ is given by the formula

$$[\rho_p(g)\, f](x) = a^{\frac{1}{p}} f\,(ax + b)\,, \qquad \text{where } g^{-1} = (a, b). \tag{4.5}$$

We consider the operators $F_\pm : L_2(\mathbb{R}) \to \mathbb{C}$ defined by:

$$F_\pm(f) = \frac{1}{2\pi i} \int_{\mathbb{R}} \frac{f(t)\,dt}{x \mp \mathrm{i}}. \tag{4.6}$$

Then the covariant transform (4.1) is the Cauchy integral from $L_p(\mathbb{R})$ to the space of functions $\hat{f}(a, b)$ such that $a^{-\frac{1}{p}}\hat{f}(a, b)$ is in the Hardy space in the upper/lower half-plane $H_p(\mathbb{R}^2_\pm)$. Although the representation (4.5) is square integrable for $p = 2$, the function $\frac{1}{x\pm\mathrm{i}}$ used in (4.6) is not an admissible vacuum vector. Thus the complex analysis becomes decoupled from the traditional wavelet theory. As a result the application of wavelet theory shall rely on extraneous mother wavelets [47].

Many important objects in complex analysis are generated by inadmissible mother wavelets like (4.6). For example, if $F : L_2(\mathbb{R}) \to \mathbb{C}$ is defined by $F : f \mapsto F_+ f + F_- f$ then the covariant transform (4.1) reduces to the *Poisson integral*. If $F : L_2(\mathbb{R}) \to \mathbb{C}^2$ is defined by $F : f \mapsto (F_+ f, F_- f)$ then the covariant transform (4.1) represents a function f on the real line as a jump:

$$f(z) = f_+(z) - f_-(z), \qquad f_\pm(z) \in H_p(\mathbb{R}^2_\pm) \tag{4.7}$$

between functions analytic in the upper and the lower half-planes. This makes a decomposition of $L_2(\mathbb{R})$ into irreducible components of the representation (4.5). Another interesting but non-admissible vector is the *Gaussian* e^{-x^2}.

Example 4.9. For the group $G = \mathrm{SL}_2(\mathbb{R})$ [97] let us consider the unitary representation ρ on the space of square integrable function $L_2(\mathbb{R}^2_+)$ on the upper half-plane through the Möbius transformations (1.1):

$$\rho(g) : f(z) \mapsto \frac{1}{(cz + d)^2}\, f\left(\frac{az + b}{cz + d}\right), \qquad g^{-1} = \begin{pmatrix} a & b \\ c & d \end{pmatrix}. \tag{4.8}$$

This is a representation from the discrete series and $L_2(\mathbb{D})$ and irreducible invariant subspaces are parametrised by integers. Let F_k be the functional $L_2(\mathbb{R}^2_+) \rightarrow \mathbb{C}$ of pairing with the lowest/highest k-weight vector in the corresponding irreducible component (Bergman space) $B_k(\mathbb{R}^2_\pm)$, $k \geq 2$ of the discrete series [97, Ch. VI]. Then we can build an operator F from various F_k similarly to the previous example. In particular, the jump representation (4.7) on the real line generalises to the representation of a square integrable function f on the upper half-plane as a sum

$$f(z) = \sum_k a_k f_k(z), \qquad f_k \in B_n(\mathbb{R}^2_\pm)$$

for prescribed coefficients a_k and analytic functions f_k in question from different irreducible subspaces.

Covariant transforms are also meaningful for principal and complementary series of representations of the group $\mathrm{SL}_2(\mathbb{R})$, which are not square integrable [64].

Example 4.10. Let $G = \mathrm{SU}(2) \times \mathrm{Aff}$ be the Cartesian product of the groups $\mathrm{SU}(2)$ of unitary rotations of $\mathbb{C}^2$ and the $ax + b$ group Aff. This group has a unitary linear representation on the space $L_2(\mathbb{R}, \mathbb{C}^2)$ of square-integrable (vector) $\mathbb{C}^2$-valued functions by the formula

$$\rho(g)\begin{pmatrix} f_1(t) \\ f_2(t) \end{pmatrix} = \begin{pmatrix} \alpha f_1(at+b) + \beta f_2(at+b) \\ \gamma f_1(at+b) + \delta f_2(at+b) \end{pmatrix},$$

where $g = \begin{pmatrix} \alpha & \beta \\ \gamma & \delta \end{pmatrix} \times (a, b) \in \mathrm{SU}(2) \times \mathrm{Aff}$. It is obvious that the vector Hardy space, that is functions with both components being analytic, is invariant under such action of G.

As a fiducial operator $F : L_2(\mathbb{R}, \mathbb{C}^2) \rightarrow \mathbb{C}$ we can take, cf. (4.6),

$$F\begin{pmatrix} f_1(t) \\ f_2(t) \end{pmatrix} = \frac{1}{2\pi i} \int_{\mathbb{R}} \frac{f_1(t)\, dt}{x - \mathrm{i}}. \tag{4.9}$$

Thus the image of the associated covariant transform is a subspace of scalar-valued bounded functions on G. In this way we can transform (without a loss of information) vector-valued problems, e.g., matrix *Wiener–Hopf factorisation* [12], to scalar questions of harmonic analysis on the group G.

Example 4.11. A straightforward generalisation of Example 4.7 is obtained if V is a Banach space and $F : V \rightarrow \mathbb{C}$ is an element of V^*. Then the covariant transform coincides with the construction of wavelets in Banach spaces [66].

Example 4.12. The next stage of generalisation is achieved if V is a Banach space and $F : V \rightarrow \mathbb{C}^n$ is a linear operator. Then the corresponding covariant transform is a map $\mathcal{W} : V \rightarrow L(G, \mathbb{C}^n)$. This is closely related to M.G. Krein's works on *directing functionals* [94], see also *multiresolution wavelet analysis* [14], Clifford-valued Fock–Segal–Bargmann spaces [20] and [2, Thm. 7.3.1].

Example 4.13. Let F be a projector $L_p(\mathbb{R}) \to L_p(\mathbb{R})$ defined by the relation $(Ff)\hat{}(\lambda) = \chi(\lambda)\hat{f}(\lambda)$, where the hat denotes the Fourier transform and $\chi(\lambda)$ is the characteristic function of the set $[-2,-1]\cup[1,2]$. Then the covariant transform $L_p(\mathbb{R}) \to C(\mathrm{Aff}, L_p(\mathbb{R}))$ generated by the representation (4.5) of the affine group from F contains all information provided by the *Littlewood–Paley operator* [34, § 5.1.1].

Example 4.14. A step in a different direction is a consideration of non-linear operators. Take again the "$ax+b$" group and its representation (4.5). We define F to be a homogeneous but non-linear functional $V \to \mathbb{R}_+$:

$$F(f) = \frac{1}{2}\int_{-1}^{1} |f(x)|\, dx.$$

The covariant transform (4.1) becomes:

$$[\mathcal{W}_p f](a,b) = F(\rho_p(a,b)f) = \frac{1}{2}\int_{-1}^{1} \left| a^{\frac{1}{p}} f\,(ax+b)\right|\, dx = a^{\frac{1}{p}}\frac{1}{2a}\int_{b-a}^{b+a} |f\,(x)|\, dx. \tag{4.10}$$

Obviously $M_f(b) = \max_a [\mathcal{W}_\infty f](a,b)$ coincides with the Hardy *maximal function*, which contains important information on the original function f. From Corollary 4.5 we deduce that the operator $M : f \mapsto M_f$ intertwines ρ_p with itself $\rho_p M = M\rho_p$.

Of course, the full covariant transform (4.10) is even more detailed than M. For example, $\|f\| = \max_b [\mathcal{W}_\infty f](\frac{1}{2}, b)$ is the shift invariant norm [48].

Example 4.15. Let $V = L_c(\mathbb{R}^2)$ be the space of compactly supported bounded functions on the plane. We take F be the linear operator $V \to \mathbb{C}$ of integration over the real line:

$$F : f(x,y) \mapsto F(f) = \int_{\mathbb{R}} f(x,0)\, dx.$$

Let G be the group of Euclidean motions of the plane represented by ρ on V by a change of variables. Then the wavelet transform $F(\rho(g)f)$ is the *Radon transform* [39].

4.3. Symbolic calculi

There is a very important class of covariant transforms which map operators to functions. Among numerous sources we wish to single out works of Berezin [10, 11]. We start from the Berezin *covariant symbol.*

Example 4.16. Let a representation ρ of a group G act on a space X. Then there is an associated representation ρ_B of G on a space $V = B(X,Y)$ of linear operators $X \to Y$ defined by the identity [11, 66]:

$$(\rho_B(g)A)x = A(\rho(g^{-1})x), \qquad x \in X,\ g \in G,\ A \in B(X,Y). \tag{4.11}$$

Following Remark 4.3 we take F to be a functional $V \to \mathbb{C}$, for example F can be defined from a pair $x \in X$, $l \in Y^*$ by the expression $F : A \mapsto \langle Ax, l\rangle$. Then the covariant transform is

$$\mathcal{W} : A \mapsto \hat{A}(g) = F(\rho_B(g)A).$$

This is an example of *covariant calculus* [10, 66].

There are several variants of the last example which are of separate interest.

Example 4.17. A modification of the previous construction is obtained if we have two groups G_1 and G_2 represented by ρ_1 and ρ_2 on X and Y^* respectively. Then we have a covariant transform $B(X,Y) \to L(G_1 \times G_2, \mathbb{C})$ defined by the formula

$$\mathcal{W} : A \mapsto \hat{A}(g_1, g_2) = \langle A\rho_1(g_1)x, \rho_2(g_2)l\rangle .$$

This generalises the above *Berezin covariant calculi* [66].

Example 4.18. Let us restrict the previous example to the case when $X = Y$ is a Hilbert space, $\rho_1 = \rho_2 = \rho$ and $x = l$ with $\|x\| = 1$. Then the range of the covariant transform,

$$\mathcal{W} : A \mapsto \hat{A}(g) = \langle A\rho(g)x, \rho(g)x\rangle ,$$

is a subset of the *numerical range* of the operator A. As a function on a group, $\hat{A}(g)$ provides a better description of A than the set of its values – numerical range.

Example 4.19. The group $\mathrm{SU}(1,1) \simeq \mathrm{SL}_2(\mathbb{R})$ consists of 2×2 matrices of the form $\begin{pmatrix}\alpha & \beta\\ \bar{\beta} & \bar{\alpha}\end{pmatrix}$ with the unit determinant [97, § IX.1]. Let T be an operator with spectral radius less than 1. Then the associated Möbius transformation

$$g : T \mapsto g\cdot T = \frac{\alpha T + \beta I}{\bar{\beta} T + \bar{\alpha} I}, \qquad \text{where} \quad g = \begin{pmatrix}\alpha & \beta\\ \bar{\beta} & \bar{\alpha}\end{pmatrix} \in \mathrm{SL}_2(\mathbb{R}), \tag{4.12}$$

produces a well-defined operator with spectral radius less than 1 as well. Thus we have a representation of $SU(1,1)$.

Let us introduce the *defect operators* $D_T = (I - T^*T)^{1/2}$ and $D_{T^*} = (I - TT^*)^{1/2}$. For the fiducial operator $F = D_{T^*}$ the covariant transform is, cf. [118, § VI.1, (1.2)],

$$[\mathcal{W}T](g) = F(g\cdot T) = -e^{\mathrm{i}\phi}\, \Theta_T(z)\, D_T, \quad \text{for } g = \begin{pmatrix} e^{\mathrm{i}\phi/2} & 0\\ 0 & e^{-\mathrm{i}\phi/2}\end{pmatrix}\begin{pmatrix}1 & -z\\ -\bar{z} & 1\end{pmatrix},$$

where the *characteristic function* $\Theta_T(z)$ [118, § VI.1, (1.1)] is

$$\Theta_T(z) = -T + D_{T^*}\,(I - zT^*)^{-1}\, z\, D_T.$$

Thus we approached the *functional model* of operators from the covariant transform. In accordance with Remark 4.3 the model is most fruitful for the case of operator $F = D_{T^*}$ being one-dimensional.

The intertwining property in the previous examples was obtained as a consequence of the general Theorem 4.4 about the covariant transform. However it may be worth selecting it as a separate definition:

Definition 4.20. A *covariant calculus*, also known as *symbolic calculus*, is a map from operators to functions, which intertwines two representations of the same group in the respective spaces.

There is a dual class of covariant transforms acting in the opposite direction: from functions to operators. The prominent examples are the Berezin *contravariant symbol* [10, 66] and symbols of a *pseudodifferential operator* (PDO) [43, 66].

Example 4.21. The classical *Riesz–Dunford functional calculus* [27, § VII.3; 106, § IV.2] maps analytical functions on the unit disk to the linear operators; it is defined through a Cauchy-type formula with a resolvent. The calculus is an intertwining operator [75] between the Möbius transformations of the unit disk, cf. (5.22), and the actions (4.12) on operators from Example 4.19. This topic will be developed in Subsection 6.1.

In line with Definition 4.20 we can directly define the corresponding calculus through the intertwining property [62, 75]:

Definition 4.22. A *contravariant calculus*, also know as *functional calculus*, is a map from functions to operators, which intertwines two representations of the same group in the respective spaces.

The duality between co- and contravariant calculi is the particular case of the duality between covariant transform and the inverse covariant transform defined in the next subsection. In many cases a proper choice of spaces makes covariant and/or contravariant calculus a bijection between functions and operators. Subsequently only one form of calculus, either co- or contravariant, is defined explicitly, although both of them are there in fact.

4.4. Inverse covariant transform

An object invariant under the left action Λ (4.2) is called *left invariant*. For example, let L and L' be two left invariant spaces of functions on G. We say that a pairing $\langle\cdot,\cdot\rangle : L \times L' \to \mathbb{C}$ is *left invariant* if

$$\langle \Lambda(g)f, \Lambda(g)f'\rangle = \langle f, f'\rangle\,, \qquad \text{for all} \quad f \in L,\ f' \in L'. \tag{4.13}$$

Remark 4.23. i. We do not require the pairing to be linear in general.

ii. If the pairing is invariant on space $L \times L'$ it is not necessarily invariant (or even defined) on the whole of $C(G) \times C(G)$.
iii. In a more general setting we shall study an invariant pairing on a homogeneous space instead of the group. However due to length constraints we cannot consider it here beyond Example 4.26.
iv. An invariant pairing on G can be obtained from an *invariant functional* l by the formula $\langle f_1, f_2\rangle = l(f_1\bar{f}_2)$.

For a representation ρ of G in V and $v_0 \in V$ we fix a function $w(g) = \rho(g)v_0$. We assume that the pairing can be extended in its second component to this V-valued function, say, in the weak sense.

Definition 4.24. Let $\langle \cdot, \cdot \rangle$ be a left invariant pairing on $L \times L'$ as above, let ρ be a representation of G in a space V; we define the function $w(g) = \rho(g)v_0$ for $v_0 \in V$. The *inverse covariant transform* $\mathcal{M}$ is a map $L \to V$ defined by the pairing:

$$\mathcal{M} : f \mapsto \langle f, w \rangle, \qquad \text{where } f \in L. \tag{4.14}$$

Example 4.25. Let G be a group with a unitary square integrable representation ρ. An invariant pairing of two square integrable functions is obviously done by integration over the Haar measure:

$$\langle f_1, f_2 \rangle = \int_G f_1(g)\bar{f}_2(g)\, dg.$$

For an admissible vector v_0 [2, Chap. 8; 26] the inverse covariant transform is known in this setup as a *reconstruction formula*.

Example 4.26. Let ρ be a square integrable representation of G modulo a subgroup $H \subset G$ and let $X = G/H$ be the corresponding homogeneous space with a quasi-invariant measure dx. Then integration over dx with an appropriate weight produces an invariant pairing. The inverse covariant transform is a more general version [2, (7.52)] of the *reconstruction formula* mentioned in the previous example.

Let ρ be not a square integrable representation (even modulo a subgroup) or let v_0 be an inadmissible vector of a square integrable representation ρ. An invariant pairing in this case is not associated with an integration over any non-singular invariant measure on G. In this case we have a *Hardy pairing*. The following example explains the name.

Example 4.27. Let G be the "$ax + b$" group and its representation ρ (4.5) from Example 4.8. An invariant pairing on G, which is not generated by the Haar measure $a^{-2}da\, db$, is

$$\langle f_1, f_2 \rangle = \lim_{a \to 0} \int_{-\infty}^{\infty} f_1(a, b)\, \bar{f}_2(a, b)\, db. \tag{4.15}$$

For this pairing we can consider functions $\frac{1}{2\pi i(x+i)}$ or e^{-x^2}, which are not admissible vectors in the sense of square integrable representations. Then the inverse covariant transform provides an *integral resolution* of the identity.

Similar pairings can be defined for other semi-direct products of two groups. We can also extend a Hardy pairing to a group, which has a subgroup with such a pairing.

Example 4.28. Let G be the group $\mathrm{SL}_2(\mathbb{R})$ from Example 4.9. Then the "$ax + b$" group is a subgroup of $\mathrm{SL}_2(\mathbb{R})$, moreover we can parametrise $\mathrm{SL}_2(\mathbb{R})$ by triples

(a, b, θ), $\theta \in (-\pi, \pi]$ with the respective Haar measure [97, III.1(3)]. Then the Hardy pairing

$$\langle f_1, f_2 \rangle = \lim_{a \to 0} \int_{-\infty}^{\infty} f_1(a, b, \theta)\, \bar{f}_2(a, b, \theta)\, db\, d\theta \tag{4.16}$$

is invariant on $SL_2(\mathbb{R})$ as well. The corresponding inverse covariant transform provides even a finer resolution of the identity which is invariant under conformal mappings of the Lobachevsky half-plane.

5. Analytic functions

We saw in the first section that an inspiring geometry of cycles can be recovered from the properties of $SL_2(\mathbb{R})$. In this section we consider a realisation of function theory within the Erlangen approach [64, 65, 68, 69]. The covariant transform will be our principal tool in this construction.

5.1. Induced covariant transform

The choice of a mother wavelet or fiducial operator F from Section 4.1 can significantly influence the behaviour of the covariant transform. Let G be a group and H be its closed subgroup with the corresponding homogeneous space $X = G/H$. Let ρ be a representation of G by operators on a space V; we denote by ρ_H the restriction of ρ to the subgroup H.

Definition 5.1. Let χ be a representation of the subgroup H in a space U and $F : V \to U$ be an intertwining operator between χ and the representation ρ_H:

$$F(\rho(h)v) = F(v)\chi(h), \qquad \text{for all } h \in H,\ v \in V. \tag{5.1}$$

Then the covariant transform (4.1) generated by F is called the *induced covariant transform.*

The following is the main motivating example.

Example 5.2. Consider the traditional wavelet transform as outlined in Example 4.7. Choose a vacuum vector v_0 to be a joint eigenvector for all operators $\rho(h)$, $h \in H$, that is $\rho(h)v_0 = \chi(h)v_0$, where $\chi(h)$ is a complex number depending of h. Then χ is obviously a character of H.

The image of wavelet transform (4.3) with such a mother wavelet will have a property:

$$\hat{v}(gh) = \langle v, \rho(gh)v_0 \rangle = \langle v, \rho(g)\chi(h)v_0 \rangle = \chi(h)\hat{v}(g).$$

Thus the wavelet transform is uniquely defined by cosets on the homogeneous space G/H. In this case we previously spoke about the *reduced wavelet transform* [65]. A representation ρ_0 is called *square integrable* mod H if the induced wavelet transform $[\mathcal{W} f_0](w)$ of the vacuum vector $f_0(x)$ is square integrable on X.

The image of an induced covariant transform has a similar property:

$$\hat{v}(gh) = F(\rho((gh)^{-1})v) = F(\rho(h^{-1})\rho(g^{-1})v) = F(\rho(g^{-1})v)\chi(h^{-1}). \qquad (5.2)$$

Thus it is enough to know the value of the covariant transform only at a single element in every coset G/H in order to reconstruct it for the entire group G by the representation χ. Since coherent states (wavelets) are now parametrised by points of homogeneous space G/H, they are referred to sometimes as coherent states which are not connected to a group [92], however this is true only in a very narrow sense as explained above.

Example 5.3. To make it more specific we can consider the representation of $\mathrm{SL}_2(\mathbb{R})$ defined on $L_2(\mathbb{R})$ by the formula, cf. (3.8):

$$\rho(g) : f(z) \mapsto \frac{1}{(cx+d)} f\left(\frac{ax+b}{cx+d}\right), \qquad g^{-1} = \begin{pmatrix} a & b \\ c & d \end{pmatrix}.$$

Let $K \subset \mathrm{SL}_2(\mathbb{R})$ be the compact subgroup of matrices $h_t = \begin{pmatrix} \cos t & \sin t \\ -\sin t & \cos t \end{pmatrix}$. Then for the fiducial operator $F_\pm$ (4.6) we have $F_\pm \circ \rho(h_t) = e^{\mp it} F_\pm$. Thus we can consider the covariant transform only for points in the homogeneous space $\mathrm{SL}_2(\mathbb{R})/K$, moreover this set can be naturally identified with the $ax+b$ group. Thus we do not obtain any advantage of extending the group in Example 4.8 from $ax+b$ to $\mathrm{SL}_2(\mathbb{R})$ if we will be still using the fiducial operator $F_\pm$ (4.6).

Functions on the group G, which have the property $\hat{v}(gh) = \hat{v}(g)\chi(h)$ (5.2), provide a space for the representation of G induced by the representation χ of the subgroup H. This explains the choice of the name for induced covariant transform.

Remark 5.4. Induced covariant transform uses the fiducial operator F which passes through the action of the subgroup H. This reduces information which we obtained from this transform in some cases.

There is also a simple connection between a covariant transform and right shifts:

Proposition 5.5. *Let G be a Lie group and ρ be a representation of G in a space V. Let $[\mathcal{W}f](g) = F(\rho(g^{-1})f)$ be a covariant transform defined by the fiducial operator $F : V \to U$. Then the right shift $[\mathcal{W}f](gg')$ by g' is the covariant transform $[\mathcal{W}'f](g) = F'(\rho(g^{-1})f)]$ defined by the fiducial operator $F' = F \circ \rho(g^{-1})$.*

In other words the covariant transform intertwines right shifts on the group G with the associated action ρ_B (4.11) on fiducial operators.

Although the above result is obvious, its infinitesimal version has interesting consequences.

Corollary 5.6 ([84]). *Let G be a Lie group with a Lie algebra $\mathfrak{g}$ and ρ be a smooth representation of G. We denote by $d\rho_B$ the derived representation of the associated representation ρ_B (4.11) on fiducial operators.*

Let a fiducial operator F be a null-solution, i.e., $AF = 0$, for the operator $A = \sum_J a_j d\rho_B^{X_j}$, where $X_j \in \mathfrak{g}$ and a_j are constants. Then the covariant transform $[\mathcal{W}f](g) = F(\rho(g^{-1})f)$ for any f satisfies

$$DF(g) = 0, \qquad \textit{where} \quad D = \sum_j \bar{a}_j \mathfrak{L}^{X_j}.$$

Here $\mathfrak{L}^{X_j}$ are the left invariant fields (Lie derivatives) on G corresponding to X_j.

Example 5.7. Consider the representation ρ (4.5) of the $ax + b$ group with the $p = 1$. Let A and N be the basis of the corresponding Lie algebra generating one-parameter subgroups $(e^t, 0)$ and $(0, t)$. Then the derived representations are

$$[d\rho^A f](x) = f(x) + xf'(x), \qquad [d\rho^N f](x) = f'(x).$$

The corresponding left invariant vector fields on $ax + b$ group are

$$\mathfrak{L}^A = a\partial_a, \qquad \mathfrak{L}^N = a\partial_b.$$

The mother wavelet $\frac{1}{x+\mathrm{i}}$ is a null solution of the operator $d\rho^A + \mathrm{i}d\rho^N = I + (x+\mathrm{i})\frac{d}{dx}$. Therefore the covariant transform with the fiducial operator F_+ (4.6) will consist with the null solutions to the operator $\mathfrak{L}^A - \mathrm{i}\mathfrak{L}^N = -\mathrm{i}a(\partial_b + \mathrm{i}\partial_a)$, that is in essence the Cauchy-Riemann operator in the upper half-plane.

There is a statement which extends the previous corollary from differential operators to integro-differential ones. We will formulate it for the wavelets setting.

Corollary 5.8. *Let G be a group and ρ be a unitary representation of G, which can be extended to a vector space V of functions or distributions on G. Let a mother wavelet $w \in V'$ satisfy the equation*

$$\int_G a(g)\, \rho(g) w \, dg = 0,$$

for a fixed distribution $a(g) \in V$ and a (not necessarily invariant) measure dg. Then any wavelet transform $F(g) = \mathcal{W}f(g) = \langle f, \rho(g) w_0 \rangle$ obeys the condition:

$$DF = 0, \qquad \textit{where} \quad D = \int_G \bar{a}(g)\, R(g)\, dg,$$

with R being the right regular representation of G.

Clearly, Corollary 5.6 is a particular case of Corollary 5.8 with a distribution a, which is a combination of derivatives of Dirac's delta functions. The last corollary will be illustrated at the end of Section 6.1.

Remark 5.9. We note that Corollaries 5.6 and 5.8 are true whenever we have an intertwining property between ρ with the right regular representation of G.

5.2. Induced wavelet transform and Cauchy integral

We again use the general scheme from Subsection 3.2. The $ax+b$ group is isomorphic to a subgroup of $\mathrm{SL}_2(\mathbb{R})$ consisting of the lower-triangular matrices:

$$F = \left\{ \frac{1}{\sqrt{a}} \begin{pmatrix} a & 0 \\ b & 1 \end{pmatrix}, \ a > 0 \right\}.$$

The corresponding homogeneous space $X = \mathrm{SL}_2(\mathbb{R})/F$ is one-dimensional and can be parametrised by a real number. The natural projection $p : \mathrm{SL}_2(\mathbb{R}) \to \mathbb{R}$ and its left inverse $s : \mathbb{R} \to \mathrm{SL}_2(\mathbb{R})$ can be defined as follows:

$$p : \begin{pmatrix} a & b \\ c & d \end{pmatrix} \mapsto \frac{b}{d}, \qquad s : u \mapsto \begin{pmatrix} 1 & u \\ 0 & 1 \end{pmatrix}. \tag{5.3}$$

Thus we calculate the corresponding map $r : \mathrm{SL}_2(\mathbb{R}) \to F$, see Subsection 2.1:

$$r : \begin{pmatrix} a & b \\ c & d \end{pmatrix} \mapsto \begin{pmatrix} d^{-1} & 0 \\ c & d \end{pmatrix}. \tag{5.4}$$

Therefore the action of $\mathrm{SL}_2(\mathbb{R})$ on the real line is exactly the Möbius map (1.1):

$$g : u \mapsto p(g^{-1} * s(u)) = \frac{au+b}{cu+d}, \qquad \text{where } g^{-1} = \begin{pmatrix} a & b \\ c & d \end{pmatrix}.$$

We also calculate that

$$r(g^{-1} * s(u)) = \begin{pmatrix} (cu+d)^{-1} & 0 \\ c & cu+d \end{pmatrix}.$$

To build an induced representation we need a character of the affine group. A generic character of F is a power of its diagonal element:

$$\rho_\kappa \begin{pmatrix} a & 0 \\ c & a^{-1} \end{pmatrix} = a^\kappa.$$

Thus the corresponding realisation of induced representation (3.6) is:

$$\rho_\kappa(g) : f(u) \mapsto \frac{1}{(cu+d)^\kappa} f\left(\frac{au+b}{cu+d} \right) \qquad \text{where } g^{-1} = \begin{pmatrix} a & b \\ c & d \end{pmatrix}. \tag{5.5}$$

The only freedom remaining by the scheme is in a choice of a value of *number* κ and the corresponding functional space where our representation acts. At this point we have a wider choice of κ than it is usually assumed: it can belong to different hypercomplex systems.

One of the important properties which would be nice to have is the unitarity of the representation (5.5) with respect to the standard inner product:

$$\langle f_1, f_2 \rangle = \int_{\mathbb{R}^2} f_1(u) \bar{f}_2(u)\, du.$$

A change of variables $x = \frac{au+b}{cu+d}$ in the integral suggests the following property is necessary and sufficient for that:

$$\kappa + \bar{\kappa} = 2. \tag{5.6}$$

A mother wavelet for an induced wavelet transform shall be an eigenvector for the action of a subgroup $\tilde{H}$ of $\mathrm{SL}_2(\mathbb{R})$, see (5.1). Let us consider the most common case of $\tilde{H} = K$ and take the infinitesimal condition with the derived representation: $d\rho_n^Z w_0 = \lambda w_0$, since Z (3.12) is the generator of the subgroup K. In other words the restriction of w_0 to a K-orbit should be given by $e^{\lambda t}$ in the exponential coordinate t along the K-orbit. However we usually need its expression in other "more natural" coordinates. For example [86], an eigenvector of the derived representation of $d\rho_n^Z$ should satisfy the differential equation in the ordinary parameter $x \in \mathbb{R}$:

$$-\kappa x f(x) - f'(x)(1+x^2) = \lambda f(x). \tag{5.7}$$

The equation does not have singular points, the general solution is globally defined (up to a constant factor) by:

$$w_{\lambda,\kappa}(x) = \frac{1}{(1+x^2)^{\kappa/2}} \left(\frac{x-\mathrm{i}}{x+\mathrm{i}}\right)^{\mathrm{i}\lambda/2} = \frac{(x-\mathrm{i})^{(\mathrm{i}\lambda-\kappa)/2}}{(x+\mathrm{i})^{(\mathrm{i}\lambda+\kappa)/2}}. \tag{5.8}$$

To avoid multivalent functions we need 2π-periodicity along the exponential coordinate on K. This implies that the parameter $m = -\mathrm{i}\lambda$ is an integer. Therefore the solution becomes:

$$w_{m,\kappa}(x) = \frac{(x+\mathrm{i})^{(m-\kappa)/2}}{(x-\mathrm{i})^{(m+\kappa)/2}}. \tag{5.9}$$

The corresponding wavelets resemble the Cauchy kernel normalised to the invariant metric in the Lobachevsky half-plane:

$$w_{m,\kappa}(u,v;x) \;=\; \rho_\kappa^F(s(u,v))w_{m,\kappa}(x) = v^{\kappa/2}\frac{(x-u+\mathrm{i}v)^{(m-\kappa)/2}}{(x-u-\mathrm{i}v)^{(m+\kappa)/2}}.$$

Therefore the wavelet transform (4.3) from a function on the real line to functions on the upper half-plane is

$$\hat{f}(u,v) \;=\; \langle f, \rho_\kappa^F(u,v)w_{m,\kappa}\rangle = v^{\bar{\kappa}/2}\int_{\mathbb{R}} f(x)\,\frac{(x-(u+\mathrm{i}v))^{(m-\kappa)/2}}{(x-(u-\mathrm{i}v))^{(m+\kappa)/2}}\,dx.$$

Introduction of a complex variable $z = u + \mathrm{i}v$ allows us to write it as

$$\hat{f}(z) = (\Im z)^{\bar{\kappa}/2}\int_{\mathbb{R}} f(x)\frac{(x-z)^{(m-\kappa)/2}}{(x-\bar{z})^{(m+\kappa)/2}}\,dx. \tag{5.10}$$

According to the general theory this wavelet transform intertwines representations ρ_κ^F (5.5) on the real line (induced by the character a^κ of the subgroup F) and ρ_m^K (3.8) on the upper half-plane (induced by the character $e^{\mathrm{i}mt}$ of the subgroup K).

5.3. The Cauchy–Riemann (Dirac) and Laplace operators

Ladder operators $L^\pm = \pm\mathrm{i}A + B$ act by raising/lowering indexes of the K-eigenfunctions $w_{m,\kappa}$ (5.8), see Subsection 3.3. More explicitly [86]:

$$d\rho_\kappa^{L^\pm} : w_{m,\kappa} \mapsto -\frac{\mathrm{i}}{2}(m \pm \kappa)w_{m\pm 2,\kappa}. \tag{5.11}$$

There are two possibilities here: $m \pm \kappa$ is zero for some m or not. In the first case the chain (5.11) of eigenfunction $w_{m,\kappa}$ terminates on one side under the transitive action (3.16) of the ladder operators; otherwise the chain is infinite in both directions. That is, the values $m = \mp\kappa$ and only those correspond to the maximal (minimal) weight function $w_{\mp\kappa,\kappa}(x) = \frac{1}{(x\pm \mathrm{i})^\kappa} \in L_2(\mathbb{R})$, which are annihilated by $L^\pm$:

$$d\rho_\kappa^{L^\pm} \, w_{\mp\kappa,\kappa} = (\pm \mathrm{i} d\rho_\kappa^A + d\rho_\kappa^B)\, w_{\mp\kappa,\kappa} = 0. \tag{5.12}$$

By Corollary 5.6 for the mother wavelets $w_{\mp\kappa,\kappa}$, which are annihilated by (5.12), the images of the respective wavelet transforms are null solutions to the left-invariant differential operator $D_\pm = \overline{\mathfrak{L}^{L^\pm}}$:

$$D_\pm = \mp \mathrm{i}\mathfrak{L}^A + \mathfrak{L}^B = -\tfrac{\mathrm{i}\kappa}{2} + v(\partial_u \pm \mathrm{i}\partial_v). \tag{5.13}$$

This is a conformal version of the Cauchy–Riemann equation. The second-order conformal Laplace-type operators $\Delta_+ = \overline{\mathfrak{L}^{L^-}\mathfrak{L}^{L^+}}$ and $\Delta_- = \overline{\mathfrak{L}^{L^+}\mathfrak{L}^{L^-}}$ are

$$\Delta_\pm = (v\partial_u - \tfrac{\mathrm{i}\kappa}{2})^2 + v^2\partial_v^2 \pm \tfrac{\kappa}{2}. \tag{5.14}$$

For the mother wavelets $w_{m,\kappa}$ in (5.12) such that $m = \mp\kappa$ the unitarity condition $\kappa + \bar\kappa = 2$, see (5.6), together with $m \in \mathbb{Z}$ implies $\kappa = \mp m = 1$. In such a case the wavelet transforms (5.10) are

$$\hat f^+(z) = (\Im z)^{\frac12} \int_{\mathbb{R}} \frac{f(x)\,dx}{x-z} \quad \text{and} \quad \hat f^-(z) = (\Im z)^{\frac12} \int_{\mathbb{R}} \frac{f(x)\,dx}{x-\bar z}, \tag{5.15}$$

for $w_{-1,1}$ and $w_{1,1}$ respectively. The first one is the Cauchy integral formula up to the factor $2\pi \mathrm{i}\sqrt{\Im z}$. Clearly, one integral is the complex conjugation of another. Moreover, the minimal/maximal weight cases can be intertwined by the following automorphism of the Lie algebra $\mathfrak{sl}_2$:

$$A \to B, \quad B \to A, \quad Z \to -Z.$$

As explained before, $\hat f^\pm(w)$ are null solutions to the operators $D_\pm$ (5.13) and $\Delta_\pm$ (5.14). These transformations intertwine unitary equivalent representations on the real line and on the upper half-plane, thus they can be made unitary for proper spaces. This is the source of two faces of the Hardy spaces: they can be defined either as square-integrable on the real line with an analytic extension to the half-plane, or analytic on the half-plane with square-integrability on an infinitesimal displacement of the real line.

For the third possibility, $m \pm \kappa \neq 0$, there is no an operator spanned by the derived representation of the Lie algebra $\mathfrak{sl}_2$ which kills the mother wavelet $w_{m,\kappa}$. However the remarkable *Casimir operator* $C = Z^2 - 2(L^-L^+ + L^+L^-)$, which spans the centre of the universal enveloping algebra of $\mathfrak{sl}_2$ [97, § X.1; 120, § 8.1], produces a second-order operator which does the job. Indeed from the identities (5.11) we get

$$d\rho_\kappa^C w_{m,\kappa} = (2\kappa - \kappa^2) w_{m,\kappa}. \tag{5.16}$$

Thus we get $d\rho^C_\kappa w_{m,\kappa} = 0$ for $\kappa = 2$ or 0. The mother wavelet $w_{0,2}$ turns out to be the *Poisson kernel* [34, Ex. 1.2.17]. The associated wavelet transform

$$\hat{f}(z) = \Im z \int_{\mathbb{R}} \frac{f(x)\,dx}{|x-z|^2} \tag{5.17}$$

consists of null solutions of the left-invariant second-order Laplacian, image of the Casimir operator, cf. (5.14):

$$\Delta (:= \mathfrak{L}^C) = v^2\partial_u^2 + v^2\partial_v^2.$$

Another integral formula producing solutions to this equation delivered by the mother wavelet $w_{m,0}$ with the value $\kappa = 0$ in (5.16):

$$\hat{f}(z) = \int_{\mathbb{R}} f(x) \left(\frac{x-z}{x-\bar{z}}\right)^{m/2} dx. \tag{5.18}$$

Furthermore, we can introduce higher-order differential operators. The functions $w_{\mp 2m+1,1}$ are annihilated by nth power of operator $d\rho^{L^\pm}_\kappa$ with $1 \le m \le n$. By Corollary 5.6 the image of wavelet transform (5.10) from a mother wavelet $\sum_1^n a_m w_{\mp 2m,1}$ will consist of null-solutions of the nth power $D^n_\pm$ of the conformal Cauchy–Riemann operator (5.13). They are a conformal flavour of *polyanalytic* functions [6].

We can similarly look for mother wavelets which are eigenvectors for other types of one-dimensional subgroups. Our consideration of subgroup K is simplified by several facts:

- The parameter κ takes only complex values.
- The derived representation does not have singular points on the real line.

For both subgroups A' and N' this will not be true. Further consideration will be given in [86].

5.4. The Taylor expansion

Consider an induced wavelet transform generated by a Lie group G, its representation ρ and a mother wavelet w which is an eigenvector of a one-dimensional subgroup $\tilde{H} \subset G$. Then by Proposition 5.5 the wavelet transform intertwines ρ with a representation $\rho^{\tilde{H}}$ induced by a character of $\tilde{H}$.

If the mother wavelet is itself in the domain of the induced wavelet transform, then the chain (3.16) of $\tilde{H}$-eigenvectors w_m will be mapped to the similar chain of their images $\hat{w}_m$. The corresponding derived induced representation $d\rho^{\tilde{H}}$ produces ladder operators with the transitive action of the ladder operators on the chain of $\hat{w}_m$. Then the vector space of "formal power series",

$$\hat{f}(z) = \sum_{m \in \mathbb{Z}} a_m \hat{w}_m(z), \tag{5.19}$$

is a module for the Lie algebra of the group G.

Coming back to the case of the group $G = \mathrm{SL}_2(\mathbb{R})$ and subgroup $\tilde{H} = K$, images $\hat{w}_{m,1}$ of the eigenfunctions (5.9) under the Cauchy integral transform (5.15) are

$$\hat{w}_{m,1}(z) = (\Im z)^{1/2} \frac{(z+\mathrm{i})^{(m-1)/2}}{(z-\mathrm{i})^{(m+1)/2}}.$$

They are eigenfunctions of the derived representation on the upper half-plane and the action of ladder operators is given by the same expressions (5.11). In particular, the $\mathfrak{sl}_2$-module generated by $\hat{w}_{1,1}$ will be one-sided since this vector is annihilated by the lowering operator. Since the Cauchy integral produces an unitary intertwining operator between two representations we get the following variant of Taylor series:

$$\hat{f}(z) = \sum_{m=0}^{\infty} c_m \hat{w}_{m,1}(z), \qquad \text{where} \qquad c_m = \langle f, w_{m,1} \rangle .$$

For two other types of subgroups, representations and mother wavelets this scheme shall be suitably adapted and detailed study will be presented elsewhere [86].

5.5. Wavelet transform in the unit disk and other domains

We can similarly construct analytic function theories in unit disks, including parabolic and hyperbolic ones [82]. This can be done simply by an application of the *Cayley transform* to the function theories in the upper half-plane. Alternatively we can apply the full procedure for properly chosen groups and subgroups. We will briefly outline such a possibility here, see also [64].

Elements of $\mathrm{SL}_2(\mathbb{R})$ can also be represented by 2×2-matrices with complex entries such that, cf. Example 4.21:

$$g = \begin{pmatrix} \alpha & \bar{\beta} \\ \beta & \bar{\alpha} \end{pmatrix}, \qquad g^{-1} = \begin{pmatrix} \bar{\alpha} & -\bar{\beta} \\ -\beta & \alpha \end{pmatrix}, \qquad |\alpha|^2 - |\beta|^2 = 1.$$

These realisations of $\mathrm{SL}_2(\mathbb{R})$ (or rather $SU(2,\mathbb{C})$) is more suitable for function theory in the unit disk. It is obtained from the form, which we used before for the upper half-plane, by means of the Cayley transform [82, § 8.1].

We may identify the *unit disk* $\mathbb{D}$ with the homogeneous space $\mathrm{SL}_2(\mathbb{R})/\mathbb{T}$ for the *unit circle* $\mathbb{T}$ through the important decomposition $\mathrm{SL}_2(\mathbb{R}) \sim \mathbb{D} \times \mathbb{T}$ with $K = \mathbb{T}$ – the compact subgroup of $\mathrm{SL}_2(\mathbb{R})$:

$$\begin{aligned} \begin{pmatrix} \alpha & \bar{\beta} \\ \beta & \bar{\alpha} \end{pmatrix} &= |\alpha| \begin{pmatrix} 1 & \bar{\beta}\bar{\alpha}^{-1} \\ \beta\alpha^{-1} & 1 \end{pmatrix} \begin{pmatrix} \frac{\alpha}{|\alpha|} & 0 \\ 0 & \frac{\bar{\alpha}}{|\alpha|} \end{pmatrix} \\ &= \frac{1}{\sqrt{1-|u|^2}} \begin{pmatrix} 1 & u \\ \bar{u} & 1 \end{pmatrix} , \begin{pmatrix} e^{ix} & 0 \\ 0 & e^{-ix} \end{pmatrix} \end{aligned} \tag{5.20}$$

where

$$x = \arg \alpha, \qquad u = \bar{\beta}\bar{\alpha}^{-1}, \qquad |u| < 1.$$

Each element $g \in \mathrm{SL}_2(\mathbb{R})$ acts by the linear-fractional transformation (the Möbius map) on $\mathbb{D}$ and $\mathbb{T}$ $H_2(\mathbb{T})$ as follows:

$$g : z \mapsto \frac{\alpha z + \beta}{\bar{\beta} z + \bar{\alpha}}, \qquad \text{where} \qquad g = \begin{pmatrix} \alpha & \beta \\ \bar{\beta} & \bar{\alpha} \end{pmatrix}. \tag{5.21}$$

In the decomposition (5.20) the first matrix on the right-hand side acts by transformation (5.21) as an orthogonal rotation of $\mathbb{T}$ or $\mathbb{D}$; and the second one – by a transitive family of maps of the unit disk onto itself.

The representation induced by a complex-valued character $\chi_k(z) = z^{-k}$ of $\mathbb{T}$ according to Section 3.2 is:

$$\rho_k(g) : f(z) \mapsto \frac{1}{(\alpha - \beta z)^k} f\left(\frac{\bar{\alpha} z - \bar{\beta}}{\alpha - \beta z}\right) \qquad \text{where} \qquad g = \begin{pmatrix} \alpha & \beta \\ \bar{\beta} & \bar{\alpha} \end{pmatrix}. \tag{5.22}$$

The representation ρ_1 is unitary on square-integrable functions and irreducible on a *Hardy space* on the unit circle.

We choose [66, 68] K-invariant function $v_0(z) \equiv 1$ to be a *vacuum vector*. Thus the associated *coherent states*

$$v(g, z) = \rho_1(g) v_0(z) = (u - z)^{-1}$$

are completely determined by the point on the unit disk $u = \bar{\beta}\bar{\alpha}^{-1}$. The family of coherent states considered as a function of both u and z is obviously the *Cauchy kernel* [64]. The *wavelet transform* [64, 66] $\mathcal{W} : L_2(\mathbb{T}) \to H_2(\mathbb{D}) : f(z) \mapsto \mathcal{W}f(g) = \langle f, v_g \rangle$ is the *Cauchy integral*

$$\mathcal{W}f(u) = \frac{1}{2\pi i} \int_{\mathbb{T}} f(z) \frac{1}{u - z}\, dz. \tag{5.23}$$

This approach can be extended to an arbitrary connected simply-connected domain. Indeed, it is known that the set of Möbius maps is the whole group of biholomorphic automorphisms of the unit disk or upper half-plane. Thus we can state the following corollary from the *Riemann mapping theorem*:

Corollary 5.10. *The group of biholomorphic automorphisms of a connected simply-connected domain with at least two points on its boundary is isomorphic to* $\mathrm{SL}_2(\mathbb{R})$.

If a domain is non-simply connected, then the group of its biholomorphic mapping can be trivial [9, 103]. However we may look for a rich group acting on function spaces rather than on geometric sets. Let a connected non-simply connected domain D be bounded by a finite collection of non-intersecting contours Γ_i, $i = 1, \ldots, n$. For each Γ_i consider the isomorphic image G_i of the $\mathrm{SL}_2(\mathbb{R})$ group which is defined by Corollary 5.10. Then define the group $G = G_1 \times G_2 \times \cdots \times G_n$ and its action on $L_2(\partial D) = L_2(\Gamma_1) \oplus L_2(\Gamma_2) \oplus \cdots \oplus L_2(\Gamma_n)$ through the Möbius action of G_i on $L_2(\Gamma_i)$.

Example 5.11. Consider an *annulus* defined by $r < |z| < R$. It is bounded by two circles: $\Gamma_1 = \{z : |z| = r\}$ and $\Gamma_2 = \{z : |z| = R\}$. For Γ_1 the Möbius action of $\mathrm{SL}_2(\mathbb{R})$ is

$$\begin{pmatrix} \alpha & \bar{\beta} \\ \beta & \bar{\alpha} \end{pmatrix} : z \mapsto \frac{\alpha z + \bar{\beta}/r}{\beta z/r + \bar{\alpha}}, \qquad \text{where} \quad |\alpha|^2 - |\beta|^2 = 1,$$

with the respective action on Γ_2. Those actions can be linearised in the spaces $L_2(\Gamma_1)$ and $L_2(\Gamma_2)$. If we consider a subrepresentation reduced to analytic function on the annulus, then one copy of $\mathrm{SL}_2(\mathbb{R})$ will act on the part of functions analytic outside of Γ_1 and another copy – on the part of functions analytic inside of Γ_2.

Thus all classical objects of complex analysis (the Cauchy-Riemann equation, the Taylor series, the Bergman space, etc.) for a rather generic domain D can also be obtained from suitable representations similarly to the case of the upper half-plane [64, 68].

6. Covariant and contravariant calculi

Functional calculus, spectrum, and the spectral mapping theorem, united in a trinity, play an exceptional rôle in functional analysis and could not be substituted by anything else. Many traditional definitions of functional calculus are covered by the following rigid template based on the *algebra homomorphism* property:

Definition 6.1. A *functional calculus* for an element $a \in \mathfrak{A}$ is a continuous linear mapping $\Phi : \mathcal{A} \to \mathfrak{A}$ such that:

i. Φ is a unital *algebra homomorphism*

$$\Phi(f \cdot g) = \Phi(f) \cdot \Phi(g).$$

ii. There is an initialisation condition: $\Phi[v_0] = a$ for a fixed function v_0, e.g., $v_0(z) = z$.

The most typical definition of a spectrum is seemingly independent and uses the important notion of resolvent:

Definition 6.2. A *resolvent* of element $a \in \mathfrak{A}$ is the function $R(\lambda) = (a - \lambda e)^{-1}$, which is the image under Φ of the Cauchy kernel $(z - \lambda)^{-1}$.

A *spectrum* of $a \in \mathfrak{A}$ is the set $\mathbf{sp}\, a$ of singular points of its resolvent $R(\lambda)$.

Then the following important theorem links spectrum and functional calculus together.

Theorem 6.3 (Spectral mapping). *For a function f suitable for the functional calculus:*

$$f(\mathbf{sp}\, a) = \mathbf{sp}\, f(a). \tag{6.1}$$

However the power of the classic spectral theory rapidly decreases if we move beyond the study of one normal operator (e.g., for quasinilpotent ones) and is virtually nil if we consider several non-commuting ones. Sometimes these severe limitations are seen to be irresistible and alternative constructions, i.e., model theory cf. Example 4.19 and [106], were developed.

Yet the spectral theory can be revived from a fresh start. While three components – functional calculus, spectrum, and spectral mapping theorem – are highly interdependent in various ways, we will nevertheless arrange them as follows:

i. Functional calculus is an *original* notion defined in some independent terms;
ii. Spectrum (or more specifically *contravariant spectrum*) (or spectral decomposition) is derived from previously defined functional calculus as its *support* (in some appropriate sense);
iii. The spectral mapping theorem then should drop out naturally in the form (6.1) or one of its variations.

Thus the entire scheme depends on the notion of the functional calculus and our ability to escape limitations of Definition 6.1. The first (known to the present author) definition of functional calculus not linked to the algebra homomorphism property was the Weyl functional calculus defined by an integral formula [3]. Then its intertwining property with affine transformations of Euclidean space was proved as a theorem. However it seems to have been the only "non-homomorphism" calculus for decades.

A different approach to the whole range of calculi was given in [62] and developed in [66, 73, 75, 84] in terms of *intertwining operators* for group representations. It was initially targeted for several non-commuting operators because no non-trivial algebra homomorphism is possible with a commutative algebra of functions in this case. However it emerged later that the new definition is a useful replacement for the classical one across all ranges of problems.

In the following subsections we will support the last claim by consideration of the simple known problem: characterisation of an $n \times n$ matrix up to similarity. Even that "freshman" question can be only sorted out by the classical spectral theory for a small set of diagonalisable matrices. Our solution in terms of new spectrum will be full and thus unavoidably coincides with one given by the Jordan normal form of matrices. Other more difficult questions are the subject of ongoing research.

6.1. Intertwining group actions on functions and operators

Any functional calculus uses properties of *functions* to model properties of *operators*. Thus changing our viewpoint on functions, as was done in Section 5, we can get another approach to operators. The two main possibilities are encoded in Definitions 4.20 and 4.22: we can assign a certain function to the given operator or *vice versa*. Here we consider the second possibility and treat the first in Subsection 6.4.

The representation ρ_1 (5.22) is unitary irreducible when it acts on the Hardy space H_2. Consequently we have one more reason to abolish the template Definition 6.1: H_2 is *not* an algebra. Instead we replace the *homomorphism property* by a *symmetric covariance*:

Definition 6.4 ([62]). A *contravariant analytic calculus* for an element $a \in \mathfrak{A}$ and an $\mathfrak{A}$-module M is a *continuous linear* mapping $\Phi : A(\mathbb{D}) \rightarrow A(\mathbb{D}, M)$ such that:

i. Φ is an *intertwining operator*

$$\Phi\rho_1 = \rho_a\Phi$$

between two representations of the $\mathrm{SL}_2(\mathbb{R})$ group ρ_1 (5.22) and ρ_a defined below in (6.4).

ii. There is an initialisation condition: $\Phi[v_0] = m$ for $v_0(z) \equiv 1$ and $m \in M$, where M is a left $\mathfrak{A}$-module.

Note that our functional calculus, released from the homomorphism condition, can take values in any left $\mathfrak{A}$-module M, which however could be $\mathfrak{A}$ itself if suitable. This adds much flexibility to our construction.

The earliest functional calculus, which is *not* an algebraic homomorphism, was the Weyl functional calculus and was defined just by an integral formula as an operator-valued distribution [3]. In that paper (joint) spectrum was defined as support of the Weyl calculus, i.e., as the set of points where this operator-valued distribution does not vanish. We also define the spectrum as a support of functional calculus, but due to our Definition 6.4 it will mean the set of non-vanishing intertwining operators with primary subrepresentations.

Definition 6.5. A corresponding *spectrum* of $a \in \mathfrak{A}$ is the *support* of the functional calculus Φ, i.e., the collection of intertwining operators of ρ_a with *primary representations* [55, § 8.3].

More variations of contravariant functional calculi are obtained from other groups and their representations [62, 66, 73, 75, 84].

A simple but important observation is that the Möbius transformations (1.1) can be easily extended to any Banach algebra. Let $\mathfrak{A}$ be a Banach algebra with the unit e, an element $a \in \mathfrak{A}$ with $\|a\| < 1$ be fixed, then

$$g : a \mapsto g \cdot a = (\bar{\alpha}a - \bar{\beta}e)(\alpha e - \beta a)^{-1}, \qquad g \in \mathrm{SL}_2(\mathbb{R}) \tag{6.2}$$

is a well-defined $\mathrm{SL}_2(\mathbb{R})$ action on a subset $\mathbb{A} = \{g \cdot a \mid g \in \mathrm{SL}_2(\mathbb{R})\} \subset \mathfrak{A}$, i.e., $\mathbb{A}$ is a $\mathrm{SL}_2(\mathbb{R})$-homogeneous space. Let us define the *resolvent* function $R(g, a) : \mathbb{A} \to \mathfrak{A}$. If

$$R(g, a) = (\alpha e - \beta a)^{-1}$$

then

$$R(g_1, \mathsf{a})R(g_2, g_1^{-1}\mathsf{a}) = R(g_1 g_2, \mathsf{a}). \tag{6.3}$$

The last identity is well known in representation theory [55, § 13.2(10)] and is a key ingredient of *induced representations*. Thus we can again linearise (6.2), cf. (5.22), in the space of continuous functions $C(\mathbb{A}, M)$ with values in a left $\mathfrak{A}$-module M, e.g., $M = \mathfrak{A}$:

$$\begin{aligned} \rho_a(g_1) : f(g^{-1} \cdot a) \quad &\mapsto \quad R(g_1^{-1}g^{-1}, a)f(g_1^{-1}g^{-1} \cdot a) \\ &= (\alpha' e - \beta' a)^{-1} f\left(\frac{\bar{\alpha}' \cdot a - \bar{\beta}' e}{\alpha' e - \beta' a}\right). \end{aligned} \tag{6.4}$$

For any $m \in M$ we can define a K-invariant *vacuum vector* as $v_m(g^{-1} \cdot a) = m \otimes v_0(g^{-1} \cdot a) \in C(\mathbb{A}, M)$. It generates the associated with v_m family of *coherent states* $v_m(u, a) = (ue - a)^{-1}m$, where $u \in \mathbb{D}$.

The *wavelet transform* defined by the same common formula based on coherent states (cf. (5.23)):

$$\mathcal{W}_m f(g) = \langle f, \rho_a(g) v_m \rangle, \tag{6.5}$$

is a version of the Cauchy integral, which maps $L_2(\mathbb{A})$ to $C(\mathrm{SL}_2(\mathbb{R}), M)$. It is closely related (but not identical!) to the Riesz-Dunford functional calculus: the traditional functional calculus is given by the case

$$\Phi : f \mapsto \mathcal{W}_m f(0) \qquad \text{for } M = \mathfrak{A} \text{ and } m = e.$$

Both conditions – the intertwining property and initial value – required by Definition 6.4 easily follow from our construction. Finally, we wish to provide an example of application of Corollary 5.8.

Example 6.6. Let a be an operator and ϕ be a function which annihilates it, i.e., $\phi(a) = 0$. For example, if a is a matrix ϕ can be its minimal polynomial. From the integral representation of the contravariant calculus on $G = \mathrm{SL}_2(\mathbb{R})$ we can rewrite the annihilation property like this:

$$\int_G \phi(g) R(g, a)\, dg = 0.$$

Then the vector-valued function $[\mathcal{W}_m f](g)$ defined by (6.5) shall satisfy the following condition:

$$\int_G \phi(g')\, [\mathcal{W}_m f](gg')\, dg' = 0$$

due to Corollary 5.8.

6.2. Jet bundles and prolongations of ρ_1

Spectrum was defined in 6.5 as the *support* of our functional calculus. To elaborate its meaning we need the notion of a *prolongation* of representations introduced by S. Lie, see [107, 108] for a detailed exposition.

Definition 6.7. [108, Chap. 4] Two holomorphic functions have nth *order contact* in a point if their value and their first n derivatives agree at that point, in other words their Taylor expansions are the same in their first $n + 1$ terms.

A point $(z, u^{(n)}) = (z, u, u_1, \ldots, u_n)$ of the *jet space* $\mathbb{J}^n \sim \mathbb{D} \times \mathbb{C}^n$ is the equivalence class of holomorphic functions having nth contact at the point z with the polynomial:

$$p_n(w) = u_n \frac{(w - z)^n}{n!} + \cdots + u_1 \frac{(w - z)}{1!} + u. \tag{6.6}$$

For a fixed n each holomorphic function $f : \mathbb{D} \to \mathbb{C}$ has nth *prolongation* (or *n-jet*) $\mathrm{j}_n f : \mathbb{D} \to \mathbb{C}^{n+1}$:

$$\mathrm{j}_n f(z) = (f(z), f'(z), \ldots, f^{(n)}(z)). \tag{6.7}$$

The graph $\Gamma_f^{(n)}$ of $\mathrm{j}_n f$ is a submanifold of $\mathbb{J}^n$ which is a section of the *jet bundle* over $\mathbb{D}$ with a fibre $\mathbb{C}^{n+1}$. We also introduce a notation J_n for the map $J_n : f \mapsto \Gamma_f^{(n)}$ of a holomorphic f to the graph $\Gamma_f^{(n)}$ of its n-jet $\mathrm{j}_n f(z)$ (6.7).

One can prolong any map of functions $\psi : f(z) \mapsto [\psi f](z)$ to a map $\psi^{(n)}$ of n-jets by the formula

$$\psi^{(n)}(J_n f) = J_n(\psi f). \tag{6.8}$$

For example such a prolongation $\rho_1^{(n)}$ of the representation ρ_1 of the group $\mathrm{SL}_2(\mathbb{R})$ in $H_2(\mathbb{D})$ (as any other representation of a Lie group [108]) will be again a representation of $\mathrm{SL}_2(\mathbb{R})$. Equivalently we can say that J_n *intertwines* ρ_1 and $\rho_1^{(n)}$:

$$J_n \rho_1(g) = \rho_1^{(n)}(g) J_n \quad \text{for all } g \in \mathrm{SL}_2(\mathbb{R}).$$

Of course, the representation $\rho_1^{(n)}$ is not irreducible: any jet subspace $\mathbb{J}^k$, $0 \leq k \leq n$ is a $\rho_1^{(n)}$-invariant subspace of $\mathbb{J}^n$. However the representations $\rho_1^{(n)}$ are *primary* [55, § 8.3] in the sense that they are not sums of two subrepresentations.

The following statement explains why jet spaces appeared in our study of functional calculus.

Proposition 6.8. *Let matrix a be a Jordan block of length k with the eigenvalue $\lambda = 0$, and m be its root vector of order k, i.e., $a^{k-1}m \neq a^k m = 0$. Then the restriction of ρ_a on the subspace generated by v_m is equivalent to the representation ρ_1^k.*

6.3. Spectrum and spectral mapping theorem

Now we are prepared to describe a spectrum of a matrix. Since the functional calculus is an intertwining operator, its support is a decomposition into intertwining operators with primary representations (we can not expect generally that these primary subrepresentations are irreducible).

Recall the transitive on $\mathbb{D}$ group of inner automorphisms of $\mathrm{SL}_2(\mathbb{R})$, which can send any $\lambda \in \mathbb{D}$ to 0 and are actually parametrised by such a λ. This group extends Proposition 6.8 to the complete characterisation of ρ_a for matrices.

Proposition 6.9. *Representation ρ_a is equivalent to a direct sum of the prolongations $\rho_1^{(k)}$ of ρ_1 in the kth jet space $\mathbb{J}^k$ intertwined with inner automorphisms. Consequently the* spectrum *of a (defined via the functional calculus $\Phi = \mathcal{W}_m$) is labelled exactly by n pairs of numbers (λ_i, k_i), $\lambda_i \in \mathbb{D}$, $k_i \in \mathbb{Z}_+$, $1 \leq i \leq n$ some of which could coincide.*

Obviously this spectral theory is a fancy restatement of the *Jordan normal form* of matrices.

Example 6.10. Let $J_k(\lambda)$ denote the Jordan block of length k for the eigenvalue λ. In Figure 11 there are two pictures of the spectrum for the matrix

$$a = J_3(\lambda_1) \oplus J_4(\lambda_2) \oplus J_1(\lambda_3) \oplus J_2(\lambda_4),$$

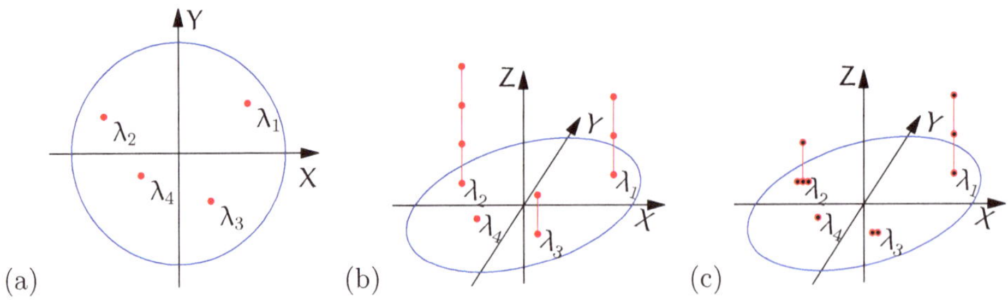

FIGURE 11. Classical spectrum of the matrix from Example 6.10 is shown at (a). Contravariant spectrum of the same matrix in the jet space is drawn at (b). The image of the contravariant spectrum under the map from Example 6.12 is presented at (c).

where

$$\lambda_1 = \frac{3}{4}e^{i\pi/4}, \quad \lambda_2 = \frac{2}{3}e^{i5\pi/6}, \quad \lambda_3 = \frac{2}{5}e^{-i3\pi/4}, \quad \lambda_4 = \frac{3}{5}e^{-i\pi/3}.$$

Part (a) represents the conventional two-dimensional image of the spectrum, i.e., eigenvalues of a, and (b) describes spectrum $\mathbf{sp}\, a$ arising from the wavelet construction. The first image did not allow one to distinguish a from many other essentially different matrices, e.g., the diagonal matrix

$$\operatorname{diag}(\lambda_1, \lambda_2, \lambda_3, \lambda_4),$$

which even have a different dimensionality. At the same time Figure 11(b) completely characterises a up to a similarity. Note that each point of $\mathbf{sp}\, a$ in Figure 11(b) corresponds to a particular root vector, which spans a primary subrepresentation.

As was mentioned in the beginning of this section, a resonable spectrum should be linked to the corresponding functional calculus by an appropriate spectral mapping theorem. The new version of spectrum is based on prolongation of ρ_1 into jet spaces (see Section 6.2). Naturally a correct version of the spectral mapping theorem should also operate in jet spaces.

Let $\phi : \mathbb{D} \to \mathbb{D}$ be a holomorphic map and let us define its action on functions $[\phi_* f](z) = f(\phi(z))$. According to the general formula (6.8) we can define the prolongation $\phi_*^{(n)}$ onto the jet space $\mathbb{J}^n$. Its associated action $\rho_1^k \phi_*^{(n)} = \phi_*^{(n)} \rho_1^n$ on the pairs (λ, k) is given by the formula

$$\phi_*^{(n)}(\lambda, k) = \left(\phi(\lambda), \left[\frac{k}{\deg_\lambda \phi}\right]\right), \tag{6.9}$$

where $\deg_\lambda \phi$ denotes the degree of zero of the function $\phi(z) - \phi(\lambda)$ at the point $z = \lambda$ and $[x]$ denotes the integer part of x.

Theorem 6.11 (Spectral mapping). *Let ϕ be a holomorphic mapping $\phi : \mathbb{D} \to \mathbb{D}$ and its prolonged action $\phi_*^{(n)}$ be defined by (6.9), then*

$$\mathbf{sp}\,\phi(a) = \phi_*^{(n)} \mathbf{sp}\, a.$$

The explicit expression of (6.9) for $\phi_*^{(n)}$, which involves derivatives of ϕ up to nth order, is known, see for example [41, Thm. 6.2.25], but was not recognised before as a form of spectral mapping.

Example 6.12. Let us continue with Example 6.10. Let ϕ map all four eigenvalues $\lambda_1, \ldots, \lambda_4$ of the matrix a into themselves. Then Figure 11(a) will represent the classical spectrum of $\phi(a)$ as well as a.

However Figure 11(c) shows that the mapping of the new spectrum for the case ϕ has *orders of zeros* at these points as follows: the order 1 at λ_1, exactly the order 3 at λ_2, an order at least 2 at λ_3, and finally any order at λ_4.

6.4. Functional model and spectral distance

Let a be a matrix and $\mu(z)$ be its *minimal polynomial*:

$$\mu_a(z) = (z - \lambda_1)^{m_1} \cdot \ldots \cdot (z - \lambda_n)^{m_n}.$$

If all eigenvalues λ_i of a (i.e., all roots of $\mu(z)$ belong to the unit disk we can consider the respective *Blaschke product*

$$B_a(z) = \prod_{i=1}^{n} \left(\frac{z - \lambda_i}{1 - \overline{\lambda_i} z} \right)^{m_i},$$

such that its numerator coincides with the minimal polynomial $\mu(z)$. Moreover, for a unimodular z we have $B_a(z) = \mu_a(z) \overline{\mu_a}^{-1}(z) z^{-m}$, where $m = m_1 + \cdots + m_n$. We also have the following covariance property:

Proposition 6.13. *The above correspondence $a \mapsto B_a$ intertwines the* $\mathrm{SL}_2(\mathbb{R})$ *action* (6.2) *on matrices with the action* (5.22) *with $k = 0$ on functions.*

The result follows from the observation that every elementary product $\frac{z-\lambda_i}{1-\overline{\lambda_i} z}$ is the Möbius transformation of z with the matrix $\begin{pmatrix} 1 & -\lambda_i \\ -\overline{\lambda_i} & 1 \end{pmatrix}$. Thus the correspondence $a \mapsto B_a(z)$ is a covariant (symbolic) calculus in the sense of Definition 4.20. See also Example 4.19.

The Jordan normal form of a matrix provides a description which is equivalent to its contravariant spectrum. From various viewpoints, e.g., numerical approximations, it is worth considering its stability under a perturbation. It is easy to see that an arbitrarily small disturbance breaks the Jordan structure of a matrix. However, the result of a random small perturbation will not be random; its nature is described by the following remarkable theorem:

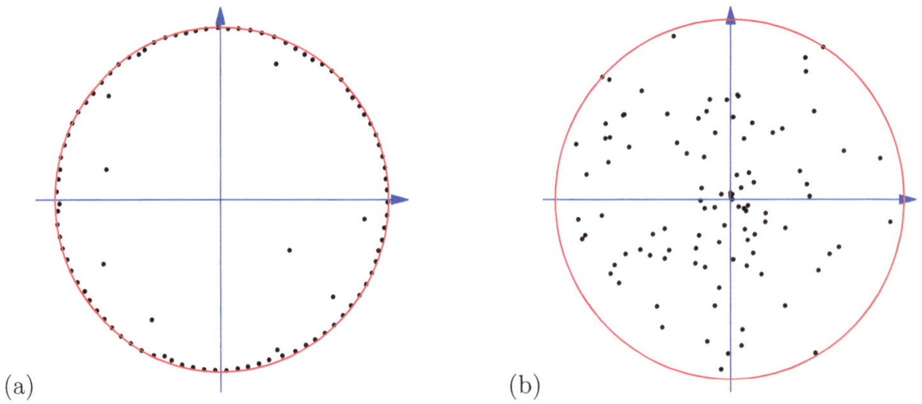

FIGURE 12. Perturbation of the Jordan block's spectrum: (a) The spectrum of the perturbation $J_{100} + \varepsilon^1 00K$ of the Jordan block J_{100} by a random matrix K. (b) The spectrum of the random matrix K.

Theorem 6.14 (Lidskii [100], see also [104]). *Let J_n be a Jordan block of a length $n > 1$ with zero eigenvalues and K be an arbitrary matrix. Then eigenvalues of the perturbed matrix $J_n + \varepsilon^n K$ admit the expansion*

$$\lambda_j = \varepsilon \xi^{1/n} + o(\varepsilon), \qquad j = 1, \ldots, n,$$

where $\xi^{1/n}$ represents all nth complex roots of certain $\xi \in \mathbb{C}$.

The left-hand picture in Figure 12 presents a perturbation of a Jordan block J_{100} by a random matrix. Perturbed eigenvalues are close to vertices of a right polygon with 100 vertices. Those regular arrangements occur despite the fact that eigenvalues of the matrix K are dispersed through the unit disk (the right-hand picture in Fig. 12). In a sense it is rather that the Jordan block regularises eigenvalues of K than that K perturbs the eigenvalue of the Jordan block.

Although the Jordan structure itself is extremely fragile, it still can be easily guessed from a perturbed eigenvalue. Thus there exists a certain characterisation of matrices which is stable under small perturbations. We will describe a sense in which the covariant spectrum of the matrix $J_n + \varepsilon^n K$ is stable for small ε. For this we introduce the covariant version of spectral distances motivated by the functional model. Our definition is different from other types known in the literature [123, Ch. 5].

Definition 6.15. Let a and b be two matrices with all their eigenvalues sitting inside of the unit disk and $B_a(z)$ and $B_b(z)$ be respective Blaschke products as defined above. The (*covariant*) *spectral distance* $d(a, b)$ between a and b is equal to the distance $\|B_a - B_b\|_2$ between $B_a(z)$ and $B_b(z)$ in the Hardy space on the unit circle.

Since the spectral distance is defined through the distance in H_2, all standard axioms of a distance are automatically satisfied. For a Blaschke product we have $|B_a(z)| = 1$ if $|z| = 1$, thus $\|B_a\|_p = 1$ in any L_p on the unit circle. Therefore an alternative expression for the spectral distance is

$$d(a,b) = 2(1 - \langle B_a, B_b \rangle).$$

In particular, we always have $0 \leq d(a,b) \leq 2$. We get an obvious consequence of Proposition 6.13, which justifies the name of the covariant spectral distance:

Corollary 6.16. *For any $g \in \mathrm{SL}_2(\mathbb{R})$ we have $d(a,b) = d(g \cdot a, g \cdot a)$, where $\cdot$ denotes the Möbius action* (6.2).

An important property of the covariant spectral distance is its stability under small perturbations.

Theorem 6.17. *For $n = 2$ let $\lambda_1(\varepsilon)$ and $\lambda_2(\varepsilon)$ be eigenvalues of the matrix $J_2 + \varepsilon^2 \cdot K$ for some matrix K. Then*

$$|\lambda_1(\varepsilon)| + |\lambda_2(\varepsilon)| = O(\varepsilon), \qquad \textit{however} \qquad |\lambda_1(\varepsilon) + \lambda_2(\varepsilon)| = O(\varepsilon^2). \tag{6.10}$$

The spectral distance from the 1-jet at 0 to two 0-jets at points λ_1 and λ_2 bounded only by the first condition in (6.10) *is $O(\varepsilon^2)$. However the spectral distance between J_2 and $J_2 + \varepsilon^2 \cdot K$ is $O(\varepsilon^4)$.*

In other words, a matrix with eigenvalues satisfying the Lisdkii condition from Theorem 6.14 is much closer to the Jordan block J_2 than a generic one with eigenvalues of the same order. Thus the covariant spectral distance is more stable under perturbation that magnitude of eigenvalues. For $n = 2$ a proof can be forced through a direct calculation. We also conjecture that a similar statement is true for any $n \geq 2$.

6.5. Covariant pencils of operators

Let H be a real Hilbert space, possibly of finite dimensionality. For bounded linear operators A and B consider the *generalised eigenvalue problem*, that is finding a scalar λ and a vector $x \in H$ such that

$$Ax = \lambda B x \qquad \text{or equivalently} \qquad (A - \lambda B)x = 0. \tag{6.11}$$

The standard eigenvalue problem corresponds to the case $B = I$, moreover for an invertible B the generalised problem can be reduced to the standard one for the operator $B^{-1}A$. Thus it is sensible to introduce an equivalence relation on the pairs of operators:

$$(A,B) \sim (DA, DB) \quad \text{for any invertible operator } D. \tag{6.12}$$

We may treat the pair (A,B) as a column vector $\begin{pmatrix} A \\ B \end{pmatrix}$. Then there is an action of the $\mathrm{SL}_2(\mathbb{R})$ group on the pairs:

$$g \cdot \begin{pmatrix} A \\ B \end{pmatrix} = \begin{pmatrix} aA + bB \\ cA + dB \end{pmatrix}, \qquad \text{where } g = \begin{pmatrix} a & b \\ c & d \end{pmatrix} \in \mathrm{SL}_2(\mathbb{R}). \tag{6.13}$$

If we consider this $\mathrm{SL}_2(\mathbb{R})$-action subject to the equivalence relation (6.12) then we will arrive at a version of the linear-fractional transformation of the operator defined in (6.2). There is a connection of the $\mathrm{SL}_2(\mathbb{R})$-action (6.13) to the problem (6.11) through the following intertwining relation:

Proposition 6.18. *Let λ and $x \in H$ solve the generalised eigenvalue problem* (6.11) *for the pair (A, B). Then the pair $(C, D) = g \cdot (A, B)$, $g \in \mathrm{SL}_2(\mathbb{R})$ has a solution μ and x, where*

$$\mu = g \cdot \lambda = \frac{a\lambda + b}{c\lambda + d}, \qquad \text{for } g = \begin{pmatrix} a & b \\ c & d \end{pmatrix} \in \mathrm{SL}_2(\mathbb{R}),$$

is defined by the Möbius transformation (1.1).

In other words the correspondence

$$(A, B) \mapsto \text{all generalised eigenvalues}$$

is another realisation of a covariant calculus in the sense of Definition 4.20. The collection of all pairs $g \cdot (A, B)$, $g \in \mathrm{SL}_2(\mathbb{R})$ is an example of a *covariant pencil* of operators. This set is an $\mathrm{SL}_2(\mathbb{R})$-homogeneous space, thus it shall be within the classification of such homogeneous spaces provided in Subsection 2.1.

Example 6.19. It is easy to demonstrate that all existing homogeneous spaces can be realised by matrix pairs.

i. Take the pair (O, I) where O and I are the zero and identity $n \times n$ matrices respectively. Then any transformation of this pair by a lower-triangular matrix from $\mathrm{SL}_2(\mathbb{R})$ is equivalent to (O, I). The respective homogeneous space is isomorphic to the real line with the Möbius transformations (1.1).
ii. Consider $H = \mathbb{R}^2$. Using the notations ι from Subsection 1.1 we define three realisations (elliptic, parabolic and hyperbolic) of an operator A_ι:

$$A_{\mathrm{i}} = \begin{pmatrix} 0 & 1 \\ -1 & 0 \end{pmatrix}, \qquad A_{\varepsilon} = \begin{pmatrix} 0 & 1 \\ 0 & 0 \end{pmatrix}, \qquad A_{\mathrm{j}} = \begin{pmatrix} 0 & 1 \\ 1 & 0 \end{pmatrix}. \tag{6.14}$$

Then for an arbitrary element h of the subgroup K, N or A the respective (in the sense of Principle 3.5) pair $h \cdot (A_\iota, I)$ is equivalent to (A_ι, I) itself. Thus those three homogeneous spaces are isomorphic to the elliptic, parabolic and hyperbolic half-planes under respective actions of $\mathrm{SL}_2(\mathbb{R})$. Note, that $A_\iota^2 = \iota^2 I$, that is A_ι is a model for hypercomplex units.
iii. Let A be a direct sum of any two different matrices out of the three A_ι from (6.14), then the fix group of the equivalence class of the pair (A, I) is the identity of $\mathrm{SL}_2(\mathbb{R})$. Thus the corresponding homogeneous space coincides with the group itself.

Having homogeneous spaces generated by pairs of operators, we can define respective functions on those spaces. Special attention is due the following paraphrase of the resolvent:

$$R_{(A,B)}(g) = (cA + dB)^{-1} \qquad \text{where} \quad g^{-1} = \begin{pmatrix} a & b \\ c & d \end{pmatrix} \in SL_2(\mathbb{R}).$$

Obviously $R_{(A,B)}(g)$ contains the essential information about the pair (A, B). Probably, the function $R_{(A,B)}(g)$ contains too much simultaneous information, so we may restrict it to get a more detailed view. For vectors u, $v \in H$ we also consider vector- and scalar-valued functions related to the generalised resolvent

$$R^u_{(A,B)}(g) = (cA + dB)^{-1}u, \qquad \text{and} \qquad R^{(u,v)}_{(A,B)}(g) = \left\langle (cA + dB)^{-1}u, v\right\rangle,$$

where $(cA+dB)^{-1}u$ is understood as a solution w of the equation $u = (cA+dB)w$ if it exists and is unique; this does not require the full invertibility of $cA + dB$.

It is easy to see that the map $(A, B) \mapsto R^{(u,v)}_{(A,B)}(g)$ is a covariant calculus as well. It is worth noticing that function $R_{(A,B)}$ can again fall into three EPH cases.

Example 6.20. For the three matrices A_ι considered in the previous example we denote by $R_\iota(g)$ the resolvent-type function of the pair (A_ι, I). Then

$$\begin{aligned} R_{\mathrm{i}}(g) &= \frac{1}{c^2+d^2}\begin{pmatrix} d & -c \\ c & d \end{pmatrix}, \\ R_{\varepsilon}(g) &= \frac{1}{d^2}\begin{pmatrix} d & -c \\ 0 & d \end{pmatrix}, \\ R_{\mathrm{j}}(g) &= \frac{1}{d^2-c^2}\begin{pmatrix} d & -c \\ -c & d \end{pmatrix}. \end{aligned}$$

Put $u = (1, 0) \in H$, then $R_\iota(g)u$ is a two-dimensional real vector-valued function with components equal to the real and imaginary parts of the hypercomplex Cauchy kernel considered in [86].

Consider the space $L(G)$ of functions spanned by all left translations of $R_{(A,B)}(g)$. As usual, a closure in a suitable metric, say L_p, can be taken. The left action $g : f(h) \mapsto f(g^{-1}h)$ of $SL_2(\mathbb{R})$ on this space is a linear representation of this group. Afterwards the representation can be decomposed into a sum of primary subrepresentations.

Example 6.21. For the matrices A_ι the irreducible components are isomorphic to analytic spaces of hypercomplex functions under the fraction-linear transformations built in Subsection 3.2.

An important observation is that a decomposition into irreducible or primary components can reveal an EPH structure even in the cases that hide it on the homogeneous space level.

Example 6.22. Take the operator $A = A_{\mathrm{i}} \oplus A_{\mathrm{j}}$ from Example 6.19(iii). The corresponding homogeneous space coincides with the entire $\mathrm{SL}_2(\mathbb{R})$. However if we take two vectors $u_{\mathrm{i}} = (1, 0) \oplus (0, 0)$ and $u_{\mathrm{j}} = (0, 0) \oplus (1, 0)$, then the respective linear spaces generated by functions $R_A(g)u_{\mathrm{i}}$ and $R_A(g)u_{\mathrm{j}}$ will be of elliptic and hyperbolic types respectively.

Let us briefly consider a *quadratic eigenvalue* problem: for given operators (matrices) A_0, A_1 and A_2 from $B(H)$, find a scalar λ and a vector $x \in H$ such that

$$Q(\lambda)x = 0, \qquad \text{where} \quad Q(\lambda) = \lambda^2 A_2 + \lambda A_1 + A_0. \tag{6.15}$$

There is a connection with our study of conic sections from Subsection 2.2 which we will only hint at now. Comparing (6.15) with the equation of the cycle (2.7) we can associate the respective Fillmore–Springer–Cnops–type matrix to $Q(\lambda)$, cf. (2.8):

$$Q(\lambda) = \lambda^2 A_2 + \lambda A_1 + A_0 \quad \longleftrightarrow \quad C_Q = \begin{pmatrix} A_1 & A_0 \\ A_2 & -A_1 \end{pmatrix}. \tag{6.16}$$

Then we can state the following analogue of Theorem 2.4 for the quadratic eigenvalues:

Proposition 6.23. *Let two quadratic matrix polynomials Q and $\tilde{Q}$ be such that their FSC matrices* (6.16) *are conjugated $C_{\tilde{Q}} = gC_Q g^{-1}$ by an element $g \in \mathrm{SL}_2(\mathbb{R})$. Then λ is a solution of the quadratic eigenvalue problem for Q and $x \in H$ if and only if $\mu = g \cdot \lambda$ is a solution of the quadratic eigenvalue problem for $\tilde{Q}$ and x. Here $\mu = g \cdot \lambda$ is the Möbius transformation* (1.1) *associated to $g \in \mathrm{SL}_2(\mathbb{R})$.*

So quadratic matrix polynomials are non-commuting analogues of the cycles and it would be exciting to extend the geometry from Section 2 to this non-commutative setting as much as possible.

Remark 6.24. It is beneficial to extend a notion of a scalar in an (generalised) eigenvalue problem to an abstract field or ring. For example, we can consider pencils of operators/matrices with polynomial coefficients. In many circumstances we may factorise the polynomial ring by an ideal generated by a collection of algebraic equations. Our work with hypercomplex units is the most elementary realisation of this setup. Indeed, the algebra of hypercomplex numbers with the hypercomplex unit ι is a realisation of the polynomial ring in a variable t factored by the single quadratic relation $t^2 + \sigma = 0$, where $\sigma = \iota^2$.

7. Quantum mechanics

Complex-valued representations of the Heisenberg group (also known as Weyl or Heisenberg-Weyl group) provide a natural framework for quantum mechanics [31, 43]. This is the most fundamental example of the Kirillov orbit method, induced representations and geometrical quantisation technique [56, 57]. Following the presentation in Section 3 we will consider representations of the Heisenberg group which are induced by hypercomplex characters of its centre: complex (which correspond to the elliptic case), dual (parabolic) and double (hyperbolic).

To describe dynamics of a physical system we use a universal equation based on inner derivations (commutator) of the convolution algebra [70, 74]. The complex-valued representations produce the standard framework for quantum mechanics with the Heisenberg dynamical equation [126].

The double number-valued representations, with the hyperbolic unit $\mathrm{j}^2 = 1$, is a natural source of hyperbolic quantum mechanics that has been developed for some time [45, 46, 49, 51, 52]. The universal dynamical equation employs a hyperbolic commutator in this case. This can be seen as a *Moyal bracket* based on the hyperbolic sine function. The hyperbolic observables act as operators on a Krein space with an indefinite inner product. Such spaces are employed in studies of $\mathcal{PT}$-symmetric Hamiltonians and the hyperbolic unit $\mathrm{j}^2 = 1$ naturally appears in this setup [38].

Representations with values in dual numbers provide a convenient description of classical mechanics. For this we do not take any sort of semiclassical limit, rather the nilpotency of the parabolic unit ($\varepsilon^2 = 0$) does the task. This removes the vicious necessity to consider the Planck *constant* tending to zero. The dynamical equation takes the Hamiltonian form. We also describe classical non-commutative representations of the Heisenberg group which acts in the first jet space.

Remark 7.1. It is worth noting that our technique is different from the contraction technique in the theory of Lie groups [37, 99]. Indeed a contraction of the Heisenberg group $\mathbb{H}^n$ is the commutative Euclidean group $\mathbb{R}^{2n}$ which does not recreate either quantum or classical mechanics.

The approach provides not only three different types of dynamics, it also generates the respective rules for addition of probabilities as well. For example, quantum interference is a consequence of the same complex-valued structure that directs the Heisenberg equation. The absence of an interference (a particle behaviour) in classical mechanics is again a consequence of the nilpotency of the parabolic unit. Double numbers created the hyperbolic law of additions of probabilities, which has been extensively investigated [49, 51]. There are still unresolved issues with positivity of the probabilistic interpretation in the hyperbolic case [45, 46].

Remark 7.2. Since Dirac's paper [25], it has been commonly accepted that the striking (or even *the only*) difference between quantum and classical mechanics is non-commutativity of observables in the first case. In particular the Heisenberg commutation relations (7.5) imply the uncertainty principle, the Heisenberg equation of motion and other quantum features. However, the entire book of Feynman on QED [30] does not contain any reference to non-commutativity. Moreover, our work shows that there is a non-commutative formulation of classical mechanics. Non-commutative representations of the Heisenberg group in dual numbers implies the Poisson dynamical equation and local addition of probabilities in Section 7.6, which are completely classical.

This entirely dispels any illusive correlation between classical/quantum and commutative/non-commutative. Instead we show that quantum mechanics is fully determined by the properties of complex numbers. In Feynman's exposition [30] complex numbers are presented by a clock, rotations of its arm encode multiplications by unimodular complex numbers. Moreover, there is no presentation of quantum mechanics that does not employ complex phases (numbers) in one or another form. Analogous parabolic and hyperbolic phases (or characters produced by associated hypercomplex numbers, see Section 3.1) lead to classical and hypercomplex mechanics respectively.

This section clarifies foundations of quantum and classical mechanics. We recovered the existence of three non-isomorphic models of mechanics from the representation theory. They were already derived in [45,46] from a translation invariant formulation, that is from the group theory as well. It was also hinted that the hyperbolic counterpart is (at least theoretically) as natural as classical and quantum mechanics are. This approach provides a framework for a description of an aggregate system which could have both quantum and classical components. Such a framework can be used to model quantum computers with classical terminals [80].

Remarkably, simultaneously with the work in [45], group-invariant axiomatics of geometry led R.I. Pimenov [114] to a description of 3^n Cayley–Klein constructions. The connections between group-invariant geometry and respective mechanics were explored in many works of N.A. Gromov, see for example [35–37]. They already highlighted the rôle of three types of hypercomplex units for the realisation of elliptic, parabolic and hyperbolic geometry and kinematics.

There is a further connection between representations of the Heisenberg group and hypercomplex numbers. The symplectomorphisms of phase space are also automorphisms of the Heisenberg group [31, § 1.2]. We recall that the symplectic group $\mathrm{Sp}(2)$ [31, § 1.2] is isomorphic to the group $\mathrm{SL}_2(\mathbb{R})$ [44,97,102] and provides linear symplectomorphisms of the two-dimensional phase space. It has three types of non-isomorphic one-dimensional continuous subgroups (2.4–2.6) with symplectic action on the phase space illustrated by Figure 9. Hamiltonians, which produce those symplectomorphisms, are of interest [121; 122; 127, § 3.8]. An analysis of those Hamiltonians from Subsection 3.3 by means of ladder operators recreates hypercomplex coefficients as well [87].

Harmonic oscillators, which we shall use as the main illustration here, are treated in most textbooks on quantum mechanics. This is efficiently done through creation/annihilation (ladder) operators, cf. § 3.3 and [13, 33]. The underlying structure is the representation theory of the Heisenberg and symplectic groups [31; 43; 97, § VI.2; 120, § 8.2]. As we will see, they are naturally connected with respective hypercomplex numbers. As a result we obtain further illustrations of the similarity and correspondence Principle 3.5.

We work with the simplest case of a particle with only one degree of freedom. Higher dimensions and the respective group of symplectomorphisms $\mathrm{Sp}(2n)$ may require consideration of Clifford algebras [20,21,38,60,115].

7.1. The Heisenberg group and its automorphisms

7.1.1. The Heisenberg group and induced representations. Let (s, x, y), where s, x, $y \in \mathbb{R}$, be an element of the one-dimensional *Heisenberg group* $\mathbb{H}^1$ [31, 43]. Consideration of the general case of $\mathbb{H}^n$ will be similar, but is beyond the scope of the present discussion. The group law on $\mathbb{H}^1$ is given as follows:

$$(s, x, y) \cdot (s', x', y') = (s + s' + \tfrac{1}{2}\omega(x, y; x', y'), x + x', y + y'), \tag{7.1}$$

where the non-commutativity is due to ω – the *symplectic form* on $\mathbb{R}^{2n}$, which is the central object of classical mechanics [4, § 37]:

$$\omega(x, y; x', y') = xy' - x'y. \tag{7.2}$$

The Heisenberg group is a non-commutative Lie group with centre

$$Z = \{(s, 0, 0) \in \mathbb{H}^1,\ s \in \mathbb{R}\}.$$

The left shifts

$$\Lambda(g) : f(g') \mapsto f(g^{-1}g') \tag{7.3}$$

act as a representation of $\mathbb{H}^1$ on a certain linear space of functions. For example, an action on $L_2(\mathbb{H}, dg)$ with respect to the Haar measure $dg = ds\,dx\,dy$ is the *left regular* representation, which is unitary.

The Lie algebra $\mathfrak{h}^n$ of $\mathbb{H}^1$ is spanned by left-(right-)invariant vector fields

$$S^{l(r)} = \pm\partial_s, \quad X^{l(r)} = \pm\partial_x - \tfrac{1}{2}y\partial_s, \quad Y^{l(r)} = \pm\partial_y + \tfrac{1}{2}x\partial_s \tag{7.4}$$

on $\mathbb{H}^1$ with the Heisenberg *commutator relation*

$$[X^{l(r)}, Y^{l(r)}] = S^{l(r)} \tag{7.5}$$

and all other commutators vanishing. We will sometimes omit the superscript l for a left-invariant field.

We can construct linear representations of $\mathbb{H}^1$ by induction [55, § 13] from a character χ of the centre Z. Here we prefer the following one, cf. § 3.2 and [55, § 13; 120, Ch. 5]. Let $F_2^\chi(\mathbb{H}^n)$ be the space of functions on $\mathbb{H}^n$ having the properties

$$f(gh) = \chi(h)f(g), \qquad \text{for all } g \in \mathbb{H}^n,\ h \in Z \tag{7.6}$$

and

$$\int_{\mathbb{R}^{2n}} |f(0, x, y)|^2\, dx\, dy < \infty. \tag{7.7}$$

Then $F_2^\chi(\mathbb{H}^n)$ is invariant under left shifts and those shifts restricted to $F_2^\chi(\mathbb{H}^n)$ make a representation ρ_χ of $\mathbb{H}^n$ induced by χ.

If the character χ is unitary, then the induced representation is unitary as well. However the representation ρ_χ is not necessarily irreducible. Indeed, left shifts commute with the right action of the group. Thus any subspace of null-solutions of a linear combination $aS + \sum_{j=1}^{n}(b_jX_j + c_jY_j)$ of left-invariant vector fields is left-invariant and we can restrict ρ_χ to this subspace. The left-invariant differential operators define analytic conditions for functions, cf. Corollary 5.6.

Example 7.3. The function $f_0(s, x, y) = e^{\mathrm{i}hs - h(x^2+y^2)/4}$, where $h = 2\pi\hbar$, belongs to $F_2^\chi(\mathbb{H}^n)$ for the character $\chi(s) = e^{\mathrm{i}hs}$. It is also a null solution for all the operators $X_j - \mathrm{i}Y_j$. The closed linear span of functions $f_g = \Lambda(g)f_0$ is invariant under left shifts and provides a model for Fock–Segal–Bargmann (FSB) type representation of the Heisenberg group, which will be considered below.

7.1.2. Symplectic automorphisms of the Heisenberg group. The group of outer automorphisms of $\mathbb{H}^1$, which trivially acts on the centre of $\mathbb{H}^1$, is the symplectic group $\mathrm{Sp}(2)$. It is the group of symmetries of the symplectic form ω in (7.1) [31, Thm. 1.22; 42, p. 830]. The symplectic group is isomorphic to $\mathrm{SL}_2(\mathbb{R})$ considered in the first half of this work. The explicit action of $\mathrm{Sp}(2)$ on the Heisenberg group is

$$g : h = (s, x, y) \mapsto g(h) = (s, x', y'), \tag{7.8}$$

where

$$g = \begin{pmatrix} a & b \\ c & d \end{pmatrix} \in \mathrm{Sp}(2), \quad \text{and} \quad \begin{pmatrix} x' \\ y' \end{pmatrix} = \begin{pmatrix} a & b \\ c & d \end{pmatrix} \begin{pmatrix} x \\ y \end{pmatrix}.$$

The Shale–Weil theorem [31, § 4.2; 42, p. 830] states that any representation $\rho_\hbar$ of the Heisenberg groups generates a unitary *oscillator* (or *metaplectic*) representation $\rho_\hbar^{\mathrm{SW}}$ of the $\widetilde{\mathrm{Sp}}(2)$, the two-fold cover of the symplectic group [31, Thm. 4.58].

We can consider the semidirect product $G = \mathbb{H}^1 \rtimes \widetilde{\mathrm{Sp}}(2)$ with the standard group law

$$(h, g) * (h', g') = (h * g(h'), g * g'), \qquad \text{where } h, h' \in \mathbb{H}^1, \quad g, g' \in \widetilde{\mathrm{Sp}}(2), \tag{7.9}$$

and the stars denote the respective group operations while the action $g(h')$ is defined as the composition of the projection map $\widetilde{\mathrm{Sp}}(2) \to \mathrm{Sp}(2)$ and the action (7.8). This group is sometimes called the *Schrödinger group* and is known as the maximal kinematical invariance group of both the free Schrödinger equation and the quantum harmonic oscillator [105]. This group is of interest not only in quantum mechanics but also in optics [121, 122]. The Shale–Weil theorem allows us to expand any representation $\rho_\hbar$ of the Heisenberg group to the representation $\rho_\hbar^2 = \rho_\hbar \oplus \rho_\hbar^{\mathrm{SW}}$ of the group G.

Consider the Lie algebra $\mathfrak{sp}_2$ of the group $\mathrm{Sp}(2)$. We again use the basis A, B, Z (3.12) with commutators (3.13). Vectors Z, $B - Z/2$ and B are generators of the one-parameter subgroups K, N' and A' (2.4–2.6) respectively. Furthermore we can consider the basis $\{S, X, Y, A, B, Z\}$ of the Lie algebra $\mathfrak{g}$ of the Lie group $G = \mathbb{H}^1 \rtimes \widetilde{\mathrm{Sp}}(2)$. All non-zero commutators besides those already listed in (7.5) and (3.13) are:

$$[A, X] = \tfrac{1}{2} X, \qquad [B, X] = -\tfrac{1}{2} Y, \qquad [Z, X] = Y; \tag{7.10}$$

$$[A, Y] = -\tfrac{1}{2} Y, \qquad [B, Y] = -\tfrac{1}{2} X, \qquad [Z, Y] = -X. \tag{7.11}$$

Of course, there is the derived form of the Shale–Weil representation for $\mathfrak{g}$. It can often be explicitly written in contrast to the Shale–Weil representation.

Example 7.4. Let $\rho_\hbar$ be the Schrödinger representation [31, § 1.3] of $\mathbb{H}^1$ in $L_2(\mathbb{R})$, that is [88, (3.5)]:

$$[\rho_\chi(s, x, y) f](q) = e^{2\pi \mathrm{i}\hbar(s - xy/2) + 2\pi \mathrm{i} x q} f(q - \hbar y).$$

Thus the action of the derived representation on the Lie algebra $\mathfrak{h}_1$ is:

$$\rho_\hbar(X) = 2\pi \mathrm{i} q, \qquad \rho_\hbar(Y) = -\hbar \frac{d}{dq}, \qquad \rho_\hbar(S) = 2\pi \mathrm{i} \hbar I. \tag{7.12}$$

Then the associated Shale–Weil representation of Sp(2) in $L_2(\mathbb{R})$ has the derived action, cf. [31, § 4.3; 121, (2.2)]:

$$\rho_\hbar^{\mathrm{SW}}(A) = -\frac{q}{2}\frac{d}{dq} - \frac{1}{4}, \quad \rho_\hbar^{\mathrm{SW}}(B) = -\frac{\hbar\mathrm{i}}{8\pi}\frac{d^2}{dq^2} - \frac{\pi\mathrm{i}q^2}{2\hbar}, \quad \rho_\hbar^{\mathrm{SW}}(Z) = \frac{\hbar\mathrm{i}}{4\pi}\frac{d^2}{dq^2} - \frac{\pi\mathrm{i}q^2}{\hbar}. \tag{7.13}$$

We can verify commutators (7.5) and (3.13), (7.11) for operators (7.12–7.13). It is also obvious that in this representation the following algebraic relations hold:

$$\rho_\hbar^{\mathrm{SW}}(A) = \frac{\mathrm{i}}{4\pi\hbar}(\rho_\hbar(X)\rho_\hbar(Y) - \tfrac{1}{2}\rho_\hbar(S)) \tag{7.14}$$
$$= \frac{\mathrm{i}}{8\pi\hbar}(\rho_\hbar(X)\rho_\hbar(Y) + \rho_\hbar(Y)\rho_\hbar(X)),$$
$$\rho_\hbar^{\mathrm{SW}}(B) = \frac{\mathrm{i}}{8\pi\hbar}(\rho_\hbar(X)^2 - \rho_\hbar(Y)^2), \tag{7.15}$$
$$\rho_\hbar^{\mathrm{SW}}(Z) = \frac{\mathrm{i}}{4\pi\hbar}(\rho_\hbar(X)^2 + \rho_\hbar(Y)^2). \tag{7.16}$$

Thus it is common in quantum optics to name $\mathfrak{g}$ as a Lie algebra with quadratic generators, see [33, § 2.2.4].

Note that $\rho_\hbar^{\mathrm{SW}}(Z)$ is the Hamiltonian of the harmonic oscillator (up to a factor). Then we can consider $\rho_\hbar^{\mathrm{SW}}(B)$ as the Hamiltonian of a repulsive (hyperbolic) oscillator. The operator $\rho_\hbar^{\mathrm{SW}}(B - Z/2) = \frac{\hbar\mathrm{i}}{4\pi}\frac{d^2}{dq^2}$ is the parabolic analog. A graphical representation of all three transformations defined by those Hamiltonians is given in Figure 9 and a further discussion of these Hamiltonians can be found in [127, § 3.8].

An important observation, which is often missed, is that the three linear symplectic transformations are unitary rotations in the corresponding hypercomplex algebra, cf. [85, § 3]. This means, that the symplectomorphisms generated by operators Z, $B - Z/2$, B within time t coincide with the multiplication of hypercomplex number $q + \iota p$ by $e^{\iota t}$, see Subsection 3.1 and Figure 9, which is just another illustration of the Similarity and Correspondence Principle 3.5.

Example 7.5. There are many advantages of considering representations of the Heisenberg group on the phase space [24; 31, § 1.6; 43, § 1.7]. A convenient expression for Fock–Segal–Bargmann (FSB) representation on the phase space is, cf. § 7.3.1 and [24, (1); 74, (2.9)]:

$$[\rho_F(s,x,y)f](q,p) = e^{-2\pi\mathrm{i}(\hbar s + qx + py)} f\left(q - \tfrac{\hbar}{2}y, p + \tfrac{\hbar}{2}x\right). \tag{7.17}$$

Then the derived representation of $\mathfrak{h}_1$ is:

$$\rho_F(X) = -2\pi\mathrm{i}q + \tfrac{\hbar}{2}\partial_p, \qquad \rho_F(Y) = -2\pi\mathrm{i}p - \tfrac{\hbar}{2}\partial_q, \qquad \rho_F(S) = -2\pi\mathrm{i}\hbar I. \tag{7.18}$$

This produces the derived form of the Shale–Weil representation:

$$\rho_F^{\mathrm{SW}}(A) = \tfrac{1}{2}\left(q\partial_q - p\partial_p\right), \quad \rho_F^{\mathrm{SW}}(B) = -\tfrac{1}{2}\left(p\partial_q + q\partial_p\right), \quad \rho_F^{\mathrm{SW}}(Z) = p\partial_q - q\partial_p. \tag{7.19}$$

Note that this representation does not contain the parameter $\hbar$ unlike the equivalent representation (7.13). Thus the FSB model explicitly shows the equivalence of $\rho^{\mathrm{SW}}_{\hbar_1}$ and $\rho^{\mathrm{SW}}_{\hbar_2}$ if $\hbar_1\hbar_2 > 0$ [31, Thm. 4.57].

As we will also see below the FSB-type representations in hypercomplex numbers produce almost the same Shale–Weil representations.

7.2. p-Mechanic formalism

Here we briefly outline a formalism [15, 63, 70, 74, 116], which allows us to unify quantum and classical mechanics.

7.2.1. Convolutions (observables) on $\mathbb{H}^n$ and commutator. Using an invariant measure $dg = ds\,dx\,dy$ on $\mathbb{H}^n$ we can define the convolution of two functions:

$$(k_1 * k_2)(g) = \int_{\mathbb{H}^n} k_1(g_1)\, k_2(g_1^{-1}g)\, dg_1. \tag{7.20}$$

This is a non-commutative operation, which is meaningful for functions from various spaces including $L_1(\mathbb{H}^n, dg)$, the Schwartz space S and many classes of distributions, which form algebras under convolutions. Convolutions on $\mathbb{H}^n$ are used as *observables* in p-mechanics [63, 74].

A unitary representation ρ of $\mathbb{H}^n$ extends to $L_1(\mathbb{H}^n, dg)$ by the formula

$$\rho(k) = \int_{\mathbb{H}^n} k(g)\rho(g)\, dg. \tag{7.21}$$

This is also an algebra homomorphism of convolutions to linear operators.

For a dynamics of observables we need inner *derivations* D_k of the convolution algebra $L_1(\mathbb{H}^n)$, which are given by the *commutator*:

$$\begin{aligned} D_k : f \mapsto [k, f] &= k * f - f * k \\ &= \int_{\mathbb{H}^n} k(g_1)\left(f(g_1^{-1}g) - f(gg_1^{-1})\right) dg_1, \quad f, k \in L_1(\mathbb{H}^n). \end{aligned} \tag{7.22}$$

To describe dynamics of a time-dependent observable $f(t, g)$ we use the universal equation, cf. [61, 63]:

$$S\dot{f} = [H, f], \tag{7.23}$$

where S is the left-invariant vector field (7.4) generated by the centre of $\mathbb{H}^n$. The presence of operator S fixes the dimensionality of both sides of the equation (7.23) if the observable H (Hamiltonian) has the dimensionality of energy [74, Rem 4.1]. If we apply a right inverse $\mathcal{A}$ of S to both sides of the equation (7.23) we obtain the equivalent equation

$$\dot{f} = \{[H, f]\}, \tag{7.24}$$

based on the universal bracket $\{[k_1, k_2]\} = k_1 * \mathcal{A}k_2 - k_2 * \mathcal{A}k_1$ [74].

Example 7.6 (Harmonic oscillator). Let $H = \frac{1}{2}(mk^2q^2 + \frac{1}{m}p^2)$ be the Hamiltonian of a one-dimensional harmonic oscillator, where k is a constant frequency and m is a constant mass. Its *p-mechanisation* will be the second-order differential operator on $\mathbb{H}^n$ [15, § 5.1]:

$$H = \tfrac{1}{2}(mk^2 X^2 + \tfrac{1}{m} Y^2),$$

where we dropped sub-indexes of vector fields (7.4) in a one-dimensional setting. We can express the commutator as a difference between the left and the right action of the vector fields:

$$[H, f] = \tfrac{1}{2}(mk^2((X^r)^2 - (X^l)^2) + \tfrac{1}{m}((Y^r)^2 - (Y^l)^2))f.$$

Thus the equation (7.23) becomes [15, (5.2)]

$$\frac{\partial}{\partial s}\dot{f} = \frac{\partial}{\partial s}\left(mk^2 y \frac{\partial}{\partial x} - \frac{1}{m} x \frac{\partial}{\partial y}\right) f. \tag{7.25}$$

Of course, the derivative $\frac{\partial}{\partial s}$ can be dropped from both sides of the equation and the general solution is found to be

$$f(t; s, x, y) = f_0\left(s, x\cos(kt) + mky\sin(kt), -\tfrac{x}{mk}\sin(kt) + y\cos(kt)\right), \tag{7.26}$$

where $f_0(s, x, y)$ is the initial value of an observable on $\mathbb{H}^n$.

Example 7.7 (Unharmonic oscillator). We consider an unharmonic oscillator with cubic potential, see [16] and references therein:

$$H = \frac{mk^2}{2} q^2 + \frac{\lambda}{6} q^3 + \frac{1}{2m} p^2. \tag{7.27}$$

Due to the absence of non-commutative products, p-mechanisation is straightforward:

$$H = \frac{mk^2}{2} X^2 + \frac{\lambda}{6} X^3 + \frac{1}{m} Y^2.$$

Similarly to the harmonic case the dynamic equation, after cancellation of $\frac{\partial}{\partial s}$ on both sides, becomes

$$\dot{f} = \left(mk^2 y \frac{\partial}{\partial x} + \frac{\lambda}{6}\left(3y \frac{\partial^2}{\partial x^2} + \frac{1}{4} y^3 \frac{\partial^2}{\partial s^2}\right) - \frac{1}{m} x \frac{\partial}{\partial y}\right) f. \tag{7.28}$$

Unfortunately, it cannot be solved analytically as easily as in the harmonic case.

7.2.2. States and probability. Let an observable $\rho(k)$ (7.21) be defined by a kernel $k(g)$ on the Heisenberg group and its representation ρ at a Hilbert space $\mathcal{H}$. A *state* on the convolution algebra is given by a vector $v \in \mathcal{H}$. A simple calculation,

$$\begin{aligned}
\langle \rho(k)v, v\rangle_{\mathcal{H}} &= \left\langle \int_{\mathbb{H}^n} k(g)\rho(g)v\,dg, v \right\rangle_{\mathcal{H}} \\
&= \int_{\mathbb{H}^n} k(g)\,\langle \rho(g)v, v\rangle_{\mathcal{H}}\,dg \\
&= \int_{\mathbb{H}^n} k(g)\overline{\langle v, \rho(g)v\rangle_{\mathcal{H}}}\,dg
\end{aligned}$$

can be restated as

$$\langle \rho(k)v, v\rangle_{\mathcal{H}} = \langle k, l\rangle, \qquad \text{where} \quad l(g) = \langle v, \rho(g)v\rangle_{\mathcal{H}}.$$

Here the left-hand side contains the inner product on $\mathcal{H}$, while the right-hand side uses a skew-linear pairing between functions on $\mathbb{H}^n$ based on the Haar measure integration. In other words we obtain, cf. [15, Thm. 3.11]:

Proposition 7.8. *A state defined by a vector $v \in \mathcal{H}$ coincides with the linear functional given by the wavelet transform*

$$l(g) = \langle v, \rho(g)v\rangle_{\mathcal{H}} \tag{7.29}$$

of v used as the mother wavelet as well.

The addition of vectors in $\mathcal{H}$ implies the following operation on states:

$$\begin{aligned}\langle v_1 + v_2, \rho(g)(v_1 + v_2)\rangle_{\mathcal{H}} &= \langle v_1, \rho(g)v_1\rangle_{\mathcal{H}} + \langle v_2, \rho(g)v_2\rangle_{\mathcal{H}} \\ &\quad + \langle v_1, \rho(g)v_2\rangle_{\mathcal{H}} + \overline{\langle v_1, \rho(g^{-1})v_2\rangle_{\mathcal{H}}}.\end{aligned} \tag{7.30}$$

The last expression can be conveniently rewritten for kernels of the functional as

$$l_{12} = l_1 + l_2 + 2A\sqrt{l_1 l_2} \tag{7.31}$$

for some real number A. This formula is behind the contextual law of addition of conditional probabilities [50] and will be illustrated below. Its physical interpretation is an interference, say, from two slits. Despite a common belief, the mechanism of such interference can be both causal and local, see [54, 72].

7.3. Elliptic characters and quantum dynamics

In this subsection we consider the representation ρ_h of $\mathbb{H}^n$ induced by the elliptic character $\chi_h(s) = e^{ihs}$ in complex numbers parametrised by $h \in \mathbb{R}$. We also use the convenient agreement $h = 2\pi\hbar$ borrowed from physical literature.

7.3.1. Fock–Segal–Bargmann and Schrödinger representations. The realisation of ρ_h by the left shifts (7.3) on $L_2^h(\mathbb{H}^n)$ is rarely used in quantum mechanics. Instead two unitary equivalent forms are more common: the Schrödinger and Fock–Segal–Bargmann (FSB) representations.

The FSB representation can be obtained from the orbit method of Kirillov [56]. It allows spatially separate irreducible components of the left regular representation, each of them become located on the orbit of the co-adjoint representation, see [56; 74, § 2.1] for details, we only present a brief summary here.

We identify $\mathbb{H}^n$ and its Lie algebra $\mathfrak{h}_n$ through the exponential map [55, § 6.4]. The dual $\mathfrak{h}_n^*$ of $\mathfrak{h}_n$ is presented by the Euclidean space $\mathbb{R}^{2n+1}$ with coordinates $(\hbar, q, p)$. The pairing $\mathfrak{h}_n^*$ and $\mathfrak{h}_n$ given by

$$\langle (s, x, y), (\hbar, q, p)\rangle = \hbar s + q \cdot x + p \cdot y.$$

This pairing defines the Fourier transform $\hat{}: L_2(\mathbb{H}^n) \to L_2(\mathfrak{h}_n^*)$ given by [57, § 2.3]:

$$\hat{\phi}(F) = \int_{\mathfrak{h}^n} \phi(\exp X) e^{-2\pi i\langle X, F\rangle}\, dX \qquad \text{where } X \in \mathfrak{h}^n,\ F \in \mathfrak{h}_n^*. \tag{7.32}$$

For a fixed $\hbar$ the left regular representation (7.3) is mapped by the Fourier transform to the FSB type representation (7.17). The collection of points $(\hbar, q, p) \in \mathfrak{h}_n^*$ for a fixed $\hbar$ is naturally identified with the *phase space* of the system.

Remark 7.9. It is possible to identify the case of $\hbar = 0$ with classical mechanics [74]. Indeed, a substitution of the zero value of $\hbar$ into (7.17) produces the commutative representation

$$\rho_0(s, x, y) : f(q, p) \mapsto e^{-2\pi \mathrm{i}(qx+py)} f(q, p). \tag{7.33}$$

It can be decomposed into the direct integral of one-dimensional representations parametrised by the points (q, p) of the phase space. The classical mechanics, including the Hamilton equation, can be recovered from those representations [74]. However the condition $\hbar = 0$ (as well as the *semiclassical limit* $\hbar \to 0$) is not completely physical. Commutativity (and subsequent relative triviality) of those representations is the main reason why they are often neglected. The commutativity can be outweighed by special arrangements, e.g., an antiderivative [74, (4.1)], but the procedure is not straightforward, see discussion in [1, 77, 79]. A direct approach using dual numbers will be shown below, cf. Remark 7.21.

To recover the Schrödinger representation we use notations and technique of induced representations from § 3.2, see also [66, Ex. 4.1]. The subgroup $H = \{(s, 0, y) \mid s \in \mathbb{R}, y \in \mathbb{R}^n\} \subset \mathbb{H}^n$ defines the homogeneous space $X = G/H$, which coincides with $\mathbb{R}^n$ as a manifold. The natural projection $\mathbf{p} : G \to X$ is $\mathbf{p}(s, x, y) = x$ and its left inverse $\mathbf{s} : X \to G$ can be as simple as $\mathbf{s}(x) = (0, x, 0)$. For the map $\mathbf{r} : G \to H$, $\mathbf{r}(s, x, y) = (s - xy/2, 0, y)$ we have the decomposition

$$(s, x, y) = \mathbf{s}(p(s, x, y)) * \mathbf{r}(s, x, y) = (0, x, 0) * (s - \tfrac{1}{2}xy, 0, y).$$

For a character $\chi_h(s, 0, y) = e^{\mathrm{i}hs}$ of H the lifting $\mathcal{L}_\chi : L_2(G/H) \to L_2^\chi(G)$ is as follows:

$$[\mathcal{L}_\chi f](s, x, y) = \chi_h(\mathbf{r}(s, x, y))\, f(\mathbf{p}(s, x, y)) = e^{\mathrm{i}h(s-xy/2)} f(x).$$

Thus the representation $\rho_\chi(g) = \mathcal{P} \circ \Lambda(g) \circ \mathcal{L}$ becomes

$$[\rho_\chi(s', x', y')f](x) = e^{-2\pi \mathrm{i}\hbar(s'+xy'-x'y'/2)} f(x - x'). \tag{7.34}$$

After the Fourier transform $x \mapsto q$ we get the Schrödinger representation on the *configuration space*

$$[\rho_\chi(s', x', y')\hat{f}\,](q) = e^{-2\pi \mathrm{i}\hbar(s'+x'y'/2)-2\pi \mathrm{i}x'q} \hat{f}(q + \hbar y'). \tag{7.35}$$

Note that this again turns into a commutative representation (multiplication by a unimodular function) if $\hbar = 0$. To get the full set of commutative representations in this way we need to use the character $\chi_{(h,p)}(s, 0, y) = e^{2\pi \mathrm{i}(\hbar+py)}$ in the above consideration.

7.3.2. Commutator and the Heisenberg equation. The property (7.6) of $F_2^\chi(\mathbb{H}^n)$ implies that the restrictions of two operators $\rho_\chi(k_1)$ and $\rho_\chi(k_2)$ to this space are equal if

$$\int_{\mathbb{R}} k_1(s,x,y)\,\chi(s)\,ds = \int_{\mathbb{R}} k_2(s,x,y)\,\chi(s)\,ds.$$

In other words, for a character $\chi(s) = e^{2\pi i\hbar s}$ the operator $\rho_\chi(k)$ depends only on

$$\hat{k}_s(\hbar,x,y) = \int_{\mathbb{R}} k(s,x,y)\,e^{-2\pi i\hbar s}\,ds,$$

which is the partial Fourier transform $s \mapsto \hbar$ of $k(s,x,y)$. The restriction to $F_2^\chi(\mathbb{H}^n)$ of the composition formula for convolutions is [74, (3.5)]:

$$(k' * k)\hat{_s} = \int_{\mathbb{R}^{2n}} e^{ih(xy'-yx')/2}\,\hat{k}'_s(\hbar,x',y')\,\hat{k}_s(\hbar,x-x',y-y')\,dx'dy'. \qquad (7.36)$$

Under the Schrödinger representation (7.35) the convolution (7.36) defines a rule for composition of two pseudo-differential operators (PDO) in the Weyl calculus [31, § 2.3; 43].

Consequently the representation (7.21) of commutator (7.22) depends only on its partial Fourier transform [74, (3.6)]

$$\begin{aligned}[k',k]\hat{_s} = 2i \int_{\mathbb{R}^{2n}} &\sin(\tfrac{h}{2}(xy'-yx')) \qquad (7.37)\\ &\times \hat{k}'_s(\hbar,x',y')\,\hat{k}_s(\hbar,x-x',y-y')\,dx'dy'.\end{aligned}$$

Under the Fourier transform (7.32) this commutator is exactly the *Moyal bracket* [129] for of $\hat{k}'$ and $\hat{k}$ on the phase space.

For observables in the space $F_2^\chi(\mathbb{H}^n)$ the action of S is reduced to multiplication $\chi(s) = e^{ihs}$ the action of S is multiplication by ih. Thus the equation (7.23) reduced to the space $F_2^\chi(\mathbb{H}^n)$ becomes the Heisenberg type equation [74, (4.4)]

$$\dot{f} = \frac{1}{ih}[H,f]\hat{_s}, \qquad (7.38)$$

based on the above bracket (7.37). The Schrödinger representation (7.35) transforms this equation into the original Heisenberg equation.

Example 7.10.

i. Under the Fourier transform $(x,y) \mapsto (q,p)$ the p-dynamic equation (7.25) of the harmonic oscillator becomes

$$\dot{f} = \left(mk^2 q\frac{\partial}{\partial p} - \frac{1}{m}p\frac{\partial}{\partial q}\right) f. \qquad (7.39)$$

The same transform creates its solution out of (7.26).

ii. Since $\frac{\partial}{\partial s}$ acts on $F_2^\chi(\mathbb{H}^n)$ as multiplication by $\mathrm{i}\hbar$, the quantum representation of unharmonic dynamics equation (7.28) is

$$\dot{f} = \left(mk^2 q \frac{\partial}{\partial p} + \frac{\lambda}{6}\left(3q^2 \frac{\partial}{\partial p} - \frac{\hbar^2}{4}\frac{\partial^3}{\partial p^3}\right) - \frac{1}{m} p \frac{\partial}{\partial q}\right) f. \tag{7.40}$$

This is exactly the equation for the Wigner function obtained in [16, (30)].

7.3.3. Quantum probabilities. For the elliptic character $\chi_h(s) = e^{\mathrm{i}hs}$ we can use the Cauchy–Schwartz inequality to demonstrate that the real number A in the identity (7.31) is between -1 and 1. Thus we can put $A = \cos\alpha$ for some angle (phase) α to get the formula for counting quantum probabilities, cf. [51, (2)]:

$$l_{12} = l_1 + l_2 + 2\cos\alpha\,\sqrt{l_1 l_2}. \tag{7.41}$$

Remark 7.11. It is interesting to note that both trigonometric functions are employed in quantum mechanics: sine is in the heart of the Moyal bracket (7.37) and cosine is responsible for the addition of probabilities (7.41). In essence the commutator and probabilities took respectively the odd and even parts of the elliptic character $e^{\mathrm{i}hs}$.

Example 7.12. Take a vector $v_{(a,b)} \in L_2^h(\mathbb{H}^n)$ defined by a Gaussian with mean value (a, b) in the phase space for a harmonic oscillator of the mass m and the frequency k:

$$v_{(a,b)}(q,p) = \exp\left(-\frac{2\pi km}{\hbar}(q-a)^2 - \frac{2\pi}{\hbar km}(p-b)^2\right). \tag{7.42}$$

A direct calculation shows that

$$\begin{aligned}\left\langle v_{(a,b)}, \rho_\hbar(s,x,y) v_{(a',b')}\right\rangle &= \frac{4}{\hbar}\exp\Big(\pi\mathrm{i}\,(2s\hbar + x(a+a') + y(b+b'))\\ &\quad - \frac{\pi}{2\hbar km}((\hbar x + b - b')^2 + (b-b')^2) - \frac{\pi km}{2\hbar}((\hbar y + a' - a)^2 + (a'-a)^2)\Big)\\ &= \frac{4}{\hbar}\exp\Big(\pi\mathrm{i}\,(2s\hbar + x(a+a') + y(b+b'))\\ &\quad - \frac{\pi}{\hbar km}((b - b' + \tfrac{\hbar x}{2})^2 + (\tfrac{\hbar x}{2})^2) - \frac{\pi km}{\hbar}((a - a' - \tfrac{\hbar y}{2})^2 + (\tfrac{\hbar y}{2})^2)\Big).\end{aligned}$$

Thus the kernel $l_{(a,b)} = \left\langle v_{(a,b)}, \rho_\hbar(s,x,y) v_{(a,b)}\right\rangle$ (7.29) for a state $v_{(a,b)}$ is

$$l_{(a,b)} = \frac{4}{\hbar}\exp\left(2\pi\mathrm{i}(s\hbar + xa + yb) - \frac{\pi\hbar}{2km}x^2 - \frac{\pi km\hbar}{2\hbar}y^2\right). \tag{7.43}$$

An observable registering a particle at a point $q = c$ of the configuration space is $\delta(q-c)$. On the Heisenberg group this observable is given by the kernel

$$X_c(s,x,y) = e^{2\pi\mathrm{i}(s\hbar + xc)}\delta(y). \tag{7.44}$$

The measurement of X_c on the state (7.42) (through the kernel (7.43)) predictably is

$$\langle X_c, l_{(a,b)} \rangle = \sqrt{\frac{2km}{\hbar}} \exp\left(-\frac{2\pi km}{\hbar}(c-a)^2\right).$$

Example 7.13. Now take two states $v_{(0,b)}$ and $v_{(0,-b)}$, where for simplicity we assume the mean values of coordinates vanish in both cases. Then the corresponding kernel (7.30) has the interference terms

$$\begin{aligned} l_i &= \langle v_{(0,b)}, \rho_\hbar(s,x,y) v_{(0,-b)} \rangle \\ &= \frac{4}{\hbar} \exp\left(2\pi i s\hbar - \frac{\pi}{2\hbar km}((\hbar x + 2b)^2 + 4b^2) - \frac{\pi\hbar km}{2} y^2\right). \end{aligned}$$

The measurement of X_c (7.44) on this term contains the oscillating part

$$\langle X_c, l_i \rangle = \sqrt{\frac{2km}{\hbar}} \exp\left(-\frac{2\pi km}{\hbar}c^2 - \frac{2\pi}{km\hbar}b^2 + \frac{4\pi i}{\hbar}cb\right).$$

Therefore on the kernel l corresponding to the state $v_{(0,b)} + v_{(0,-b)}$ the measurement is

$$\langle X_c, l \rangle = 2\sqrt{\frac{2km}{\hbar}} \exp\left(-\frac{2\pi km}{\hbar}c^2\right)\left(1 + \exp\left(-\frac{2\pi}{km\hbar}b^2\right)\cos\left(\frac{4\pi}{\hbar}cb\right)\right).$$

The presence of the cosine term in the last expression can generate an interference picture. In practise it does not happen for the minimal uncertainty state (7.42) which we are using here: it rapidly vanishes outside of the neighbourhood of zero, where oscillations of the cosine occurs, see Figure 13(a).

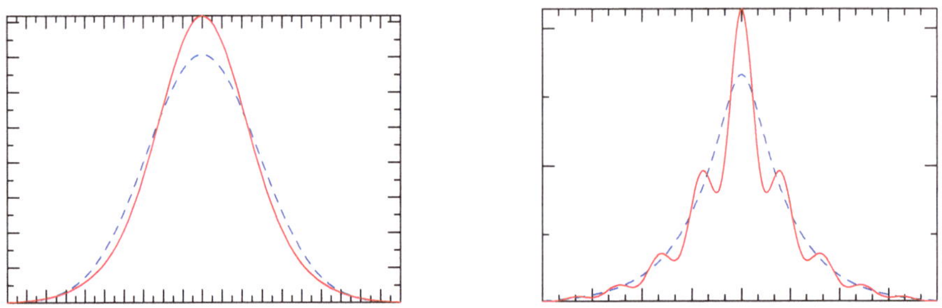

(a) (b)

FIGURE 13. Quantum probabilities: the blue (dashed) graph shows the addition of probabilities without interaction, the red (solid) graph presents the quantum interference. The left picture shows the Gaussian state (7.42), the right – the rational state (7.45)

Example 7.14. To see a traditional interference pattern one can use a state which is far from the minimal uncertainty. For example, we can consider the state

$$u_{(a,b)}(q,p) = \frac{\hbar^2}{((q-a)^2+\hbar/km)((p-b)^2+\hbar km)}. \qquad (7.45)$$

To evaluate the observable X_c (7.44) on the state $l(g) = \langle u_1, \rho_h(g)u_2\rangle$ (7.29) we use the following formula:

$$\langle X_c, l\rangle = \frac{2}{\hbar}\int_{\mathbb{R}^n} \hat{u}_1(q, 2(q-c)/\hbar)\,\overline{\hat{u}_2(q, 2(q-c)/\hbar)}\,dq,$$

where $\hat{u}_i(q,x)$ denotes the partial Fourier transform $p \mapsto x$ of $u_i(q,p)$. The formula is obtained by swapping order of integrations. The numerical evaluation of the state obtained by the addition $u_{(0,b)} + u_{(0,-b)}$ is plotted on Figure 13(b); the red curve shows the canonical interference pattern.

7.4. Ladder operators and harmonic oscillator

Let ρ be a representation of the Schrödinger group $G = \mathbb{H}^1 \rtimes \widetilde{\mathrm{Sp}}(2)$ (7.9) in a space V. Consider the derived representation of the Lie algebra $\mathfrak{g}$ [97, § VI.1] and write $\tilde{X} = \rho(X)$ for $X \in \mathfrak{g}$. To see the structure of the representation ρ we can decompose the space V into eigenspaces of the operator $\tilde{X}$ for some $X \in \mathfrak{g}$. The canonical example is the Taylor series in complex analysis.

We are going to consider three cases corresponding to three non-isomorphic subgroups (2.4–2.6) of Sp(2) starting from the compact case. Let $H = Z$ be a generator of the compact subgroup K. Corresponding symplectomorphisms (7.8) of the phase space are given by orthogonal rotations with matrices $\begin{pmatrix} \cos t & \sin t \\ -\sin t & \cos t\end{pmatrix}$. The Shale–Weil representation (7.13) coincides with the Hamiltonian of the harmonic oscillator in Schrödinger representation.

Since $\widetilde{\mathrm{Sp}}(2)$ is a two-fold cover the corresponding eigenspaces of a compact group $\tilde{Z}v_k = ikv_k$ are parametrised by a half-integer $k \in \mathbb{Z}/2$. Explicitly for a half-integer k eigenvectors are

$$v_k(q) = H_{k+\frac{1}{2}}\left(\sqrt{\frac{2\pi}{\hbar}}q\right)e^{-\frac{\pi}{\hbar}q^2}, \qquad (7.46)$$

where H_k is the *Hermite polynomial* [28, 8.2(9); 31, § 1.7].

From the point of view of quantum mechanics as well as the representation theory it is beneficial to introduce the ladder operators $L^{\pm}$ (3.14), known also as *creation/annihilation* in quantum mechanics [13; 31, p. 49]. There are two ways to search for ladder operators: in (complexified) Lie algebras $\mathfrak{h}_1$ and $\mathfrak{sp}_2$. The latter in essence coincides with our consideration in Section 3.3.

7.4.1. Ladder operators from the Heisenberg group. Assuming $L^+ = a\tilde{X} + b\tilde{Y}$ we obtain from the relations (7.10–7.11) and (3.14) the linear equations with unknown a and b:

$$a = \lambda_+ b, \qquad -b = \lambda_+ a.$$

The equations have a solution if and only if $\lambda_+^2 + 1 = 0$, and the raising/lowering operators are $L^\pm = \tilde{X} \mp \mathrm{i}\tilde{Y}$.

Remark 7.15. Here we have an interesting asymmetric response: due to the structure of the semidirect product $\mathbb{H}^1 \rtimes \widetilde{\mathrm{Sp}}(2)$ it is the symplectic group which acts on $\mathbb{H}^1$, not vise versa. However the Heisenberg group has a weak action in the opposite direction; it shifts eigenfunctions of $\mathrm{Sp}(2)$.

In the Schrödinger representation (7.12) the ladder operators are

$$\rho_\hbar(L^\pm) = 2\pi \mathrm{i} q \pm \mathrm{i}\hbar \frac{d}{dq}. \tag{7.47}$$

The standard treatment of the harmonic oscillator in quantum mechanics, which can be found in many textbooks, e.g., [31, § 1.7; 33, § 2.2.3], is as follows. The vector $v_{-1/2}(q) = e^{-\pi q^2/\hbar}$ is an eigenvector of $\tilde{Z}$ with the eigenvalue $-\frac{\mathrm{i}}{2}$. In addition $v_{-1/2}$ is annihilated by L^+. Thus the chain (3.16) terminates to the right and the complete set of eigenvectors of the harmonic oscillator Hamiltonian is presented by $(L^-)^k v_{-1/2}$ with $k = 0, 1, 2, \ldots$.

We can make a wavelet transform generated by the Heisenberg group with the mother wavelet $v_{-1/2}$, and the image will be the Fock–Segal–Bargmann (FSB) space [31, § 1.6; 43]. Since $v_{-1/2}$ is the null solution of $L^+ = \tilde{X} - \mathrm{i}\tilde{Y}$, then by Corollary 5.6 the image of the wavelet transform will be null-solutions of the corresponding linear combination of the Lie derivatives (7.4):

$$D = \overline{X^r - \mathrm{i}Y^r} = (\partial_x + \mathrm{i}\partial_y) - \pi\hbar(x - \mathrm{i}y), \tag{7.48}$$

which turns out to be the Cauchy–Riemann equation on a weighted FSB-type space.

7.4.2. Symplectic ladder operators. We can also look for ladder operators within the Lie algebra $\mathfrak{sp}_2$, see § 3.3.1 and [85, § 8]. Assuming $L_2^+ = a\tilde{A} + b\tilde{B} + c\tilde{Z}$ from the relations (3.13) and defining condition (3.14) we obtain the following linear equations with unknown a, b and c:

$$c = 0, \qquad 2a = \lambda_+ b, \qquad -2b = \lambda_+ a.$$

The equations have a solution if and only if $\lambda_+^2 + 4 = 0$, and the raising/lowering operators are $L_2^\pm = \pm\mathrm{i}\tilde{A} + \tilde{B}$. In the Shale–Weil representation (7.13) they turn out to be

$$L_2^\pm = \pm\mathrm{i}\left(\frac{q}{2}\frac{d}{dq} + \frac{1}{4}\right) - \frac{\hbar\mathrm{i}}{8\pi}\frac{d^2}{dq^2} - \frac{\pi\mathrm{i}q^2}{2\hbar} = -\frac{\mathrm{i}}{8\pi\hbar}\left(\mp 2\pi q + \hbar\frac{d}{dq}\right)^2. \tag{7.49}$$

Since this time $\lambda_+ = 2\mathrm{i}$, the ladder operators $L_2^\pm$ produce a shift on the diagram (3.16) twice bigger than the operators $L^\pm$ from the Heisenberg group. After all, this is not surprising since from the explicit representations (7.47) and (7.49) we get

$$L_2^\pm = -\frac{\mathrm{i}}{8\pi\hbar}(L^\pm)^2.$$

7.5. Hyperbolic quantum mechanics

Now we turn to double numbers also known as hyperbolic, split-complex, etc. numbers [53; 124; 128, App. C]. They form a two-dimensional algebra $\mathbb{O}$ spanned by 1 and j with the property $\mathrm{j}^2 = 1$. There are zero divisors:

$$\mathrm{j}_\pm = \tfrac{1}{\sqrt{2}}(1 \pm j), \qquad \text{such that} \quad \mathrm{j}_+\mathrm{j}_- = 0 \quad \text{and} \quad \mathrm{j}_\pm^2 = \mathrm{j}_\pm.$$

Thus double numbers are algebraically isomorphic to two copies of $\mathbb{R}$ spanned by $\mathrm{j}_\pm$. Being algebraically dull, double numbers are nevertheless interesting as a homogeneous space [82, 85] and they are relevant in physics [49, 124, 125]. The combination of the p-mechanical approach with hyperbolic quantum mechanics was already discussed in [15, § 6].

For the hyperbolic character $\chi_{\mathrm{j}h}(s) = e^{\mathrm{j}hs} = \cosh hs + \mathrm{j}\sinh hs$ of $\mathbb{R}$ one can define the hyperbolic Fourier-type transform

$$\hat{k}(q) = \int_{\mathbb{R}} k(x)\, e^{-\mathrm{j}qx}\, dx.$$

It can be understood in the sense of distributions on the space dual to the set of analytic functions [52, § 3]. A hyperbolic Fourier transform intertwines the derivative $\frac{d}{dx}$ and multiplication by $\mathrm{j}q$ [52, Prop. 1].

Example 7.16. For the Gaussian the hyperbolic Fourier transform is the ordinary function (note the sign difference!)

$$\int_{\mathbb{R}} e^{-x^2/2} e^{-\mathrm{j}qx}\, dx = \sqrt{2\pi}\, e^{q^2/2}.$$

However the opposite identity

$$\int_{\mathbb{R}} e^{x^2/2} e^{-\mathrm{j}qx}\, dx = \sqrt{2\pi}\, e^{-q^2/2}$$

is true only in a suitable distributional sense. To this end we may note that $e^{x^2/2}$ and $e^{-q^2/2}$ are null solutions to the differential operators $\frac{d}{dx} - x$ and $\frac{d}{dq} + q$ respectively, which are intertwined (up to the factor j) by the hyperbolic Fourier transform. The above differential operators $\frac{d}{dx} - x$ and $\frac{d}{dq} + q$ are images of the ladder operators (7.47) in the Lie algebra of the Heisenberg group. They are intertwining by the Fourier transform, since this is an automorphism of the Heisenberg group [42].

An elegant theory of hyperbolic Fourier transform may be achieved by a suitable adaptation of [42], which uses representation theory of the Heisenberg group.

7.5.1. Hyperbolic representations of the Heisenberg group. Consider the space $F_h^{\mathrm{j}}(\mathbb{H}^n)$ of $\mathbb{O}$-valued functions on $\mathbb{H}^n$ with the property

$$f(s + s', h, y) = e^{\mathrm{j}hs'} f(s, x, y), \qquad \text{for all } (s, x, y) \in \mathbb{H}^n,\ s' \in \mathbb{R}, \tag{7.50}$$

and the square integrability condition (7.7). Then the hyperbolic representation is obtained by the restriction of the left shifts to $F_h^{\mathrm{j}}(\mathbb{H}^n)$. To obtain an equivalent representation on the phase space we take an $\mathbb{O}$-valued functional of the Lie algebra $\mathfrak{h}_n$:

$$\chi^j_{(h,q,p)}(s,x,y) = e^{\mathrm{j}(hs+qx+py)} = \cosh(hs+qx+py) + \mathrm{j}\sinh(hs+qx+py). \quad (7.51)$$

The hyperbolic Fock–Segal–Bargmann type representation is intertwined with the left group action by means of a Fourier transform (7.32) with the hyperbolic functional (7.51). Explicitly this representation is

$$\rho_\hbar(s,x,y) : f(q,p) \mapsto e^{-\mathrm{j}(hs+qx+py)} f\left(q - \tfrac{h}{2}y, p + \tfrac{h}{2}x\right). \quad (7.52)$$

For a hyperbolic Schrödinger type representation we again use the scheme described in § 3.2. Similarly to the elliptic case one obtains the formula, resembling (7.34),

$$[\rho^{\mathrm{j}}_\chi(s',x',y')f](x) = e^{-\mathrm{j}h(s'+xy'-x'y'/2)} f(x-x'). \quad (7.53)$$

Application of the hyperbolic Fourier transform produces a Schrödinger type representation on the configuration space, cf. (7.35):

$$[\rho^{\mathrm{j}}_\chi(s',x',y')\hat{f}\,](q) = e^{-\mathrm{j}h(s'+x'y'/2)-\mathrm{j}x'q}\,\hat{f}(q+hy').$$

The extension of this representation to kernels according to (7.21) generates hyperbolic pseudodifferential operators introduced in [52, (3.4)].

7.5.2. Hyperbolic dynamics. Similarly to the elliptic (quantum) case we consider a convolution of two kernels on $\mathbb{H}^n$ restricted to $F_h^{\mathrm{j}}(\mathbb{H}^n)$. The composition law becomes, cf. (7.36),

$$(k' * k)\hat{_s} = \int_{\mathbb{R}^{2n}} e^{\mathrm{j}h(xy'-yx')}\,\hat{k}'_s(h,x',y')\,\hat{k}_s(h,x-x',y-y')\,dx'dy'. \quad (7.54)$$

This is close to the calculus of hyperbolic PDO obtained in [52, Thm. 2]. Respectively for the commutator of two convolutions we get, cf. (7.37),

$$[k',k]\hat{_s} = \int_{\mathbb{R}^{2n}} \sinh(h(xy'-yx'))\,\hat{k}'_s(h,x',y')\,\hat{k}_s(h,x-x',y-y')\,dx'dy'. \quad (7.55)$$

This the hyperbolic version of the Moyal bracket, cf. [52, p. 849], which generates the corresponding image of the dynamic equation (7.23).

Example 7.17.

i. For a quadratic Hamiltonian, e.g., harmonic oscillator from Example 7.6, the hyperbolic equation and respective dynamics are identical to the quantum considered before.
ii. Since $\frac{\partial}{\partial s}$ acts on $F_2^{\mathrm{j}}(\mathbb{H}^n)$ as multiplication by $\mathrm{j}h$ and $\mathrm{j}^2 = 1$, the hyperbolic image of the unharmonic equation (7.28) becomes

$$\dot{f} = \left(mk^2 q\frac{\partial}{\partial p} + \frac{\lambda}{6}\left(3q^2\frac{\partial}{\partial p} + \frac{\hbar^2}{4}\frac{\partial^3}{\partial p^3}\right) - \frac{1}{m}p\frac{\partial}{\partial q}\right) f.$$

The difference with quantum mechanical equation (7.40) is in the sign of the cubic derivative.

Notably, the hyperbolic setup allows us to linearise many non-linear problems of classical mechanics. It will be interesting to realise new hyperbolic coordinates introduced to this end in [112, 113] as a hyperbolic phase space.

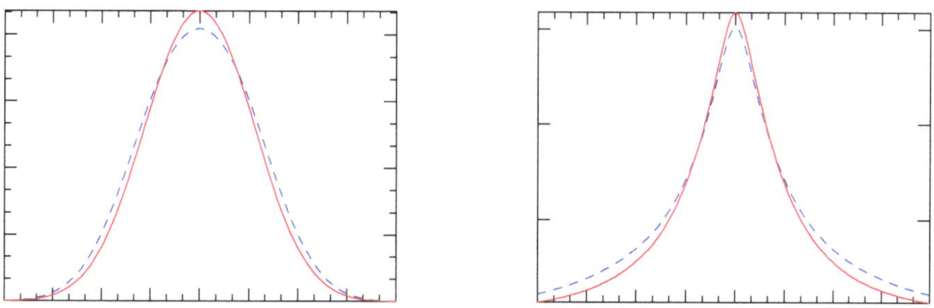

(a) (b)

FIGURE 14. Hyperbolic probabilities: the blue (dashed) graph shows the addition of probabilities without interaction, the red (solid) graph presents the quantum interference. The left picture shows the Gaussian state (7.42), with the same distribution as in quantum mechanics, cf. Figure 13(a). The right picture shows the rational state (7.45); note the absence of interference oscillations in comparison with the quantum state on Figure 13(b).

7.5.3. Hyperbolic probabilities. To calculate probability distribution generated by a hyperbolic state we will use the general procedure from Section 7.2.2. The main differences with the quantum case are as follows:

i. The real number A in the expression (7.31) for the addition of probabilities is bigger than 1 in absolute value. Thus it can be associated with the hyperbolic cosine, $\cosh\alpha$, cf. Remark 7.11, for a certain phase $\alpha \in \mathbb{R}$ [52].
ii. The nature of hyperbolic interference on two slits is affected by the fact that $e^{\mathrm{j}hs}$ is not periodic and the hyperbolic exponent $e^{\mathrm{j}t}$ and cosine $\cosh t$ do not oscillate. It is worth noticing that for Gaussian states the hyperbolic interference is exactly the same as quantum one, cf. Figs. 13(a) and 14(a). This is similar to coincidence of quantum and hyperbolic dynamics of a harmonic oscillator.

    The contrast between two types of interference is prominent for the rational state (7.45), which is far from the minimal uncertainty, see the different patterns on Figsure 13(b) and 14(b).

7.5.4. Ladder operators for the hyperbolic subgroup. Consider the case of the Hamiltonian $H = 2B$, which is a repulsive (hyperbolic) harmonic oscillator [127, § 3.8]. The corresponding one-dimensional subgroup of symplectomorphisms produces hyperbolic rotations of the phase space, see Figure 9. The eigenvectors v_μ of the operator

$$\rho_\hbar^{\mathrm{SW}}(2B)v_\nu = -\mathrm{i}\left(\frac{\hbar}{4\pi}\frac{d^2}{dq^2} + \frac{\pi q^2}{\hbar}\right)v_\nu = \mathrm{i}\nu v_\nu,$$

are *Weber–Hermite* (or *parabolic cylinder*) functions $v_\nu = D_{\nu-\frac{1}{2}}\left(\pm 2e^{\mathrm{i}\frac{\pi}{4}}\sqrt{\frac{\pi}{\hbar}}q\right)$, see [28, § 8.2; 117] for fundamentals of Weber–Hermite functions and [121] for further illustrations and applications in optics.

The corresponding one-parameter group is not compact and the eigenvalues of the operator $2\tilde{B}$ are not restricted by any integrality condition, but the raising/lowering operators are still important [44, § II.1; 102, § 1.1]. We again seek solutions in two subalgebras $\mathfrak{h}_1$ and $\mathfrak{sp}_2$ separately. However the additional options will be provided by a choice of the number system: either complex or double.

Example 7.18 (Complex ladder operators). Assuming $L_h^+ = a\tilde{X} + b\tilde{Y}$ from the commutators (7.10–7.11) we obtain the linear equations

$$-a = \lambda_+ b, \qquad -b = \lambda_+ a. \tag{7.56}$$

The equations have a solution if and only if $\lambda_+^2 - 1 = 0$. Taking the real roots $\lambda = \pm 1$ we obtain that the raising/lowering operators are $L_h^\pm = \tilde{X} \mp \tilde{Y}$. In the Schrödinger representation (7.12) the ladder operators are

$$L_h^\pm = 2\pi \mathrm{i} q \pm \hbar \frac{d}{dq}. \tag{7.57}$$

The null solutions $v_{\pm\frac{1}{2}}(q) = e^{\pm\frac{\pi \mathrm{i}}{\hbar}q^2}$ to operators $\rho_\hbar(L^\pm)$ are also eigenvectors of the Hamiltonian $\rho_\hbar^{\mathrm{SW}}(2B)$ with the eigenvalue $\pm\frac{1}{2}$. However the important distinction from the elliptic case is, that they are not square-integrable on the real line anymore.

We can also look for ladder operators within the $\mathfrak{sp}_2$, that is in the form $L_{2h}^+ = a\tilde{A} + b\tilde{B} + c\tilde{Z}$ for the commutator $[2\tilde{B}, L_h^+] = \lambda L_h^+$, see § 3.3.2. Within complex numbers we get only the values $\lambda = \pm 2$ with the ladder operators $L_{2h}^\pm = \pm 2\tilde{A} + \tilde{Z}/2$, see [44, § II.1; 102, § 1.1]. Each indecomposable $\mathfrak{h}_1$- or $\mathfrak{sp}_2$-module is formed by a one-dimensional chain of eigenvalues with a transitive action of ladder operators $L_h^\pm$ or $L_{2h}^\pm$ respectively. And we again have a quadratic relation between the ladder operators:

$$L_{2h}^\pm = \frac{\mathrm{i}}{4\pi\hbar}(L_h^\pm)^2.$$

7.5.5. Double ladder operators. There are extra possibilities in the context of hyperbolic quantum mechanics [49, 51, 52]. Here we use the representation of $\mathbb{H}^1$

induced by a hyperbolic character $e^{\mathrm{j}ht} = \cosh(ht) + \mathrm{j}\sinh(ht)$, see [88, (4.5)], and obtain the hyperbolic representation of $\mathbb{H}^1$, cf. (7.35),

$$[\rho_h^{\mathrm{j}}(s',x',y')\hat{f}\,](q) = e^{\mathrm{j}h(s'-x'y'/2)+\mathrm{j}x'q}\,\hat{f}(q-hy'). \tag{7.58}$$

The corresponding derived representation is

$$\rho_h^{\mathrm{j}}(X) = \mathrm{j}q, \qquad \rho_h^{\mathrm{j}}(Y) = -h\frac{d}{dq}, \qquad \rho_h^{\mathrm{j}}(S) = \mathrm{j}hI. \tag{7.59}$$

Then the associated Shale–Weil derived representation of $\mathfrak{sp}_2$ in the Schwartz space $S(\mathbb{R})$ is, cf. (7.13),

$$\rho_h^{\mathrm{SW}}(A) = -\frac{q}{2}\frac{d}{dq} - \frac{1}{4}, \quad \rho_h^{\mathrm{SW}}(B) = \frac{\mathrm{j}h}{4}\frac{d^2}{dq^2} - \frac{\mathrm{j}q^2}{4h}, \quad \rho_h^{\mathrm{SW}}(Z) = -\frac{\mathrm{j}h}{2}\frac{d^2}{dq^2} - \frac{\mathrm{j}q^2}{2h}. \tag{7.60}$$

Note that $\rho_h^{\mathrm{SW}}(B)$ now generates a usual harmonic oscillator, not the repulsive one like $\rho_\hbar^{\mathrm{SW}}(B)$ in (7.13). However the expressions in the quadratic algebra are still the same (up to a factor), cf. (7.14–7.16):

$$\begin{aligned}\rho_h^{\mathrm{SW}}(A) &= -\frac{\mathrm{j}}{2h}(\rho_h^{\mathrm{j}}(X)\rho_h^{\mathrm{j}}(Y) - \tfrac{1}{2}\rho_h^{\mathrm{j}}(S)) &(7.61)\\ &= -\frac{\mathrm{j}}{4h}(\rho_h^{\mathrm{j}}(X)\rho_h^{\mathrm{j}}(Y) + \rho_h^{\mathrm{j}}(Y)\rho_h^{\mathrm{j}}(X)),\\ \rho_h^{\mathrm{SW}}(B) &= \frac{\mathrm{j}}{4h}(\rho_h^{\mathrm{j}}(X)^2 - \rho_h^{\mathrm{j}}(Y)^2), &(7.62)\\ \rho_h^{\mathrm{SW}}(Z) &= -\frac{\mathrm{j}}{2h}(\rho_h^{\mathrm{j}}(X)^2 + \rho_h^{\mathrm{j}}(Y)^2). &(7.63)\end{aligned}$$

This is due to Principle 3.5 of similarity and correspondence: we can swap operators Z and B with simultaneous replacement of hypercomplex units i and j.

The eigenspace of the operator $2\rho_h^{\mathrm{SW}}(B)$ with an eigenvalue $\mathrm{j}\nu$ are spanned by the Weber–Hermite functions $D_{-\nu-\frac{1}{2}}\left(\pm\sqrt{\frac{2}{h}}x\right)$, see [28, § 8.2]. Functions D_ν are generalisations of the Hermite functions (7.46).

The compatibility condition for a ladder operator within the Lie algebra $\mathfrak{h}_1$ will be (7.56) as before, since it depends only on the commutators (7.10–7.11). Thus we still have the set of ladder operators corresponding to values $\lambda = \pm 1$:

$$L_h^{\pm} = \tilde{X} \mp \tilde{Y} = \mathrm{j}q \pm h\frac{d}{dq}.$$

Admitting double numbers we have an extra way to satisfy $\lambda^2 = 1$ in (7.56) with values $\lambda = \pm\mathrm{j}$. Then there is an additional pair of hyperbolic ladder operators, which are identical (up to factors) to (7.47):

$$L_{\mathrm{j}}^{\pm} = \tilde{X} \mp \mathrm{j}\tilde{Y} = \mathrm{j}q \pm \mathrm{j}h\frac{d}{dq}.$$

Pairs $L_h^{\pm}$ and $L_{\mathrm{j}}^{\pm}$ shift eigenvectors in the "orthogonal" directions changing their eigenvalues by ± 1 and $\pm\mathrm{j}$. Therefore an indecomposable $\mathfrak{sp}_2$-module can be parametrised by a two-dimensional lattice of eigenvalues in double numbers, see Figure 10.

The functions

$$v_{\frac{1}{2}}^{\pm h}(q) = e^{\mp \mathrm{j} q^2/(2h)} = \cosh\frac{q^2}{2h} \mp \mathrm{j}\sinh\frac{q^2}{2h}, \qquad v_{\frac{1}{2}}^{\pm \mathrm{j}}(q) = e^{\mp q^2/(2h)}$$

are null solutions to the operators $L_h^{\pm}$ and $L_{\mathrm{j}}^{\pm}$ respectively. They are also eigenvectors of $2\rho_h^{\mathrm{SW}}(B)$ with eigenvalues $\mp\frac{\mathrm{j}}{2}$ and $\mp\frac{1}{2}$ respectively. If these functions are used as mother wavelets for the wavelet transforms generated by the Heisenberg group, then the image space will consist of the null-solutions of the following differential operators, see Corollary 5.6:

$$D_h = \overline{X^r - Y^r} = (\partial_x - \partial_y) + \tfrac{h}{2}(x+y), \quad D_{\mathrm{j}} = \overline{X^r - \mathrm{j}Y^r} = (\partial_x + \mathrm{j}\partial_y) - \tfrac{h}{2}(x - \mathrm{j}y),$$

for $v_{\frac{1}{2}}^{\pm h}$ and $v_{\frac{1}{2}}^{\pm \mathrm{j}}$ respectively. This is again in line with the classical result (7.48). However annihilation of the eigenvector by a ladder operator does not mean that the part of the 2D-lattice becomes void since it can be reached via alternative routes on this lattice. Instead of multiplication by a zero, as it happens in the elliptic case, a half-plane of eigenvalues will be multiplied by the divisors of zero $1 \pm \mathrm{j}$.

We can also search ladder operators within the algebra $\mathfrak{sp}_2$ and admitting double numbers we will again find two sets of them, cf. § 3.3.2:

$$L_{2h}^{\pm} = \pm\tilde{A} + \tilde{Z}/2 = \mp\frac{q}{2}\frac{d}{dq} \mp \frac{1}{4} - \frac{\mathrm{j}h}{4}\frac{d^2}{dq^2} - \frac{\mathrm{j}q^2}{4h} = -\frac{\mathrm{j}}{4h}(L_h^{\pm})^2,$$

$$L_{2\mathrm{j}}^{\pm} = \pm\mathrm{j}\tilde{A} + \tilde{Z}/2 = \mp\frac{\mathrm{j}q}{2}\frac{d}{dq} \mp \frac{\mathrm{j}}{4} - \frac{\mathrm{j}h}{4}\frac{d^2}{dq^2} - \frac{\mathrm{j}q^2}{4h} = -\frac{\mathrm{j}}{4h}(L_{\mathrm{j}}^{\pm})^2.$$

Again the operators $L_{2h}^{\pm}$ and $L_{2h}^{\pm}$ produce double shifts in the orthogonal directions on the same two-dimensional lattice in Figure 10.

7.6. Parabolic (classical) representations on the phase space

After the previous two cases it is natural to link classical mechanics with dual numbers generated by the parabolic unit $\varepsilon^2 = 0$. Connection of the parabolic unit ε with the Galilean group of symmetries of classical mechanics is around for a while [128, App. C].

However the nilpotency of the parabolic unit ε make it difficult if we are working with dual number-valued functions only. To overcome this issue we consider a commutative real algebra $\mathfrak{C}$ spanned by 1, i, ε and iε with identities $\mathrm{i}^2 = -1$ and $\varepsilon^2 = 0$. A seminorm on $\mathfrak{C}$ is defined as

$$|a + b\mathrm{i} + c\varepsilon + d\mathrm{i}\varepsilon|^2 = a^2 + b^2.$$

7.6.1. Classical non-commutative representations. We wish to build a representation of the Heisenberg group which will be a classical analog of the Fock–Segal–

Bargmann representation (7.17). To this end we introduce the space $F^\varepsilon_h(\mathbb{H}^n)$ of $\mathfrak{C}$-valued functions on $\mathbb{H}^n$ with the property

$$f(s+s',h,y) = e^{\varepsilon h s'} f(s,x,y), \qquad \text{for all } (s,x,y)\in\mathbb{H}^n,\ s'\in\mathbb{R}, \tag{7.64}$$

and the square integrability condition (7.7). It is invariant under the left shifts and we restrict the left group action to $F^\varepsilon_h(\mathbb{H}^n)$.

There is a unimodular $\mathfrak{C}$-valued function on the Heisenberg group parametrised by a point $(h,q,p)\in\mathbb{R}^{2n+1}$:

$$E_{(h,q,p)}(s,x,y) = e^{2\pi(\varepsilon s\hbar + \mathrm{i}xq + \mathrm{i}yp)} = e^{2\pi\mathrm{i}(xq+yp)}(1+\varepsilon sh).$$

This function, if used instead of the ordinary exponent, produces a modification $\mathcal{F}_c$ of the Fourier transform (7.32). The transform intertwines the left regular representation with the following action on $\mathfrak{C}$-valued functions on the phase space:

$$\begin{aligned}\rho^\varepsilon_h(s,x,y): f(q,p) \quad\mapsto\quad & e^{-2\pi\mathrm{i}(xq+yp)}(f(q,p) \\ & +\varepsilon h(sf(q,p) + \frac{y}{2\pi\mathrm{i}} f'_q(q,p) - \frac{x}{2\pi\mathrm{i}} f'_p(q,p))).\end{aligned} \tag{7.65}$$

Remark 7.19. Comparing the traditional infinite-dimensional (7.17) and one-dimensional (7.33) representations of $\mathbb{H}^n$ we can note that the properties of the representation (7.65) are a non-trivial mixture of the former:

i. The action (7.65) is non-commutative, similarly to the quantum representation (7.17) and unlike the classical one (7.33). This non-commutativity will produce the Hamilton equations below in a way very similar to the Heisenberg equation, see Remark 7.21.
ii. The representation (7.65) does not change the support of a function f on the phase space, similarly to the classical representation (7.33) and unlike the quantum one (7.17). Such a localised action will be responsible later for the absence of an interference in classical probabilities.
iii. The parabolic representation (7.65) can not be derived from either the elliptic (7.17) or hyperbolic (7.52) by the plain substitution $h=0$.

We may also write a classical Schrödinger type representation. According to § 3.2 we get a representation formally very similar to the elliptic (7.34) and hyperbolic versions (7.53):

$$\begin{aligned}[= [\rho^\varepsilon_\chi(s',x',y')ft(x) &= e^{-\varepsilon h(s'+xy'-x'y'/2)} f(x-x') \\ &= (1-\varepsilon h(s'+xy'-\tfrac{1}{2}x'y'))f(x-x').\end{aligned} \tag{7.66}$$

However due to nilpotency of ε the (complex) Fourier transform $x\mapsto q$ produces a different formula for parabolic Schrödinger type representation in the configuration space, cf. (7.35) and (7.58):

$$[\rho^\varepsilon_\chi(s',x',y')\hat{f}](q) = e^{2\pi\mathrm{i}x'q}\left(\left(1-\varepsilon h(s'-\tfrac{1}{2}x'y')\right)\hat{f}(q) + \frac{\varepsilon h y'}{2\pi\mathrm{i}}\hat{f}'(q)\right).$$

This representation shares all properties mentioned in Remark 7.19 as well.

7.6.2. Hamilton equation. The identity $e^{\varepsilon t} - e^{-\varepsilon t} = 2\varepsilon t$ can be interpreted as a parabolic version of the sine function, while the parabolic cosine is identically equal to 1, cf. § 3.1 and [40, 81]. From this we obtain the parabolic version of the commutator (7.37),

$$[k', k]\hat{}_s(\varepsilon h, x, y) = \varepsilon h \int_{\mathbb{R}^{2n}} (xy' - yx') \\ \times \hat{k}'_s(\varepsilon h, x', y')\, \hat{k}_s(\varepsilon h, x - x', y - y')\, dx'dy',$$

for the partial parabolic Fourier-type transform $\hat{k}_s$ of the kernels. Thus the parabolic representation of the dynamical equation (7.23) becomes

$$\varepsilon h \frac{d\hat{f}_s}{dt}(\varepsilon h, x, y; t) = \varepsilon h \int_{\mathbb{R}^{2n}} (xy' - yx')\, \hat{H}_s(\varepsilon h, x', y')\, \hat{f}_s(\varepsilon h, x - x', y - y'; t)\, dx'dy'. \tag{7.67}$$

Although there is no possibility to divide by ε (since it is a zero divisor) we can obviously eliminate εh from both sides if the rest of the expressions are real. Moreover this can be done "in advance" through a kind of the antiderivative operator considered in [74, (4.1)]. This will prevent "imaginary parts" of the remaining expressions (which contain the factor ε) from vanishing.

Remark 7.20. It is noteworthy that the Planck constant completely disappeared from the dynamical equation. Thus the only prediction about it following from our construction is $h \neq 0$, which was confirmed by experiments, of course.

Using the duality between the Lie algebra of $\mathbb{H}^n$ and the phase space we can find an adjoint equation for observables on the phase space. To this end we apply the usual Fourier transform $(x, y) \mapsto (q, p)$. It turns to be the Hamilton equation [74, (4.7)]. However the transition to the phase space is more a custom rather than a necessity and in many cases we can efficiently work on the Heisenberg group itself.

Remark 7.21. It is noteworthy, that the non-commutative representation (7.65) allows one to obtain the Hamilton equation directly from the commutator $[\rho^{\varepsilon}_{h}(k_1), \rho^{\varepsilon}_{h}(k_2)]$. Indeed its straightforward evaluation will produce exactly the above expression. On the contrary such a commutator for the commutative representation (7.33) is zero and to obtain the Hamilton equation we have to work with an additional tool, e.g., an anti-derivative [74, (4.1)].

Example 7.22.

i. For the harmonic oscillator in Example 7.6 the equation (7.67) again reduces to the form (7.25) with the solution given by (7.26). The adjoint equation of the harmonic oscillator on the phase space is not different from the quantum written in Example 7.10(i). This is true for any Hamiltonian of at most quadratic order.

ii. For non-quadratic Hamiltonians, classical and quantum dynamics are different, of course. For example, the cubic term of ∂_s in the equation (7.28) will generate the factor $\varepsilon^3 = 0$ and thus vanish. Thus the equation (7.67) of the unharmonic oscillator on $\mathbb{H}^n$ becomes

$$\dot{f} = \left(mk^2 y \frac{\partial}{\partial x} + \frac{\lambda y}{2} \frac{\partial^2}{\partial x^2} - \frac{1}{m} x \frac{\partial}{\partial y} \right) f.$$

The adjoint equation on the phase space is

$$\dot{f} = \left(\left(mk^2 q + \frac{\lambda}{2} q^2 \right) \frac{\partial}{\partial p} - \frac{1}{m} p \frac{\partial}{\partial q} \right) f.$$

The last equation is the classical Hamilton equation generated by the cubic potential (7.27). Qualitative analysis of its dynamics can be found in many textbooks [4, § 4.C, Pic. 12; 110, § 4.4].

Remark 7.23. We have obtained the *Poisson bracket* from the commutator of convolutions on $\mathbb{H}^n$ without any quasiclassical limit $h \to 0$. This has a common source with the deduction of main calculus theorems in [17] based on dual numbers. As explained in [82, Rem. 6.9] this is due to the similarity between the parabolic unit ε and the infinitesimal number used in non-standard analysis [23]. In other words, we never need to take care about terms of order $O(h^2)$ because they will be wiped out by $\varepsilon^2 = 0$.

An alternative derivation of classical dynamics from the Heisenberg group is given in the recent paper [101].

7.6.3. Classical probabilities. It is worth noticing that dual numbers are not only helpful in reproducing classical Hamiltonian dynamics, they also provide the classic rule for addition of probabilities. We use the same formula (7.29) to calculate kernels of the states. The important difference is now that the representation (7.65) does not change the support of functions. Thus if we calculate the correlation term $\langle v_1, \rho(g)v_2 \rangle$ in (7.30), then it will be zero for every two vectors v_1 and v_2 which have disjoint supports in the phase space. Thus no interference similar to quantum or hyperbolic cases (Subsection 7.3.3) is possible.

7.6.4. Ladder operator for the nilpotent subgroup. Finally we look for ladder operators for the Hamiltonian $\tilde{B} + \tilde{Z}/2$ or, equivalently, $-\tilde{B} + \tilde{Z}/2$. It can be identified with a free particle [127, § 3.8].

We can look for ladder operators in the representation (7.12–7.13) within the Lie algebra $\mathfrak{h}_1$ in the form $L_\varepsilon^\pm = a\tilde{X} + b\tilde{Y}$. This is possible if and only if

$$-b = \lambda a, \quad 0 = \lambda b. \tag{7.68}$$

The compatibility condition $\lambda^2 = 0$ implies $\lambda = 0$ within complex numbers. However such a "ladder" operator produces only the zero shift on the eigenvectors, cf. (3.15).

Another possibility appears if we consider the representation of the Heisenberg group induced by dual-valued characters. On the configuration space such a representation is [88, (4.11)]:

$$[\rho^{\varepsilon}_{\chi}(s,x,y)f](q) = e^{2\pi \mathrm{i} xq}\left(\left(1-\varepsilon h(s-\tfrac{1}{2}xy)\right)f(q)+\frac{\varepsilon hy}{2\pi\mathrm{i}}f'(q)\right). \tag{7.69}$$

The corresponding derived representation of $\mathfrak{h}_1$ is

$$\rho^{p}_{h}(X)=2\pi \mathrm{i}q,\qquad \rho^{p}_{h}(Y)=\frac{\varepsilon h}{2\pi\mathrm{i}}\frac{d}{dq},\qquad \rho^{p}_{h}(S)=-\varepsilon hI. \tag{7.70}$$

However the Shale–Weil extension generated by this representation is inconvenient. It is better to consider the FSB–type parabolic representation (7.65) on the phase space induced by the same dual-valued character. Then the derived representation of $\mathfrak{h}_1$ is

$$\rho^{p}_{h}(X)=-2\pi\mathrm{i}q-\frac{\varepsilon h}{4\pi\mathrm{i}}\partial_p,\qquad \rho^{p}_{h}(Y)=-2\pi\mathrm{i}p+\frac{\varepsilon h}{4\pi\mathrm{i}}\partial_q,\qquad \rho^{p}_{h}(S)=\varepsilon hI. \tag{7.71}$$

An advantage of the FSB representation is that the derived form of the parabolic Shale–Weil representation coincides with the elliptic one (7.19).

Eigenfunctions with the eigenvalue μ of the parabolic Hamiltonian $\tilde{B}+\tilde{Z}/2 = q\partial_p$ have the form

$$v_\mu(q,p)=e^{\mu p/q}f(q),\ \text{with an arbitrary function } f(q). \tag{7.72}$$

The linear equations defining the corresponding ladder operator $L^{\pm}_{\varepsilon} = a\tilde{X} + b\tilde{Y}$ in the algebra $\mathfrak{h}_1$ are (7.68). The compatibility condition $\lambda^2=0$ implies $\lambda=0$ within complex numbers again. Admitting dual numbers we have additional values $\lambda=\pm\varepsilon\lambda_1$ with $\lambda_1\in\mathbb{C}$ with the corresponding ladder operators

$$L^{\pm}_{\varepsilon}=\tilde{X}\mp\varepsilon\lambda_1\tilde{Y}=-2\pi\mathrm{i}q-\frac{\varepsilon h}{4\pi\mathrm{i}}\partial_p\pm 2\pi\varepsilon\lambda_1\mathrm{i}p=-2\pi\mathrm{i}q+\varepsilon\mathrm{i}(\pm 2\pi\lambda_1 p+\frac{h}{4\pi}\partial_p).$$

For the eigenvalue $\mu=\mu_0+\varepsilon\mu_1$ with μ_0, $\mu_1\in\mathbb{C}$ the eigenfunction (7.72) can be rewritten as

$$v_\mu(q,p)=e^{\mu p/q}f(q)=e^{\mu_0 p/q}\left(1+\varepsilon\mu_1\frac{p}{q}\right)f(q) \tag{7.73}$$

due to the nilpotency of ε. Then the ladder action of $L^{\pm}_{\varepsilon}$ is $\mu_0+\varepsilon\mu_1\mapsto\mu_0+\varepsilon(\mu_1\pm\lambda_1)$. Therefore these operators are suitable for building $\mathfrak{sp}_2$-modules with a one-dimensional chain of eigenvalues.

Finally, consider the ladder operator for the same element $B+Z/2$ within the Lie algebra $\mathfrak{sp}_2$, cf. § 3.3.3. There is the only operator $L^{\pm}_{p}=\tilde{B}+\tilde{Z}/2$ corresponding to complex coefficients, which does not affect the eigenvalues. However the dual numbers lead to the operators

$$L^{\pm}_{\varepsilon}=\pm\varepsilon\lambda_2\tilde{A}+\tilde{B}+\tilde{Z}/2=\pm\frac{\varepsilon\lambda_2}{2}(q\partial_q-p\partial_p)+q\partial_p,\qquad \lambda_2\in\mathbb{C}.$$

These operators act on eigenvalues in a non-trivial way.

7.6.5. Similarity and correspondence. We wish to summarise our findings. Firstly, the appearance of hypercomplex numbers in ladder operators for $\mathfrak{h}_1$ follows exactly the same pattern as was already noted for $\mathfrak{sp}_2$, see Remark 3.8:

- the introduction of complex numbers is a necessity for the *existence* of ladder operators in the elliptic case;
- in the parabolic case we need dual numbers to make ladder operators *useful*;
- in the hyperbolic case double numbers are not required, neither for the existence nor for the usability of ladder operators, but they do provide an enhancement.

In the spirit of the similarity and correspondence Principle 3.5 we have the following extension of Proposition 3.9:

Proposition 7.24. *Let a vector $H \in \mathfrak{sp}_2$ generate the subgroup K, N' or A', that is $H = Z$, $B + Z/2$, or $2B$ respectively. Let ι be the respective hypercomplex unit. Then the ladder operators $L^{\pm}$ satisfying the commutation relation*

$$[H, L_2^{\pm}] = \pm \iota L^{\pm}$$

are given by:

i. *Within the Lie algebra $\mathfrak{h}_1$: $L^{\pm} = \tilde{X} \mp \iota \tilde{Y}$.*
ii. *Within the Lie algebra $\mathfrak{sp}_2$: $L_2^{\pm} = \pm \iota \tilde{A} + \tilde{E}$. Here $E \in \mathfrak{sp}_2$ is a linear combination of B and Z with the properties:*
    - *$E = [A, H]$.*
    - *$H = [A, E]$.*
    - *Killings form $K(H, E)$ [55, § 6.2] vanishes.*

    Any of the above properties defines the vector $E \in \operatorname{span}\{B, Z\}$ up to a real constant factor.

It is worth continuing this investigation and describing in detail hyperbolic and parabolic versions of FSB spaces.

8. Open problems

The reader may already note numerous objects and results that deserve further consideration. It may also be useful to state some open problems explicitly. In this section we indicate several directions for further work, which go through four main areas described in the paper.

8.1. Geometry

Geometry is the most elaborated area so far, yet many directions are waiting for further exploration.

i. Möbius transformations (1.1) with three types of hypercomplex units appear from the action of the group $SL_2(\mathbb{R})$ on the homogeneous space $SL_2(\mathbb{R})/H$ [85], where H is any subgroup A, N, K from the Iwasawa decomposition (1.3). Which other actions and hypercomplex numbers can be obtained from other Lie groups and their subgroups?

ii. Lobachevsky geometry of the upper half-plane is an extremely beautiful and well-developed subject [8, 22]. However the traditional study is limited to one subtype out of nine possible: with the complex numbers for Möbius transformation and the complex imaginary unit used in FSCc (2.8). The remaining eight cases should be explored in various directions, notably in the context of discrete subgroups [7].
iii. The Fillmore-Springer-Cnops construction, see Subsection 2.2, is closely related to the *orbit method* [57] applied to $SL_2(\mathbb{R})$. An extension of the orbit method from the Lie algebra dual to matrices representing cycles may be fruitful for semisimple Lie groups.
iv. A development of a discrete version of the geometrical notions can be derived from suitable discrete groups. A natural first example is the group $SL_2(\mathbb{F})$, where $\mathbb{F}$ is a finite field, e.g., $\mathbb{Z}_p$ the field of integers modulo a prime p.

8.2. Analytic functions

It is known that in several dimensions there are different notions of analyticity, e.g., several complex variables and Clifford analysis. However, analytic functions of a complex variable are usually thought to be the only options in a plane domain. The following seem to be promising:

i. Development of the basic components of analytic function theory (the Cauchy integral, the Taylor expansion, the Cauchy-Riemann and Laplace equations, etc.) from the same construction and principles in the elliptic, parabolic and hyperbolic cases and respective subcases.
ii. Identification of Hilbert spaces of analytic functions of Hardy and Bergman types, investigation of their properties. Consideration of the corresponding Toeplitz operators and algebras generated by them.
iii. Application of analytic methods to elliptic, parabolic and hyperbolic equations and corresponding boundary and initial value problems.
iv. Generalisation of the results obtained to higher-dimensional spaces. Detailed investigation of physically significant cases of three and four dimensions.
v. There is a current interest in construction of analytic function theory on discrete sets. Our approach is ready for application to analytic functions in a discrete geometric set-up as outlined in item 8.1.iv above.

8.3. Functional calculus

The functional calculus of a finite-dimensional operator considered in Section 6 is elementary but provides a coherent and comprehensive treatment. It should be extended to further cases where other approaches seem to be rather limited.

i. Nilpotent and quasinilpotent operators have the most trivial spectrum possible (the single point $\{0\}$) while their structure can be highly non-trivial. Thus the standard spectrum is insufficient for this class of operators. In contrast, the covariant calculus and the spectrum give complete descriptions of nilpotent operators – the basic prototypes of quasi-nilpotent ones. For quasi-nilpotent operators the construction will be more complicated and should use analytic functions as mentioned in item 8.2.i.

ii. The version of covariant calculus described above is based on the *discrete series* representations of the $SL_2(\mathbb{R})$ group and is particularly suitable for the description of the *discrete spectrum* (note the remarkable coincidence in the names).

It would be interesting to develop similar covariant calculi based on the two other representation series of $SL_2(\mathbb{R})$: *principal* and *complementary* [97]. The corresponding versions of analytic function theories for principal [64] and complementary series [82] have been initiated within a unifying framework. The classification of analytic function theories into elliptic, parabolic, hyperbolic [78, 82] should be also compared with discrete, continuous and residual spectrum of an operator.

iii. Let a be an operator with $\mathbf{sp}\, a \in \bar{\mathbb{D}}$ and $\|a^k\| < Ck^p$. It is typical to consider instead of a the *power bounded* operator ra, where $0 < r < 1$, and consequently develop its H_∞ calculus. However such a regularisation is very rough and hides the nature of extreme points of $\mathbf{sp}\, a$. To restore full information a subsequent limit transition $r \to 1$ of the regularisation parameter r is required. This make the entire technique rather cumbersome and many results have an indirect nature.

The regularisation $a^k \to a^k/k^p$ is more natural and accurate for polynomially bounded operators. However it cannot be achieved within the homomorphic calculus Definition 6.1 because it is not compatible with any algebra homomorphism. Albeit this may be achieved within the covariant calculus Defninition 6.4 and Bergman type space from item 8.2.ii.

iv. Several non-commuting operators are especially difficult to treat with functional calculus Definition 6.1 or a joint spectrum. For example, deep insights on joint spectrum of commuting tuples [119] refused to be generalised to the non-commuting case so far. The covariant calculus was initiated [62] as a new approach to this hard problem and was later found useful elsewhere as well. Multidimensional covariant calculus [73] should use analytic functions described in item 8.2.iv.

v. As we noted above there is a duality between the co- and contravariant calculi from Definitions. 4.20 and 4.22. We also saw in Section 6 that functional calculus is an example of contravariant calculus and the functional model is a case of a covariant one. It would be interesting to explore the duality between them further.

8.4. Quantum mechanics

Due to space restrictions we only touched upon quantum mechanics; further details can be found in [63, 74, 76, 77, 79, 88]. In general, the Erlangen approach is much more popular among physicists than among mathematicians. Nevertheless its potential is not exhausted even there.

i. There is a possibility to build representation of the Heisenberg group using characters of its centre with values in dual and double numbers rather than in complex ones. This would naturally unify classical mechanics, traditional QM and hyperbolic QM [52]. In particular, a full construction of the corresponding Fock–Segal–Bargmann spaces would be of interest.
ii. Representations of nilpotent Lie groups with multidimensional centres in Clifford algebras as a framework for consistent quantum field theories based on De Donder–Weyl formalism [76].

Remark 8.1. This work has been done within the "Erlangen program at large" framework [78, 82], thus it would be suitable to explain the numbering of various papers. Since the logical order may be different from the chronological one the following numbering scheme is used:

Prefix	Branch description
"0" or no prefix	Mainly geometrical works, within the classical field of the Erlangen program by F. Klein, see [82, 85]
"1"	Papers on analytical functions theories and wavelets, e.g., [64]
"2"	Papers on operator theory, functional calculi and spectra, e.g., [75]
"3"	Papers on mathematical physics, e.g., [88]

For example, [88] is the first paper in the mathematical physics area. The present paper [89] outlines the whole framework and thus does not carry a subdivision number. The on-line version of this paper may be updated in due course to reflect the achieved progress.

Acknowledgement

The material of these notes was derived from lectures given at various summer/winter schools and advanced courses. Those presentations helped me to clarify ideas and improve my understanding of the subject. In particular, I am grateful to Prof. S.V. Rogosin and Dr. A.A. Koroleva for a kind invitation to the Minsk Winter School in 2010, which was an exciting event. I would like also to acknowledge the support of MAGIC group during my work on those notes.

References

[1] F. Agostini, S. Caprara, and G. Ciccotti, *Do we have a consistent non-adiabatic quantum-classical mechanics?*, Europhys. Lett. EPL **78** (2007), no. 3, Art. 30001, 6. doi: 10.1209/0295-5075/78/30001. MR2366698 (2008k:81004) ↑59

[2] Syed Twareque Ali, Jean-Pierre Antoine, and Jean-Pierre Gazeau, *Coherent states, wavelets and their generalizations*, Graduate Texts in Contemporary Physics, Springer-Verlag, New York, 2000. MR2002m:81092 ↑22, 23, 24, 25, 29

[3] Robert F.V. Anderson, *The Weyl functional calculus*, J. Functional Analysis **4** (1969), 240–267. MR58#30405 ↑40, 41

[4] V.I. Arnol'd, *Mathematical methods of classical mechanics*, Graduate Texts in Mathematics, vol. 60, Springer-Verlag, New York, 1991. Translated from the 1974 Russian original by K. Vogtmann and A. Weinstein, corrected reprint of the second (1989) edition. MR96c:70001 ↑53, 73

[5] Damir Z. Arov and Harry Dym, *J-contractive matrix valued functions and related topics*, Encyclopedia of Mathematics and its Applications, vol. 116, Cambridge University Press, Cambridge, 2008. MR2474532 ↑18

[6] M.B. Balk, *Polyanalytic functions and their generalizations* [*MR*1155418 (93*f*: 30050)], Complex analysis, I, 1997, pp. 195–253. MR1464199 ↑36

[7] Alan F. Beardon, *The geometry of discrete groups*, Graduate Texts in Mathematics, vol. 91, Springer-Verlag, New York, 1995. Corrected reprint of the 1983 original. MR1393195 (97d:22011) ↑76

[8] ———, *Algebra and geometry*, Cambridge University Press, Cambridge, 2005. MR2153234 (2006a:00001) ↑2, 76

[9] Alan F. Beardon and Ian Short, *Conformal symmetries of regions*, Irish Math. Soc. Bull. **59** (2007), 49–60. MR2353408 (2008j:30013) ↑38

[10] F.A. Berezin, *Covariant and contravariant symbols of operators*, Izv. Akad. Nauk SSSR Ser. Mat. **36** (1972), 1134–1167. Reprinted in [11, pp. 228–261]. MR50#2996 ↑26, 27, 28

[11] ———, *Metod vtorichnogo kvantovaniya*, Second, "Nauka", Moscow, 1986. Edited and with a preface by M.K. Polivanov. MR89c:81001 ↑26, 79

[12] Albrecht Böttcher, Yuri I. Karlovich, and Ilya M. Spitkovsky, *Convolution operators and factorization of almost periodic matrix functions*, Operator Theory: Advances and Applications, vol. 131, Birkhäuser Verlag, Basel, 2002. MR1898405 (2003c:47047) ↑25

[13] Charles P. Boyer and Willard Miller Jr., *A classification of second-order raising operators for Hamiltonians in two variables*, J. Mathematical Phys. **15** (1974), 1484–1489. MR0345542 (49 #10278) ↑52, 63

[14] Ola Bratteli and Palle E.T. Jorgensen, *Isometries, shifts, Cuntz algebras and multiresolution wavelet analysis of scale N*, Integral Equations Operator Theory **28** (1997), no. 4, 382–443. `arXiv:funct-an/9612003`. ↑25

[15] Alastair Brodlie and Vladimir V. Kisil, *Observables and states in p-mechanics*, Advances in mathematics research. vol. 5, 2003, pp. 101–136. `arXiv:quant-ph/0304023`. MR2117375 ↑56, 57, 58, 65

[16] Esteban Calzetta and Enric Verdaguer, *Real-time approach to tunnelling in open quantum systems: decoherence and anomalous diffusion*, J. Phys. A **39** (2006), no. 30, 9503–9532. MR2246702 (2007f:82059) ↑57, 61

[17] Francesco Catoni, Roberto Cannata, and Enrico Nichelatti, *The parabolic analytic functions and the derivative of real functions*, Adv. Appl. Clifford Algebras **14** (2004), no. 2, 185–190. ↑73

[18] Jens Gerlach Christensen and Gestur Ólafsson, *Examples of coorbit spaces for dual pairs*, Acta Appl. Math. **107** (2009), no. 1-3, 25–48. MR2520008 ↑22, 24

[19] Jan Cnops, *An introduction to Dirac operators on manifolds*, Progress in Mathematical Physics, vol. 24, Birkhäuser Boston Inc., Boston, MA, 2002. MR1917 405 ↑7

[20] Jan Cnops and Vladimir V. Kisil, *Monogenic functions and representations of nilpotent Lie groups in quantum mechanics*, Math. Methods Appl. Sci. **22** (1999), no. 4, 353–373. arXiv:math/9806150. **Zbl** # 1005.22003. MR1671449 (2000b:81044) ↑25, 52

[21] Denis Constales, Nelson Faustino, and Rolf Sören Kraußhar, *Fock spaces, Landau operators and the time-harmonic Maxwell equations*, Journal of Physics A: Mathematical and Theoretical **44** (2011), no. 13, 135303. ↑52

[22] H.S.M. Coxeter and S.L. Greitzer, *Geometry revisited.*, Random House, New York, 1967 (English). **Zbl** # 0166.16402. ↑76

[23] Martin Davis, *Applied nonstandard analysis*, Wiley-Interscience [John Wiley & Sons], New York, 1977. MR0505473 (58 #21590) ↑73

[24] Maurice A. de Gosson, *Spectral properties of a class of generalized Landau operators*, Comm. Partial Differential Equations **33** (2008), no. 10-12, 2096–2104. MR2475331 (2010b:47128) ↑55

[25] P.A.M. Dirac, *Quantum mechanics and a preliminary investigation of the hydrogen atom*, Proceedings of the Royal Society of London. Series A, Containing Papers of a Mathematical and Physical Character **110** (1926), no. 755, pp. 561–579 (English). ↑51

[26] M. Duflo and Calvin C. Moore, *On the regular representation of a nonunimodular locally compact group*, J. Functional Analysis **21** (1976), no. 2, 209–243. MR52#14145 ↑24, 29

[27] Nelson Dunford and Jacob T. Schwartz, *Linear operators. Part i: General theory*, Pure and Applied Mathematics, vol. VII, John Wiley & Sons, Inc., New York, 1957. ↑28

[28] Arthur Erdélyi, Wilhelm Magnus, Fritz Oberhettinger, and Francesco G. Tricomi, *Higher transcendental functions. Vol. II*, Robert E. Krieger Publishing Co. Inc., Melbourne, Fla., 1981. Based on notes left by Harry Bateman, Reprint of the 1953 original. MR698780 (84h:33001b) ↑63, 68, 69

[29] Feichtinger, Hans G. and Groechenig, K.H., *Banach spaces related to integrable group representations and their atomic decompositions, I*, J. Funct. Anal. **86** (1989), no. 2, 307–340. **Zbl** # 691.46011. ↑22, 24

[30] R.P. Feynman, *Qed: the strange theory of light and matter*, Penguin Press Science Series, Penguin, 1990. ↑51

[31] Gerald B. Folland, *Harmonic analysis in phase space*, Annals of Mathematics Studies, vol. 122, Princeton University Press, Princeton, NJ, 1989. MR92k:22017 ↑50, 52, 54, 55, 56, 60, 63, 64

[32] Hartmut Führ, *Abstract harmonic analysis of continuous wavelet transforms*, Lecture Notes in Mathematics, vol. 1863, Springer-Verlag, Berlin, 2005. MR2130226 (2006m:43003) ↑22, 24

[33] Jean-Pierre Gazeau, *Coherent States in Quantum Physics*, Wiley-VCH Verlag, 2009. ↑52, 55, 64

[34] Loukas Grafakos, *Classical Fourier analysis*, Second, Graduate Texts in Mathematics, vol. 249, Springer, New York, 2008. MR2445437 ↑26, 36

[35] N.A. Gromov, *Контракции и аналитические продолжения классических групп. Единыи подход.* (*Russian*) [*Contractions and analytic extensions of classical groups. Unified approach*], Akad. Nauk SSSR Ural. Otdel. Komi Nauchn. Tsentr, Syktyvkar, 1990. MR1092760 (91m:81078) ↑52

[36] ———, *Transitions: Contractions and analytical continuations of the Cayley–Klein groups*, Int. J. Theor. Phys. **29** (1990), 607–620. ↑52

[37] N.A. Gromov and V.V. Kuratov, *All possible Cayley-Klein contractions of quantum orthogonal groups*, Yadernaya Fiz. **68** (2005), no. 10, 1752–1762. MR2189521 (2006g:81101) ↑51, 52

[38] Uwe Günther and Sergii Kuzhel, *$\mathcal{PT}$–symmetry, Cartan decompositions, Lie triple systems and Krein space-related Clifford algebras*, Journal of Physics A: Mathematical and Theoretical **43** (2010), no. 39, 392002. ↑51, 52

[39] Sigurdur Helgason, *Integral geometry and Radon transforms*, Springer, New York, 2011. MR2743116 ↑26

[40] Francisco J. Herranz, Ramón Ortega, and Mariano Santander, *Trigonometry of spacetimes: a new self-dual approach to a curvature/signature (in)dependent trigonometry*, J. Phys. A **33** (2000), no. 24, 4525–4551. `arXiv:math-ph/9910041`. MR1768742 (2001k:53099) ↑16, 72

[41] Roger A. Horn and Charles R. Johnson, *Topics in matrix analysis*, Cambridge University Press, Cambridge, 1994. Corrected reprint of the 1991 original. MR95c: 15001 ↑45

[42] Roger Howe, *On the role of the Heisenberg group in harmonic analysis*, Bull. Amer. Math. Soc. (N.S.) **3** (1980), no. 2, 821–843. MR81h:22010 ↑2, 54, 65

[43] ———, *Quantum mechanics and partial differential equations*, J. Funct. Anal. **38** (1980), no. 2, 188–254. MR83b:35166 ↑28, 50, 52, 55, 60, 64

[44] Roger Howe and Eng-Chye Tan, *Nonabelian harmonic analysis. Applications of $SL(2,\mathbf{R})$*, Springer-Verlag, New York, 1992. MR1151617 (93f:22009) ↑20, 52, 68

[45] Robin Hudson, *Generalised translation-invariant mechanics*, D. Phil. thesis, Bodleian Library, Oxford, 1966. ↑51, 52

[46] ———, *Translation invariant phase space mechanics*, Quantum theory: reconsideration of foundations – 2, 2004, pp. 301–314. MR2111131 (2006e:81134) ↑51, 52

[47] Ondrej Hutník, *On Toeplitz-type operators related to wavelets*, Integral Equations Operator Theory **63** (2009), no. 1, 29–46. MR2480637 ↑24

[48] Andreas Johansson, *Shift-invariant signal norms for fault detection and control*, Systems Control Lett. **57** (2008), no. 2, 105–111. MR2378755 (2009d:93035) ↑26

[49] A.Yu. Khrennikov, *Hyperbolic quantum mechanics*, Dokl. Akad. Nauk **402** (2005), no. 2, 170–172. MR2162434 (2006d:81118) ↑51, 65, 68

[50] Andrei Khrennikov, *'Quantum probabilities' as context depending probabilities* (2001). `arXiv:quant-ph/0106073`. ↑58

[51] ———, *Hyperbolic Quantum Mechanics*, Adv. Appl. Clifford Algebras **13** (2003), no. 1, 1–9 (English). `arXiv:quant-ph/0101002`. ↑51, 61, 68

[52] ______, *Hyperbolic quantization*, Adv. Appl. Clifford Algebr. **18** (2008), no. 3-4, 843–852. MR2490591 ↑51, 65, 66, 67, 68, 78

[53] Andrei Khrennikov and Gavriel Segre, *Hyperbolic quantization*, Quantum probability and infinite dimensional analysis, 2007, pp. 282–287. MR2359402 ↑65

[54] A.Yu. Khrennikov and Ya.I. Volovich, *Numerical experiment on interference for macroscopic particles* (2001). `arXiv:quant-ph/0111159`. ↑58

[55] A.A. Kirillov, *Elements of the theory of representations*, Springer-Verlag, Berlin, 1976. Translated from the Russian by Edwin Hewitt, Grundlehren der Mathematischen Wissenschaften, Band 220. MR54#447 ↑5, 14, 16, 18, 21, 41, 43, 53, 58, 75

[56] ______, *Introduction to the theory of representations and noncommutative harmonic analysis* [*MR*90*a*:22005], Representation theory and noncommutative harmonic analysis, i, 1994, pp. 1–156, 227–234. MR1311488. MR1311 488 ↑50, 58

[57] ______, *Merits and demerits of the orbit method*, Bull. Amer. Math. Soc. (N.S.) **36** (1999), no. 4, 433–488. MR2000h:22001 ↑50, 58, 76

[58] A.A. Kirillov, *A tale on two fractals*, 2010. URL: `http://www.math.upenn.edu/~kirillov/MATH480-F07/tf.pdf`, in publication. ↑9

[59] Anastasia V. Kisil, *Isometric action of* $SL_2(\mathbb{R})$ *on homogeneous spaces*, Adv. App. Clifford Algebras **20** (2010), no. 2, 299–312. `arXiv:0810.0368`. ↑14

[60] Vladimir V. Kisil, *Clifford valued convolution operator algebras on the Heisenberg group. A quantum field theory model*, Clifford algebras and their applications in mathematical physics, proceedings of the third international conference held in Deinze, 1993, pp. 287–294. MR1266878. ↑52

[61] ______, *Quantum probabilities and non-commutative Fourier transform on the Heisenberg group*, Interaction between functional analysis, harmonic analysis and probability (Columbia, MO, 1994), 1995, pp. 255–266. MR97b:81060. ↑56

[62] ______, *Möbius transformations and monogenic functional calculus*, Electron. Res. Announc. Amer. Math. Soc. **2** (1996), no. 1, 26–33. On-line. MR1405966 (98a:47018) ↑28, 40, 41, 77

[63] ______, *Plain mechanics: classical and quantum*, J. Natur. Geom. **9** (1996), no. 1, 1–14. `arXiv:funct-an/9405002`. MR1374912 (96m:81112) ↑56, 77

[64] ______, *Analysis in* $\mathbf{R}^{1,1}$ *or the principal function theory*, Complex Variables Theory Appl. **40** (1999), no. 2, 93–118. `arXiv:funct-an/9712003`. MR1744876 (2000k:30078) ↑5, 14, 22, 25, 30, 37, 38, 39, 77, 78

[65] ______, *Two approaches to non-commutative geometry*, Complex methods for partial differential equations (Ankara, 1998), 1999, pp. 215–244. `arXiv:funct-an/9703001`. MR1744440 (2001a:01002) ↑30

[66] ______, *Wavelets in Banach spaces*, Acta Appl. Math. **59** (1999), no. 1, 79–109. `arXiv:math/9807141`, On-line. MR1740458 (2001c:43013) ↑22, 23, 25, 26, 27, 28, 38, 40, 41, 59

[67] ______, *Umbral calculus and cancellative semigroup algebras*, Z. Anal. Anwendungen **19** (2000), no. 2, 315–338. `arXiv:funct-an/9704001`. **Zbl** # 0959.43004. MR1768995 (2001g:05017) ↑23

[68] ———, *Spaces of analytical functions and wavelets – Lecture notes*, 2000. 92 pp. arXiv:math.CV/0204018. ↑30, 38, 39

[69] ———, *Meeting Descartes and Klein somewhere in a noncommutative space*, Highlights of mathematical physics (London, 2000), 2002, pp. 165–189. arXiv: math-ph/0112059. MR2001578 (2005b:43015) ↑4, 30

[70] ———, *Quantum and classical brackets*, Internat. J. Theoret. Phys. **41** (2002), no. 1, 63–77. arXiv:math-ph/0007030. On-line. MR2003b:81105 ↑50, 56

[71] ———, *Tokens: an algebraic construction common in combinatorics, analysis, and physics*, Ukrainian mathematics congress – 2001 (Ukrainian), 2002, pp. 146–155. arXiv:math.FA/0201012. MR2228860 (2007d:05010) ↑23

[72] ———, *Two slits interference is compatible with particles' trajectories*, Quantum theory: Reconsideration of foundations, 2002, pp. 215–226. arXiv:quant-ph/0111094. ↑58

[73] ———, *Monogenic calculus as an intertwining operator*, Bull. Belg. Math. Soc. Simon Stevin **11** (2004), no. 5, 739–757. arXiv:math.FA/0311285, On-line. MR2130636 (2006a:47025) ↑40, 41, 77

[74] ———, *p-Mechanics as a physical theory: an introduction*, J. Phys. A **37** (2004), no. 1, 183–204. arXiv:quant-ph/0212101, On-line. **Zbl** # 1045.81032. MR2044764 (2005c:81078) ↑50, 55, 56, 58, 59, 60, 72, 77

[75] ———, *Spectrum as the support of functional calculus*, Functional analysis and its applications, 2004, pp. 133–141. arXiv:math.FA/0208249. MR2098877 ↑4, 28, 40, 41, 78

[76] ———, *p-mechanics and field theory*, Rep. Math. Phys. **56** (2005), no. 2, 161–174. arXiv:quant-ph/0402035, On-line. MR2176789 (2006h:53104) ↑77, 78

[77] ———, *A quantum-classical bracket from p-mechanics*, Europhys. Lett. **72** (2005), no. 6, 873–879. arXiv:quant-ph/0506122, On-line. MR2213328 (2006k:81134) ↑59, 77

[78] ———, *Erlangen program at large – 0: Starting with the group* $SL_2(\mathbf{R})$, Notices Amer. Math. Soc. **54** (2007), no. 11, 1458–1465. arXiv:math/0607387, On-line. MR2361159 ↑77, 78

[79] ———, *Comment on "Do we have a consistent non-adiabatic quantum-classical mechanics?" by Agostini F. et al.*, Europhys. Lett. EPL **89** (2010), 50005. arXiv:0907.0855. ↑16, 59, 77

[80] ———, *Computation and dynamics: Classical and quantum*, AIP Conference Proceedings **1232** (2010), no. 1, 306–312. arXiv:0909.1594. ↑52

[81] ———, *Erlangen program at large – 2: Inventing a wheel. The parabolic one*, Trans. Inst. Math. of the NAS of Ukraine, 2010, pp. 89–98. arXiv:0707.4024. ↑72

[82] ———, *Erlangen program at large – 1: Geometry of invariants*, SIGMA, Symmetry Integrability Geom. Methods Appl. **6** (2010), no. 076, 45. arXiv:math.CV/0512416. ↑4, 7, 11, 14, 37, 65, 73, 77, 78

[83] ———, *Wavelets beyond admissibility*, Proceedings of the 10th ISAAC Congress, London 2009, 2010, pp. 6 pp. arXiv:0911.4701. ↑22

[84] ———, *Covariant transform*, Journal of Physics: Conference Series **284** (2011), no. 1, 012038. arXiv:1011.3947. ↑31, 40, 41

[85] ______, *Erlangen program at large – 2 1/2: Induced representations and hypercomplex numbers*, *Известия Коми научного центра УрО РАН* [Izvestiya Komi nauchnogo centra UrO RAN] **5** (2011), no. 1, 4–10. arXiv:0909.4464. ↑5, 14, 18, 19, 55, 64, 65, 75, 78

[86] ______, *Erlangen Programme at Large* 1.1*: Integral transforms and differential operators*, in preparation (2011). ↑34, 36, 37, 49

[87] ______, *Erlangen Programme at Large* 3.2*: Ladder operators in hypercomplex mechanics*, Acta Polytechnica **51** (2011), no. 4, 44–53. arXiv:1103.1120. ↑16, 52

[88] ______, *Hypercomplex representations of the Heisenberg group and mechanics*, Internat. J. Theoret. Phys. **51** (2012), no. 3, 964–984. arXiv:1005.5057. ↑16, 54, 69, 74, 77, 78

[89] ______, *Erlangen program at large: an Overview*, Advances in applied analysis, 2012, pp. 1–65. arXiv:1106.1686. ↑78

[90] ______, *Geometry of Möbius transformations: Elliptic, parabolic and hyperbolic actions of* $SL_2(\mathbf{R})$, Imperial College Press, 2012. Includes a live DVD. ↑11, 14

[91] ______, *Operator covariant transform and local principle*, J. Phys. A: Math. Theor. **45** (2012), 244022. arXiv:1201.1749. ↑22

[92] John R. Klauder, *Coherent states for the hydrogen atom*, J. Phys. A **29** (1996), no. 12, L293–L298. MR1398598 (97f:81052) ↑31

[93] John R. Klauder and Bo-Sture Skagerstam (eds.), *Coherent states. Applications in physics and mathematical physics*, World Scientific Publishing Co., Singapur, 1985. ↑22

[94] M.G. Kreĭn, *On Hermitian operators with directed functionals*, Akad. Nauk Ukrain. RSR. Zbirnik Prac′ Inst. Mat. **1948** (1948), no. 10, 83–106. MR14:56c, reprinted in [95]. ↑25

[95] ______, *Избранные Труды. II*, Akad. Nauk Ukrainy Inst. Mat., Kiev, 1997. MR96m:01030. ↑84

[96] Joseph P.S. Kung (ed.), *Gian-Carlo Rota on combinatorics: Introductory papers and commentaries*, Contemporary Mathematicians, vol. 1, Birkhäuser Verlag, Boston, 1995. ↑23

[97] Serge Lang, $SL_2(\mathbf{R})$, Graduate Texts in Mathematics, vol. 105, Springer-Verlag, New York, 1985. Reprint of the 1975 edition. MR803508 (86j:22018) ↑2, 3, 4, 17, 19, 22, 24, 25, 27, 30, 35, 52, 63, 77

[98] Lavrent′ev, M.A. and Shabat, B.V., *Проблемы гидродинамики и их математические модели. (Russian) [Problems of hydrodynamics and their mathematical models]*, Second, Izdat. "Nauka", Moscow, 1977. MR56#17392 ↑16

[99] Jean-Marc Lévy-Leblond, *Une nouvelle limite non-relativiste du groupe de Poincaré*, Ann. Inst. H. Poincaré Sect. A (N.S.) **3** (1965), 1–12. MR0192900 (33 #1125) ↑51

[100] V.B. Lidskiĭ, *On the theory of perturbations of nonselfadjoint operators*, Ž. Vyčisl. Mat. i Mat. Fiz. **6** (1966), no. 1, 52–60. MR0196930 (33 #5114) ↑46

[101] S.G. Low, *Noninertial Symmetry Group of Hamilton's Mechanics* (March 2009), available at 0903.4397. arXiv:0903.4397. ↑73

[102] Volodymyr Mazorchuk, *Lectures on* sl_2*-modules*, World Scientific, 2009. ↑20, 52, 68

[103] Vladimir V. Mityushev and Sergei V. Rogosin, *Constructive methods for linear and nonlinear boundary value problems for analytic functions*, Chapman & Hall/CRC Monographs and Surveys in Pure and Applied Mathematics, vol. 108, Chapman & Hall/CRC, Boca Raton, FL, 2000. Theory and applications. MR1739063 (2001d:30075) ↑38

[104] Julio Moro, James V. Burke, and Michael L. Overton, *On the Lidskii-Vishik-Lyusternik perturbation theory for eigenvalues of matrices with arbitrary Jordan structure*, SIAM J. Matrix Anal. Appl. **18** (1997), no. 4, 793–817. MR1471994 (98k:15014) ↑46

[105] U. Niederer, *The maximal kinematical invariance group of the free Schrödinger equation*, Helv. Phys. Acta **45** (1972/73), no. 5, 802–810. MR0400948 (53 #4778) ↑54

[106] N. K. Nikol′skiĭ, *Treatise on the shift operator*, Springer-Verlag, Berlin, 1986. Spectral function theory, with an appendix by S. V. Hruščev [S. V. Khrushchëv] and V. V. Peller, translated from the Russian by Jaak Peetre. MR87i:47042 ↑28, 39

[107] Peter J. Olver, *Applications of Lie groups to differential equations*, Second, Springer-Verlag, New York, 1993. MR94g:58260 ↑42

[108] ———, *Equivalence, invariants, and symmetry*, Cambridge University Press, Cambridge, 1995. MR96i:58005 ↑42, 43

[109] John R. Parker, *Hyperbolic spaces*, University of Durham, 2007. URL: `http://maths.dur.ac.uk/~dma0jrp/img/HSjyvaskyla.pdf`. ↑14

[110] Ian Percival and Derek Richards, *Introduction to Dynamics.*, Cambridge etc.: Cambridge University Press. VIII, 228 p., 1982 (English). ↑73

[111] A. Perelomov, *Generalized coherent states and their applications*, Texts and Monographs in Physics, Springer-Verlag, Berlin, 1986. MR87m:22035 ↑22, 24

[112] Valery N. Pilipchuk, *Nonlinear dynamics. Between linear and impact limits.*, Lecture Notes in Applied and Computational Mechanics, vol. 52, Springer, Berlin, 2010 (English). ↑67

[113] ———, *Non-Smooth Spatio-Temporal Coordinates in Nonlinear Dynamics* (January 2011), available at `1101.4597`. `arXiv:1101.4597`. ↑67

[114] R.I. Pimenov, *Unified axiomatics of spaces with maximal movement group*, Litov. Mat. Sb. **5** (1965), 457–486 (Russian). **Zbl** # 0139.37806. ↑52

[115] Ian R. Porteous, *Clifford algebras and the classical groups*, Cambridge Studies in Advanced Mathematics, vol. 50, Cambridge University Press, Cambridge, 1995. MR1369094 (97c:15046) ↑52

[116] Oleg V. Prezhdo and Vladimir V. Kisil, *Mixing quantum and classical mechanics*, Phys. Rev. A (3) **56** (1997), no. 1, 162–175. `arXiv:quant-ph/9610016`. MR1459700 (99j:81010) ↑56

[117] H.M. Srivastava, Vu Kim Tuan, and S.B. Yakubovich, *The Cherry transform and its relationship with a singular Sturm-Liouville problem*, Q. J. Math. **51** (2000), no. 3, 371–383. MR1782100 (2001g:44010) ↑68

[118] Béla Sz.-Nagy and Ciprian Foiaş, *Harmonic analysis of operators on Hilbert space*, North-Holland Publishing Company, Amsterdam, 1970. ↑27

[119] Joseph L. Taylor, *A general framework for a multi-operator functional calculus*, Advances in Math. **9** (1972), 183–252. MR0328625 (48 #6967) ↑77

[120] Michael E. Taylor, *Noncommutative harmonic analysis*, Mathematical Surveys and Monographs, vol. 22, American Mathematical Society, Providence, RI, 1986. MR88a:22021 ↑19, 22, 35, 52, 53

[121] A. Torre, *A note on the general solution of the paraxial wave equation: a Lie algebra view*, Journal of Optics A: Pure and Applied Optics **10** (2008), no. 5, 055006 (14pp). ↑52, 54, 55, 68

[122] ———, *Linear and quadratic exponential modulation of the solutions of the paraxial wave equation*, Journal of Optics A: Pure and Applied Optics **12** (2010), no. 3, 035701 (11pp). ↑52, 54

[123] Eugene E. Tyrtyshnikov, *A brief introduction to numerical analysis*, Birkhäuser Boston Inc., Boston, MA, 1997. MR1442956 (97m:65005) ↑46

[124] S. Ulrych, *Relativistic quantum physics with hyperbolic numbers*, Phys. Lett. B **625** (2005), no. 3-4, 313–323. MR2170329 (2006e:81103a) ↑65

[125] ———, *Representations of Clifford algebras with hyperbolic numbers*, Adv. Appl. Clifford Algebr. **18** (2008), no. 1, 93–114. MR2377525 (2009d:81139) ↑65

[126] A. Vourdas, *Analytic representations in quantum mechanics*, J. Phys. A **39** (2006), no. 7, R65–R141. MR2210163 (2007g:81069) ↑50

[127] Carl E. Wulfman, *Dynamical Symmetry*, World Scientific, 2010. ↑52, 55, 68, 73

[128] I. M. Yaglom, *A simple non-Euclidean geometry and its physical basis*, Heidelberg Science Library, Springer-Verlag, New York, 1979. Translated from the Russian by Abe Shenitzer, with the editorial assistance of Basil Gordon. MR520230 (80c:51007) ↑2, 6, 13, 16, 65, 70

[129] Cosmas Zachos, *Deformation quantization: quantum mechanics lives and works in phase-space*, Internat. J. Modern Phys. A **17** (2002), no. 3, 297–316. `arXiv:hep-th/0110114`. MR1888 937 ↑60

Index

Vladimir V. Kisil
School of Mathematics
University of Leeds
Leeds, LS2 9JT, UK
e-mail: kisilv@maths.leeds.ac.uk
URL: http://www.maths.leeds.ac.uk/~kisilv/

Advances in Applied Analysis
Trends in Mathematics, 95–114

The Riemann Zeta-function: Approximation of Analytic Functions

Antanas Laurinčikas

Abstract. In the paper, a short survey on the theory of the Riemann zeta-function is given. The main attention is given to universality-approximation of analytic functions by shifts of the Riemann zeta-function. This includes the effectivization problem, generalization for other zeta-functions, joint universality as well as some applications.

Mathematics Subject Classification (2010). 11M06, 11M41.

Keywords. Approximation of analytic functions, joint universality, Riemann zeta-function, universality.

1. Introduction

We recall that the function $f(z)$ is called analytic at the point z_0 if $f(z)$ has a power series expansion

$$f(z) = \sum_{m=0}^{\infty} a_m (z - z_0)^m$$

which is convergent in some neighbourhood of the point z_0. The function $f(z)$ is analytic in the set D if it is analytic in each point of D.

A set of points (x, y), where

$$x = f_1(t), \quad y = f_2(t), \quad 0 \le t \le 1, \tag{1.1}$$

and $f_1(t)$ and $f_2(t)$ are continuous functions such that, for given x and y the system (1.1) has no more than one solution, is called the Jordan arc.

A set D is said to be connected if any two of its points, arbitrarily chosen, can be connected by a Jordan arc lying in D. A domain is an open connected set.

It is well known that analytic functions can be approximated by polynomials. The ultimate result in this field belongs to S.N. Mergelyan [S.N. Mergelyan (1951)], [S.N. Mergelyan (1952)], see also [J.L. Walsh (1960)].

Theorem 1.1. *Suppose that K is a compact subset on the complex plane with connected complement, and the function $f(z)$ is continuous on K and analytic in the interior of K. Then, for every $\epsilon > 0$, there exists a polynomial $P(z)$ such that*

$$\sup_{z \in K} |f(z) - P(z)| < \epsilon.$$

Note that conditions on the set K and function $f(z)$ are necessary.

In Theorem 1.1, the approximating polynomial depends on the function $f(z)$. It turns out that there exist functions $F(z)$ such that their shifts $F(z+i\tau)$ approximate any analytic function. The simplest of $F(z)$ is the Riemann zeta-function.

In this survey, we give an introduction to the theory of the Riemann zeta-function, state the universality theorem, discuss the effectivization problem of this theorem and present recent results on universality of zeta-functions.

2. The Riemann zeta-function

Let $s = \sigma + it$ be a complex variable. The Riemann zeta-function $\zeta(s)$ is defined, in the half-plane $\sigma > 1$, by the series

$$\zeta(s) = \sum_{m=1}^{\infty} \frac{1}{m^s}.$$

We recall that the series of the form

$$\sum_{m=1}^{\infty} a_m \mathrm{e}^{-\lambda_m s},$$

where $a_m \in \mathbb{C}$ and $\{\lambda_m\}$ is an increasing sequence of real numbers such that $\lim_{m\to\infty} \lambda_m = +\infty$, are called general Dirichlet series. If $\lambda_m = \log m$, then we have an ordinary Dirichlet series

$$\sum_{m=1}^{\infty} \frac{a_m}{m^s}.$$

The region of convergence as well as of absolute convergence of Dirichlet series is a half-plane. Thus, the Riemann zeta-function is given, for $\sigma > 1$, by ordinary Dirichlet series with coefficients $a_m \equiv 1$.

The function $\zeta(s)$, as a function of a complex variable, was introduced by B. Riemann in 1859. However, the function $\zeta(s)$ with real s earlier was studied by L. Euler.

Denote by $[u]$ the integer part of u. Summing by parts, it it easy to obtain that, for $\sigma > 1$,

$$\zeta(s) = \frac{1}{s-1} + \frac{1}{2} + s \int_1^{\infty} \frac{[x] - x + \frac{1}{2}}{x^{s+1}} \, \mathrm{d}x. \tag{2.1}$$

Clearly, the integral converges absolutely for $\sigma > 0$, and uniformly for $\sigma \geq \varepsilon$ with arbitrary $\varepsilon > 0$. Therefore, it defines a function analytic for $\sigma > 0$. Hence, (2.1) gives analytic continuation for $\zeta(s)$ to the region $\sigma > 0$, except for a simple pole at the point $s = 1$ with residue 1.

Denote, as usual, by $\Gamma(s)$ the Euler gamma-function which is defined, for $\sigma > 0$, by

$$\Gamma(s) = \int_0^\infty e^{-u} u^{s-1} du.$$

Moreover, the function $\Gamma(s)$ is meromorphically continuable over the whole complex plane, the points $s = -m$, $m = 0, 1, 2, \dots$, are simple poles, and

$$\mathrm{Res}_{s=-m}\Gamma(s) = \frac{(-1)^m}{m!}.$$

The Euler gamma function is involved in the functional equation of the Riemann zeta function

$$\pi^{-\frac{s}{2}}\Gamma\left(\frac{s}{2}\right)\zeta(s) = \pi^{-\frac{1-s}{2}}\Gamma\left(\frac{1-s}{2}\right)\zeta(1-s) \tag{2.2}$$

which implies analytic continuation for $\zeta(s)$ to the region $\sigma < \frac{1}{2}$.

Riemann began to study the function $\zeta(s)$ for needs of the distribution of prime numbers, i.e., for the asymptotics for the function

$$\pi(x) = \sum_{p \le x} 1, \quad p \text{ is prime},$$

as $x \to \infty$. A relation of $\zeta(s)$ with prime numbers is clearly seen from the Euler identity

$$\zeta(s) = \prod_p \left(1 - \frac{1}{p^s}\right)^{-1}, \quad \sigma > 1, \tag{2.3}$$

which is a simple consequence of the principal theorem of arithmetics and definition of $\zeta(s)$ by Dirichlet series. Riemann proposed [B. Riemann (1859)] an original way to obtain the asymptotic formula

$$\pi(x) \sim \int_2^x \frac{du}{\log u}, \quad x \to \infty,$$

however, his work was not completely correct. Riemann's ideas were realized probably 50 years later independently by C.J. de la Vallée Poussin [C.J. de la Vallée-Poussin (1896)] and J. Hadamard [J. Hadamard (1896)].

It turned out that a problem of the asymptotics for $\pi(x)$ is closely connected to zeros of the function $\zeta(s)$. From the functional equation (2.2) it follows that $\zeta(s) = 0$ for $s = -2m$, $m \in \mathbb{N}$. These zeros of $\zeta(s)$ are called trivial and, in general, are not interesting. The Euler identity (2.3) shows that $\zeta(s) \neq 0$ for $\sigma > 1$. It is not difficult to show that $\zeta(s) \neq 0$ on the line $\sigma = 1$, and this is already sufficient to obtain the asymptotics for $\pi(x)$. This and (2.2) imply that $\zeta(s) \neq 0$ for $\sigma \leq 0$. Application of elements of the theory of entire functions of order 1 for the function

$$\xi(s) = \frac{1}{2}s(s-1)\pi^{-\frac{s}{2}}\Gamma\left(\frac{s}{2}\right)\zeta(s)$$

allows us to prove that the function $\zeta(s)$ has infinitely many zeros in the strip $0 < \sigma < 1$. These zeros of $\zeta(s)$ are called non-trivial, and play an important role not only in analytic number theory but in mathematics in general. The famous Riemann hypothesis (RH) says that all non-trivial zeros of $\zeta(s)$ lie on the critical line $\sigma = \frac{1}{2}$. RH is included in the list of seven Millennium Prize Problems, for each of which a solution carries a prize of US $ 1 million, set up by the Clay Mathematics Institute [The Millennium Prize problems (2006)].

There exist results favorable for RH, however some facts do not support it. There are many computations on the location of zeros of $\zeta(s)$. For example, in [J. van de Lune, H.J.J. te Riele, D.T. Winter (1986)] the first 1500000001 zeros of $\zeta(s)$ were found, all lying on the critical line; moreover, they all are simple. Of course, calculations can not prove RH, they can only disprove it.

In applications, it is important to know regions where $\zeta(s) \neq 0$. The best result in this direction is of the form: there exists an absolute constant $c > 0$ such that $\zeta(s) \neq 0$ in the region

$$\sigma \geq 1 - \frac{c}{(\log t)^{2/3}(\log\log t)^{1/3}}, \quad t \geq t_0 > 0.$$

This result is due to H.-E. Richert who never published its proof.

For $T > 0$, let $N(T)$ denote the number of non-trivial zeros of $\zeta(s)$ lying in the rectangle $0 < \sigma < 1$, $0 < t \leq T$. Then the von Mangoldt formula

$$N(T) = \frac{T}{2\pi}\log\frac{T}{2\pi} - \frac{T}{2\pi} + O(\log T), \quad T \to \infty,$$

is true. This formula was conjectured by Riemann and proved in [H. von Mangoldt (1895)] by H. von Mangoldt.

Let $N_0(T)$ denote the number of zeros of $\zeta(s)$ of the form $s = \frac{1}{2}+it, 0 < t \leq T$. Then RH is equivalent to the assertion that $N_0(T) = N(T)$ for all $T > 0$. We recall some results on the relation between $N(T)$ and $N_0(T)$.

In 1914, G.H. Hardy proved [G.H. Hardy (1914)] that $N_0(T) \to \infty$ as $T \to \infty$. More precisely, he obtained that $N_0(T) > cT$, $c > 0$, for $T \geq T_0$.

In 1942, A. Selberg found [A. Selberg (1942)] that $N_0(T) > cT\log T$, $c > 0$, for $T \geq T_0$, that is, that a positive proportion of non-trivial zeros lies on the critical line.

A very important result belongs to N. Levinson. In 1974, he proved [N. Levinson (1974)] that

$$N_0(T) \geq \frac{1}{3}N(T).$$

In 1983, J.B. Conrey improved [J.B. Conrey (1989)] this result till $N_0(T) \geq \frac{2}{5}N(T)$.

One more important conjecture in the theory of the Riemann zeta-function is the Lindelöf hypothesis (LH). LH asserts that, with arbitrary $\varepsilon > 0$,

$$\zeta\left(\frac{1}{2}+it\right) = \mathrm{O}_\varepsilon(t^\varepsilon), \quad t \geq t_0,$$

or equivalently

$$\zeta(\sigma+it) = \mathrm{O}_\varepsilon(t^\varepsilon), \quad t \geq t_0,$$

for all $\sigma > \frac{1}{2}$. The classical estimate says that

$$\zeta\left(\frac{1}{2}+it\right) = \mathrm{O}\left(t^{\frac{1}{6}}\right).$$

The best result in this direction belongs to M.N. Huxley [M.N. Huxley (2005)], and is of the form

$$\zeta\left(\frac{1}{2}+it\right) = \mathrm{O}\left(t^{\frac{32}{205}+\varepsilon}\right).$$

It is well known that RH implies LH.

There are several equivalents of LH. One of them is related to the moments of $\zeta(s)$. Namely, LH is equivalent to the estimates: for arbitrary $\varepsilon > 0$

$$\int_1^T \left|\zeta\left(\frac{1}{2}+it\right)\right|^{2k} dt = \mathrm{O}_\varepsilon(T^{1+\varepsilon}), \quad k \in \mathbb{N},$$

or

$$\int_1^T |\zeta(\sigma+it)|^{2k} dt = \mathrm{O}_\varepsilon(T^{1+\varepsilon}), \quad k \in \mathbb{N},$$

for all $\sigma > \frac{1}{2}$.

In general, the moment problem is a very important and difficult one in the theory of the Riemann zeta-function.In some applications, individual values of $\zeta(s)$ can be replaced by its mean-value estimates. There exists a conjecture that, as $T \to \infty$,

$$\int_1^T \left|\zeta\left(\frac{1}{2}+it\right)\right|^{2k} dt \sim c_k T(\log T)^{k^2}, \quad k > 0.$$

This is proved only for three values of k:

$c_1 = 1$, Hardy-Littlewood (1918);

$c_2 = \frac{1}{2\pi^2}$, Ingham (1926).

From a probabilistic limit theorem for $\left|\zeta\left(\frac{1}{2}+it\right)\right|$ it follows [A. Laurinčikas (1996)] that $c_k = 1$ for $k = c(\log\log T)^{-\frac{1}{2}}$, $c > 0$.

3. Universality

We understand the universal mathematical object as an object having influence for a wide class of other objects. In analysis, this influence often is related with a certain approximation.

The first universal object in analysis was found by Fekete in 1914. He proved that there exists a real power series

$$\sum_{m=1}^{\infty} a_m x^m$$

which is divergent for all $x \neq 0$. Moreover, this divergence is so extreme that,for every continuous function f on $[-1,1]$, $f(0)=0$, there exists a sequence $\{n_k\} \subset \mathbb{N}$ such that

$$\lim_{k\to\infty} \sum_{m=1}^{n_k} a_m x^m = f(x)$$

uniformly on $[-1,1]$.

After Fekete's result, many universal objects were found. We recall one theorem of Birkhoff. He proved [G.D. Birkhoff (1929)] that there exists an entire function $f(z)$ such that, for every entire function $g(z)$, there exists a sequence of complex numbers $\{a_m\}$ such that

$$\lim_{m\to\infty} f(z+a_m) = g(z)$$

uniformly on compact subsets of the complex plane.

The term of universality was used for the first time by J. Marcinkiewicz in [J. Marcinkiewicz (1935)]. He obtained the following result. Let $\{h_n\}$ be a sequence of real numbers and $\lim\limits_{n\to\infty} h_n = 0$. Then he proved that there exists a continuous function $f \in C[0,1]$ such that, for every continuous function $g \in C[0,1]$, there exists an increasing sequence $\{n_k\} \subset \mathbb{N}$ such that

$$\lim_{k\to\infty} \frac{f(x+h_{n_k}) - f(x)}{h_{n_k}} = g(x)$$

almost everywhere on $[0,1]$. Marcinkiewicz called the function f a primitive universal.

However, all of the above and other known universal objects, were not explicitly given, only their existence was proved. As recently as 1975, S.M. Voronin [S.M. Voronin (1975)] found the first explicitly given universal (in a certain sense object). It was not very strange that this object is the famous Riemann zeta-function $\zeta(s)$.

The first version of the Voronin theorem is as follows.

Theorem 3.1 [S.M. Voronin (1975)]. *Let $0 < r < \frac{1}{4}$. Suppose that $f(s)$ is a continuous non-vanishing function on the disc $|s| \leq r$ which is analytic in the interior* *of this disc. Then, for every $\varepsilon > 0$, there exists a real number $\tau = \tau(\varepsilon)$ such that*

$$\max_{|s|\le r}\left|\zeta\left(s+\frac{3}{4}+i\tau\right)-f(s)\right|<\varepsilon.$$

Roughly speaking, the Voronin theorem asserts that any analytic function is approximated with desired accuracy uniformly on the disc by shifts of the Riemann zeta-function. Voronin himself called his theorem "theorem o kruzhochkakh". Nowadays its name is the Voronin universality theorem.

A modern version of the Voronin theorem has a bit more general form. Denote by meas$\{A\}$ the Lebesgue measure of a measurable set $A \subset \mathbb{R}$, see, for example, [A. Laurinčikas (1996)].

Theorem 3.2. *Let K be a compact subset of the strip $D = \{s \in \mathbb{C} : \frac{1}{2} < \sigma < 1\}$ with connected complement. Let $f(s)$ be a continuous and non-vanishing on K function which is analytic in the interior of K. Then, for every $\varepsilon > 0$,*

$$\liminf_{T\to\infty}\frac{1}{T}\text{meas}\left\{\tau\in[0,T]:\sup_{s\in K}|\zeta(s+i\tau)-f(s)|<\varepsilon\right\}>0.$$

The latter theorem shows that the set of shifts $\zeta(s+i\tau)$ whose approximate uniformly on K a given analytic function $f(s)$ is sufficiently rich, it has a positive lower density.

The universality in the Voronin sense of $\zeta(s)$ has a direct connection to RH.It is known, see, for example, [J. Steuding (2007)] that RH is equivalent to the following statement: for every $\varepsilon > 0$,

$$\liminf_{T\to\infty}\frac{1}{T}\text{meas}\left\{\tau\in[0,T]:\sup_{s\in K}|\zeta(s+i\tau)-\zeta(s)|<\varepsilon\right\}>0,$$

where the set K is the same as in Theorem 3.2.

In Theorem 3.2, the shifts $\zeta(s+i\tau)$ occur, where τ varies continuously in the interval $[0,T]$. Therefore, the universality of $\zeta(s)$ in Theorem 3.2 is called continuous. Also, a discrete universality of $\zeta(s)$ is known. It is included in the following theorem.

Theorem 3.3. *Let $h > 0$ be a fixed number, and K and $f(s)$ be the same as in Theorem 3.2. Then, for every $\varepsilon > 0$,*

$$\liminf_{N\to\infty}\frac{1}{N+1}\sharp\left\{0\le m\le N:\sup_{s\in K}|\zeta(s+imh)-f(s)|<\varepsilon\right\}>0.$$

4. Effectivization problem

The universality theorem for $\zeta(s)$ has one very important shortcoming. It is not effective in the sense that we do not know any concrete value $\tau \in \mathbb{R}$ for which

$$\sup_{s\in K}|\zeta(s+i\tau)-f(s)|<\varepsilon,$$

or $m \in \mathbb{N}$ with

$$\sup_{s\in K} |\zeta(s+imh) - f(s)| < \varepsilon.$$

For applications of the universality theorem, it is sufficient to know at least an interval $[0, T_0]$ containing τ with an approximating property. However, in our opinion, the latter problem is very difficult. The first attempt in this direction was made in [A. Good (1981)] by A. Good, however, his results are too complicated to be given here. Interesting results were obtained by my student R. Garunkštis. Suppose that the function $f(s)$ is analytic on the disc $|s| \le 0.05$ and $\max_{|s|\le 0.05} |f(s)| < 1$. Then Garunkštis proved [R. Garunkštis (2003)] that, for every $0 < \varepsilon < \frac{1}{2}$, there exists τ,

$$0 \le \tau \le \exp\left\{\exp\{10\varepsilon^{-13}\}\right\},$$

such that

$$\max_{|s|\le 0.0001} \left|\log\zeta\left(s + \frac{3}{4} + i\tau\right) - f(s)\right| < \varepsilon,$$

and

$$\liminf_{T\to\infty} \frac{1}{T}\text{meas}\left\{\tau \in [0,T] : \max_{|s|\le 0.0001} \left|\log\zeta\left(s+\frac{3}{4}+i\tau\right) - f(s)\right| < \varepsilon\right\}$$
$$\ge \exp\left\{-\varepsilon^{-13}\right\}.$$

Also, an estimate for the upper universality density is known [J. Steuding (2003)]. Suppose that $r \in \left(0, \frac{1}{4}\right)$ and the function $f(s)$ is non-vanishing and analytic on the disc $|s| \le r$. Then, for every $\varepsilon \ge 0$,

$$\limsup_{T\to\infty} \frac{1}{T}\text{meas}\left\{\tau \in [0,T] : \max_{|s|\le r} \left|\zeta\left(s+\frac{3}{4}+i\tau\right) - f(s)\right| < \varepsilon\right\} = \mathrm{O}(\varepsilon).$$

The last result in this direction is as follows. For $\underline{b} = (b_0, b_1, \dots, b_{n-1}) \in \mathbb{C}^n$, let

$$\|\underline{b}\| = \sum_{k=0}^{n-1} |b_n|,$$

and

$$A(n, \underline{b}, \varepsilon) = |\log|b_0|| + \left(\frac{\|\underline{b}\|}{\varepsilon}\right)^{n^2}.$$

Then we have the following statement.

Theorem 4.1. [R. Garunkštis, A. Laurinčikas, K. Matsumoto, J. Steuding, R. Steuding (2010)] *Let* $s_0 = \sigma_0 + it_0$, $\sigma_0 \in \left(\frac{1}{2}, 1\right)$ *and* $K = \{s \in \mathbb{C} : |s - s_0| \le r\}$. *Moreover, let* $g : K \to \mathbb{C}$ *be a continuous function,* $g(s_0) \ne 0$, *which is analytic on the disc* $|s - s_0| \le r$. *Then, for every* $\varepsilon \in (0, |g(s_0)|)$, *there exist real numbers* $\tau \in [T, 2T]$ *and* $\delta = \delta(\varepsilon, g, \tau) > 0$, *connected by the equality*

$$M(\tau)\frac{\delta^n}{1-\delta} = \frac{\varepsilon}{3}\left(2 - \mathrm{e}^{\delta r}\right)$$

with

$$M(\tau) = \max_{|s-s_0|=r} |\zeta(s+i\tau)|,$$

such that

$$\max_{|s-s_0|\leq\delta r} |\zeta(s+i\tau) - g(s)| < \varepsilon.$$

Here $T = T(g, \varepsilon, \sigma_0) > r$ *satisfies the inequality*

$$T \geq C(n, \sigma_0)\exp\left\{\exp\left\{5A\left(n, \underline{g}, \frac{\varepsilon}{3}\right)^{\frac{8}{1-\sigma_0}+\frac{8}{\sigma_0-\frac{1}{2}}}\right\}\right\},$$

where

$$\underline{g} = \left(g(s_0), g'((s_0), \ldots, g^{(n-1)}(s_0)\right),$$

and $C(n, \sigma_0)$ *is an effective computable constant depending on* n *and* σ_0.

Remark. *The requirement* $g(s_0) \neq 0$ *can be removed if* $A(n, \underline{g}, \frac{\varepsilon}{3})$ *is changed by* $A(n, \underline{g}_{\varepsilon}, \frac{\varepsilon}{3})$, *where*

$$\underline{g}_{\varepsilon} = \left(\frac{\varepsilon}{2}, g'(s_0), \ldots, g^{(n-1)}(s_0)\right).$$

We note that Theorem 4.1 gives only an approximation to the effectivization problem of the universality theorem, and is far from the full solution of the problem.

5. Other zeta-functions

The Riemann zeta-function is not unique in having the above universality property. There exists the Linnik-Ibragimov conjecture that all functions $Z(s)$ in some half-plane given by Dirichlet series

$$Z(s) = \sum_{m=1}^{\infty} \frac{a_m}{m^s}, \quad \sigma > \sigma_0,$$

analytically continuable to the half-plane $\sigma > \sigma_1$, $\sigma_1 < \sigma_0$, and satisfying certain natural growth conditions, are universal in the Voronin sense. Usually, in the proof of universality the estimates, for $\sigma_1 < \sigma < \sigma_0$,

$$\int_0^T |Z(\sigma+it)|^2 \mathrm{d}t \ll T$$

and

$$Z(\sigma+it) \ll t^a, \quad a > 0, \quad t > t_0 > 0,$$

are applied. In our opinion, the latter conjecture is very difficult, however, the majority of classical zeta-functions are universal.

On the other hand, there exist non-universal functions given by Dirichlet series. For example, suppose that

$$a_m = \begin{cases} 1 \text{ if } m = m_0^k, \ k \in \mathbb{N}, \\ 0 \text{ if } m \neq m_0^k, \end{cases}$$

where $m_0 \in \mathbb{N}\backslash\{1\}$. Then we have that, for $\sigma > 0$,

$$\sum_{m=1}^{\infty} \frac{a_m}{m^s} = \sum_{k=1}^{\infty} \frac{1}{m_0^{ks}} = \frac{1}{m_0^s - 1}.$$

The function $(m_0^s - 1)^{-1}$ is analytic in the whole complex plane except, for simple poles on the line $\sigma = 0$; it however, obviously, is non-universal. In this section, we discuss some examples of other universal zeta-functions.

Let $0 < \alpha \leq 1$ be a fixed parameter. The Hurwitz zeta-function $\zeta(s, \alpha)$ is defined, for $\sigma > 1$, by

$$\zeta(s, \alpha) = \sum_{m=0}^{\infty} \frac{1}{(m + \alpha)^s},$$

and can be meromorphically continued to the whole complex plane. The point $s = 1$ is a simple pole with residue 1. Obviously, $\zeta(s, 1) = \zeta(s)$, so $\zeta(s, \alpha)$ is a generalization of the Riemann zeta-function. On the other hand, its properties are governed by the arithmetical nature of the parameter α. The simplest case is of transcendental α, i.e., when α is not a root of any polynomial with rational coefficients. In this case, the set

$$\{\log(m + \alpha) : m \in \mathbb{N}_0\}, \quad \mathbb{N}_0 = \mathbb{N} \cup \{0\},$$

is linearly independent over the field $\mathbb{Q}$ of rational numbers. We observe that $\zeta(s, \alpha)$ with transcendental α has no Euler product over primes, therefore, its universality differs from that of the Riemann zeta-function: the approximated function can be not necessarily non-vanishing, see, for example, [A. Laurinčikas, R. Garunkštis (2002)].

Theorem 5.1. *Suppose that α is transcendental. Let K be a compact subset of the strip D with connected complement, and $f(s)$ be a continuous function on K and analytic in the interior of K. Then, for every $\varepsilon > 0$,*

$$\liminf_{T \to \infty} \frac{1}{T} \text{meas} \left\{ \tau \in [0, T] : \sup_{s \in K} |\zeta(s + i\tau, \alpha) - f(s)| < \varepsilon \right\} > 0.$$

If α is rational, then the function $\zeta(s, \alpha)$ is also universal. However, the case of algebraic irrational α is an open problem.

Some periodic generalizations of the Riemann and Hurwitz zeta-functions are known. Let $\mathfrak{a} = a_m : m \in \mathbb{N}$ be a periodic with a minimal period $k \in \mathbb{N}$ sequence of complex numbers. Then the function

$$\zeta(s;\mathfrak{a}) = \sum_{m=1}^{\infty} \frac{a_m}{m^s}, \qquad \sigma > 1,$$

is called a periodic zeta-function. The periodicity of $\mathfrak{a}$ implies, for $\sigma > 1$, the equality

$$\zeta(s;\mathfrak{a}) = \frac{1}{k^s} \sum_{l=1}^{k} a_l \zeta\left(s, \frac{l}{k}\right).$$

Hence, by virtue of the well-known properties of $\zeta(s,\alpha)$, we have that the function $\zeta(s;\mathfrak{a})$ has analytic continuation to the whole complex plane. If

$$a = \frac{1}{k} \sum_{l=1}^{k} a_l \neq 0,$$

then the point $s = 1$ is a simple pole of $\zeta(s;\mathfrak{a})$ with residue a,while if $a \neq 0$, then $\zeta(s;\mathfrak{a})$ is an entire function.

If $\mathfrak{a}$ is a multiplicative sequence, i.e., $a_1 = 1$ and $a_{mn} = a_m a_n$ for all coprimes $m, n \in \mathbb{N}$, then an analogue of Theorem 3.2 is true [A. Laurinčikas, D. Šiaučiūnas (2006)] for the function $\zeta(\alpha, \alpha)$. In the general case, the following result is known [J. Kaczorowski (2009)].

Theorem 5.2. *For every non-zero periodic sequence $\mathfrak{a}$ of complex numbers with period k, there exists a positive constant $c_0 = c_0(\mathfrak{a})$ such that, for every compact subset $K \subset D$ with connected complement,*

$$\max_{s\in K} \mathrm{Im} s - \min_{s \in K} \mathrm{Im} s \leq c_0,$$

every continuous non-vanishing function $f(s)$ on K which is analytic in the interior of K, and every $\varepsilon > 0$,

$$\liminf_{T\to\infty} \frac{1}{T} \mathrm{meas}\left\{\tau \in [0,T] : \sup_{s\in K} |\zeta(s+i\tau,\mathfrak{a}) - f(s)| < \varepsilon\right\} > 0.$$

Now let $\mathfrak{b} = \{b_m : m \in \mathbb{N}_0\}$ be another periodic with a minimal period $l \in \mathbb{N}$ sequence of complex numbers, and α, $0 < \alpha \leq 1$, be a fixed parameter. The periodic Hurwitz zeta-function $\zeta(s, \alpha, \mathfrak{b})$ is defined, for $\sigma > 1$, by

$$\zeta(s,\alpha;\mathfrak{b}) = \sum_{m=0}^{\infty} \frac{b_m}{(m+\alpha)^s}.$$

In view of periodicity of $\mathfrak{b}$,for $\sigma > 1$,

$$\zeta(s,\alpha;\mathfrak{b}) = \frac{1}{l^s} \sum_{k=0}^{l-1} a_k \zeta\left(s, \frac{k+l}{l}\right),$$

and this equality gives analytic continuation for $\zeta(s,\alpha;\mathfrak{b})$ to the whole complex plane. The function $\zeta(s,\alpha;\mathfrak{b})$ is entire, if

$$b = \frac{1}{l}\sum_{k=0}^{l-1} a_k = 0,$$

and has a simple pole with residue b at $s=1$ if $b \neq 0$.

If α is transcendental, then an analogue of Theorem 5.1 is true [A. Javtokas, A. Laurinčikas (2006)] for the function $\zeta(s,\alpha;\mathfrak{b})$.

We will present one more example of universal zeta-functions with the Euler product. Let $SL(2,\mathbb{Z})$ denote the full modular group, i.e.,

$$SL(2,\mathbb{Z}) = \left\{ \begin{pmatrix} a & b \\ c & d \end{pmatrix} : a,b,c,d \in \mathbb{Z}, ad - bc = 1 \right\}.$$

Suppose that the function $F(z)$ is analytic in the upper half-plane $\text{Im} z > 0$ and, for all $\begin{pmatrix} a & b \\ c & d \end{pmatrix} \in SL(2,\mathbb{Z})$, the functional equation

$$F\left(\frac{az+b}{cz+d}\right) = (cz+d)^{\kappa} F(z)$$

with a certain $\kappa \in 2\mathbb{N}$ is satisfied. Then $F(s)$ has the Fourier series expansion

$$F(z) = \sum_{m=-\infty}^{\infty} c_m e^{2\pi imz}.$$

In the case when $c_m = 0$ for $m \leq 0$, the function is called a cusp form of weight κ. Additionally, suppose that $F(z)$ is an eigenform of all Hecke operators

$$(T_n f)(z) = n^{\kappa-1} \sum_{d|n} d^{-\kappa} \sum_{b=0}^{d-1} f\left(\frac{nz+bd}{d^2}\right).$$

Then it is proved that $c_m \neq 0$, and, after normalization, we have that

$$F(z) = \sum_{m=1}^{\infty} c_m e^{2\pi imz} \quad \text{with } c_1 = 1. \tag{5.1}$$

To the cusp form (5.1), we attach the zeta-function

$$\varphi(s,F) = \sum_{m=1}^{\infty} \frac{c_m}{m^s}, \quad \sigma > \frac{\kappa+1}{2}.$$

Since the coefficients c_m are multiplicative, $\varphi(s,F)$ has the Euler product representation

$$\varphi(s,F) = \prod_p \left(1 - \frac{\alpha(p)}{p^s}\right)^{-1} \left(1 - \frac{\beta(p)}{p^s}\right)^{-1}, \quad \sigma > \frac{\kappa+1}{2}.$$

Here $\alpha(p)$ and $\beta(p)$ are complex numbers, $\beta(p) = \overline{\alpha}(p)$, $\alpha(p)\beta(p) = 1$ and $\alpha(p) + \beta(p) = c(p)$. Moreover, the function $\varphi(s,F)$ is analytically continued to an entire function. The universality of $\varphi(s,F)$ has been proved in [A. Laurinčikas, K. Matsumoto (2001)].

Theorem 5.3 [A. Laurinčikas, K. Matsumoto (2001)]. *Let K be a compact subset of the strip $\{s\in\mathbb{C}: \frac{\kappa}{2}<\sigma<\frac{\kappa+1}{2}\}$ with connected complement, and $f(s)$ be a continuous non-vanishing on K function which is analytic in the interior of K. Then, for every $\varepsilon>0$,*

$$\liminf_{T\to\infty}\frac{1}{T}\mathrm{meas}\left\{\tau\in[0,T]:\sup_{s\in K}|\varphi(s+i\tau,F)-f(s)|<\varepsilon\right\}>0.$$

6. Joint universality

A more complicated and interesting problem is a simultaneous approximation of a collection of analytic functions by shifts of zeta-functions. The first result in this direction also belongs to S.M. Voronin: in [S.M. Voronin (1975)], he obtained the joint universality for Dirichlet L-functions. For the definition of Dirichlet L-functions, a notion of a Dirichlet character is needed. A full definition is rather complicated, therefore, we only observe that every arithmetical function $g(m)\neq 0$ satisfies the following conditions:

1° $g(m)$ is a completely multiplicative function ($g(mn)=g(m)g(n)$) for all $m,n\in\mathbb{N}$;
2° $g(m)$ is periodic with period k;
3° $g(m)=0$ if $(m,k)>1$, and $g(m)\neq 0$ if $(m,k)=1$ coincides with one of the Dirichlet characters modulo k.

Let χ be a Dirichlet character. Then the corresponding Dirichlet L-function $L(s,\chi)$ is defined, for $\sigma>1$, by

$$L(s,\chi)=\sum_{m=1}^{\infty}\frac{\chi(m)}{m^s}=\prod_p\left(1-\frac{\chi(p)}{p^s}\right)^{-1}.$$

A character

$$\chi_0(m)=\begin{cases}1, & \text{if } (m,k)=1,\\ 0, & \text{if } (m,k)>0,\end{cases}$$

is called the principal character modulo k. It is not difficult to see that, for $\sigma>1$,

$$L(s,\chi_0)=\zeta(s)\prod_{p|k}\left(1-\frac{1}{p^s}\right),$$

thus $L(s,\chi_0)$ has a simple pole at $s=1$ with residue $\prod\limits_{p|k}\left(1-\frac{1}{p^s}\right)$. If $\chi\neq\chi_0$, then the function $L(s,\chi)$ is entire.

Let $l,k\in\mathbb{N}$, $(l,k)=1$. Define

$$\pi(x,k,l)=\sum_{\substack{p\le x\\ p\equiv l(\mathrm{mod}k)}}1.$$

Dirichlet L-functions are applied for the investigation of prime numbers in arithmetical progressions, i.e., for the asymptotics of the function $\pi(x,k,l)$ as $x \to \infty$. It has been proved that

$$\pi(x,k,l) \sim \frac{x}{\varphi(k)\log x}, \quad x \to \infty,$$

where $\varphi(k)$ is the Euler function: $\varphi(k) = \sharp\{1 \le m \le k : (m,k) = 1\}$.

Let $\chi_1(\mathrm{mod} k_1)$ and $\chi_2(\mathrm{mod} k_2)$ be two Dirichlet characters, and $k = [k_1, k_2]$ denote the least common multiple. The characters χ_1 and χ_2 are called equivalent if, for $(m,k) = 1$,

$$\chi_1(m) = \chi_2(m).$$

Each Dirichlet L-function is also universal in the Voronin sense. Moreover, the first example of the joint universality is related to Dirichlet L-functions.

Theorem 6.1 [S.M. Voronin (1975)]. *Suppose that $\chi_1, \dots, \chi_n$ are pairwise non-equivalent Dirichlet characters, and $L(s,\chi_1), \dots, L(s,\chi_n)$ are the corresponding Dirichlet L-functions. Let $K_1, \dots, K_n$ be compact subsets of the strip $\{s \in \mathbb{C} : 1/2 < \sigma < 1\}$ with connected complements, and let functions $f_1(s), \dots, f_n(s)$ be continuous non-vanishing on $K_1, \dots, K_n$ and analytic in the interior of $K_1, \dots, K_n$, respectively. Then, for every $\varepsilon > 0$,*

$$\liminf_{T\to\infty} \frac{1}{T} \mathrm{meas} \left\{ \tau \in [0,T] : \sup_{1<j\le n} \sup_{s\in K_j} |L(s+i\tau, \chi_j) - f(s)| < \varepsilon \right\} > 0.$$

Now let $\alpha_j \in \mathbb{R}$, $0 < \alpha_j \le 1$, $j = 1, \dots, r$, and

$$L(\alpha_1, \dots, \alpha_r) = \{\log(m + x_j) : m \in \mathbb{N}_0,\ j = 1, \dots, r\}.$$

A joint universality theorem for Hurwitz zeta-functions is of the form.

Theorem 6.2. *Suppose that the set $L(\alpha_1, \dots, \alpha_r)$ is linearly independent over $\mathbb{Q}$. For $j = 1, \dots, r$, let K_j be a compact subset of the strip D with connected complement, and let $f_j(s)$ be a continuous function on K_j which is analytic in the interior of K_j. Then, for every $\varepsilon > 0$,*

$$\liminf_{T\to\infty} \frac{1}{T} \mathrm{meas} \left\{ \tau \in [0,T] : \sup_{1<j\le r} \sup_{s\in K_j} |\zeta(s+i\tau, \alpha_j) - f_j(s)| < \varepsilon \right\} > 0.$$

Note that, differently from Theorem 6.1, the approximated analytic functions in Theorem 6.2 can have zeros on the set K_j.

There exist other results on joint universality, however, some of them are conditional. We present a recent theorem on joint universality of zeta-functions with periodic coefficients.

Let $\mathfrak{a}_j = \{a_{jm} : m \in \mathbb{N}\}$ be a periodic sequence of complex numbers with minimal period $k_j \in \mathbb{N}$, and $\zeta(s; \mathfrak{a}_j)$ denote the corresponding periodic zeta-function, $j = 1, \dots, r_1$, $r_1 > 1$. Let $\mathfrak{b}_j = \{b_{jm} : m \in \mathbb{N}_0\}$ be another periodic sequence of complex numbers with minimal period $l_j \in \mathbb{N}$, $0 < \alpha_j \le 1$, and $\zeta(s, \alpha_j; \mathfrak{b}_j)$ be the corresponding periodic Hurwitz zeta-function, $j = 1, \dots, r_2$, $r_2 > 1$.

Denote by $k = [k_1, \dots, k_2]$ the least common multiple of the periods $k_1, \dots, k_r$, and let $\eta_1, \dots, \eta_{\varphi(k)}$ be the reduced residue system modulo k. Define the matrix

$$A = \begin{pmatrix} a_{1\eta_1} & a_{2\eta_1} & \dots & a_{r_1\eta_1} \\ a_{1\eta_2} & a_{2\eta_2} & \dots & a_{r_1\eta_2} \\ \dots & \dots & \dots & \dots \\ a_{1\eta_{\varphi(k)}} & a_{2\eta_{\varphi(k)}} & \dots & a_{r_1\eta_{\varphi(k)}} \end{pmatrix}.$$

Theorem 6.3 [A. Laurinčikas (2010)]. *Suppose that $\mathfrak{a}_1, \dots, \mathfrak{a}_{r_1}$ are multiplicative,* rank$(A) = r_1$*, and the numbers $\alpha_1, \dots, \alpha_{r_2}$ are algebraically independent over $\mathbb{Q}$. Let $K_1, \dots, K_{r_1}$ be compact subsets of the strip $D = \{s \in \mathbb{C} : 1/2 < \sigma < 1\}$ with connected complements, and let the functions $f_1(s), \dots, f_r(s)$ be continuous non-vanishing on $K_1, \dots, K_{r_1}$ and analytic in the interior of $K_1, \dots, K_{r_1}$, respectively. Let $\hat{K}_1, \dots, \hat{K}_{r_1}$ also be compact subsets of D with connected complements, and let the functions $\hat{f}_1(s), \dots, \hat{f}_{r_1}(s)$ be continuous on $\hat{K}_1, \dots, \hat{K}_{r_2}$ and analytic in the interior of $\hat{K}_1, \dots, \hat{K}_{r_2}$, respectively. Then, for every $\varepsilon > 0$,*

$$\liminf_{T\to\infty} \frac{1}{T} \text{meas}\Big\{\tau \in [0,T] : \sup_{1<j\le r_1} \sup_{s\in K_j} |\zeta(s+i\tau, \mathfrak{a}_j) - f_j(s)| < \varepsilon,$$
$$\sup_{1<j\le r_2} \sup_{s\in \hat{K}_j} |\zeta(s+i\tau, \alpha_j; \mathfrak{b}_j) - \hat{f}_j(s)| < \varepsilon\Big\} > 0.$$

7. Proof of universality theorems

The original proof of the Voronin universality theorem is based on an analogue of the Riemann theorem on rearrangement of terms of series in Hilbert spaces. However, a more convenient and universal approach uses probabilistic limit theorems in the sense of weak convergence of probability measures in the space of analytic functions.

Let S be a metric space, and let $\mathcal{B}(S)$ denote the class of Borel sets of the space S, i.e., the σ-field generated by open sets of S. Let P_n, $n \in \mathbb{N}$, and P be probability measures on $(S, \mathcal{B}(S))$. We recall that P_n converges weakly to P as $n \to \infty$ if

$$\lim_{n\to\infty} \int_S f \mathrm{d}P_n = \int_S f \mathrm{d}P$$

for every real continuous bounded function f on S.

Denote by $H(D)$ the space of analytic functions on D equipped with the topology of uniform convergence on compacta. On $(H(D), \mathcal{B}(H(D)))$, define the probability measure

$$P_T(A) = \frac{1}{T} \text{meas}\{\tau \in [0,T] : \zeta(s+i\tau) \in A\}.$$

To state a limit theorem for the measure P_T, we need some notation. Let

$$\Omega = \prod_p \gamma_p,$$

where $\gamma_p = \{s \in \mathbb{C} : |s| = 1\}$ for each prime p. With the product topology and pointwise multiplication, the infinite-dimensional torus Ω is a compact topological Abelian group. Therefore, on $(\Omega, \mathcal{B}(\Omega))$, the probability Haar measure m_H can be defined, and this gives the probability space $(\Omega, \mathcal{B}(\Omega), m_H)$. Denote by $\omega(p)$ the projection of $\omega \in \Omega$ to the coordinate space γ_p, and on the probability space $(\Omega, \mathcal{B}(\Omega), m_H)$, define the $H(D)$-valued random element $\zeta(s, \omega)$ by the formula

$$\zeta(s, \omega) = \prod_p \left(1 - \frac{\omega(p)}{p^s}\right)^{-1}.$$

Note that the latter product converges uniformly on compact subset of the half-plane $\sigma > \frac{1}{2}$ for almost all $\omega \in \Omega$.Denote by P_ζ the distribution of the random element $\zeta(s, \omega)$, i.e.,

$$P_\zeta(A) = m_H(\omega \in \Omega : \zeta(s, \omega) \in A), \quad A \in \mathcal{B}(H(D)).$$

Theorem 7.1. *The probability measure P_T converges weakly to P_ζ as $T \to \infty$.*

The next ingredient of the proof of universality for $\zeta(s)$ is the support of the measure P_ζ.We recall that the support of P_ζ is a minimal closed set S_ζ such that $P_\zeta(S_\zeta) = 1$. The support S_ζ consists of elements $x \in H(D)$ such that, for every neighbourhood G of x, the inequality $P_\zeta(G) > 0$ is satisfied.

Theorem 7.2. *The support of the measure P_ζ is the set*

$$\{g \in H(D) : g(s) \neq 0 \text{ or } g(s) \equiv 0\}.$$

Proof of Theorem 3.2. First suppose that the function $f(s)$ has a non-vanishing analytic continuation to the strip D. Then, by Theorem 7.2, $f(s) \in S_\zeta$, therefore, defining an open set G by

$$G = \{g \in H(D) : \sup_{s \in K} |g(s) - f(s)| < \varepsilon\},$$

we obtain that

$$\liminf_{T \to \infty} \frac{1}{T} \text{meas} \left\{ \tau \in [0, T] : \sup_{s \in K} |\zeta(s + i\tau) - f(s)| < \varepsilon \right\} \geq P_\zeta(G) > 0.$$

Now let $f(s)$ be as in Theorem 3.2. Then, by Theorem 1.1, there exists a polynomial $p(s)$ such that

$$\sup_{s \in K} |f(s) - p(s)| < \frac{\varepsilon}{4}.$$

We have that $p(s) \neq 0$ on K. Therefore, we can choose a continuous branch of $\log p(s)$ which is analytic in some region containing K. By Theorem 1.1 again, we can find a polynomial $q(s)$ such that

$$\sup_{s\in K} |p(s) - \mathrm{e}^{q(s)}| < \frac{\varepsilon}{4}. \tag{7.1}$$

This and (7.1) show that

$$\sup_{s\in K} |f(s) - \mathrm{e}^{q(s)}| < \frac{\varepsilon}{2}. \tag{7.2}$$

However, $e^{q(s)}$ is a non-vanishing analytic function on D. Thus, by the first part of the proof

$$\liminf_{T\to\infty} \frac{1}{T} \mathrm{meas}\left\{\tau \in [0,T] : \sup_{s\in K} |\zeta(s+i\tau) - \mathrm{e}^{q(s)}| < \frac{\varepsilon}{2}\right\} > 0.$$

In view of (7.2),

$$\left\{\tau\in[0,T]:\sup_{s\in K}|\zeta(s+i\tau)-f(s)|<\varepsilon\right\}\supset\left\{\tau\in[0,T]:\sup_{s\in K}|\zeta(s+i\tau)-\mathrm{e}^{q(s)}|<\frac{\varepsilon}{2}\right\},$$

hence the theorem follows.

8. Some applications

Universality theorems for zeta- and L-functions have theoretical and practical applications. One of the theoretical applications is related to the functional independence of functions.

In 1887, O. Hölder proved [O. Hölder (1887)] that the Euler gamma-function $\Gamma(s)$ does not satisfy any algebraic-differential equation, i.e., there are no polynomials $P \not\equiv$ such that

$$P(\Gamma(s), \Gamma'(s), \dots, \Gamma^{(n-1)}) \equiv 0.$$

In 1900, Hilbert observed that the algebraic-differential independence of the Riemann zeta-function can be proved by using the above Hölder's result and the functional equation for $\zeta(s)$. S.M. Voronin, using a universality theorem, obtained [S.M. Voronin (1973)] the functional independence of $\zeta(s)$.

Theorem 8.1. *Suppose that the functions $F_j : \mathbb{C}^N \to \mathbb{C}$ are continuous, $j = 0, \dots, r$, and*

$$\sum_{j=0}^{r} s^j F_j(\zeta(s), \dots, \zeta^{(N-1)}(s)) \equiv 0.$$

Then $F_j \equiv 0$ for $j = 0, \dots, r$.

The functional independence also follows for other zeta- and L-functions that are universal in the above sense.

The universality also can be used for approximate computations with analytic functions. Usually, zeta-functions satisfy approximate functional equations.

For example, for the function $\zeta(s)$, the following equation is true [A. Ivič(1985)]. Suppose that $0 \leq \sigma \leq 1$, $x, y, t \geq c > 0$ and $2\pi xy = t$. Then uniformly in σ,

$$\zeta(s) = \sum_{m \leq x} \frac{1}{m^s} + \chi(s) \sum_{m \leq y} \frac{1}{m^{1-s}} + \mathrm{O}(x^{-\sigma}) + \mathrm{O}\left(t^{1/2-\sigma} y^{\sigma-1}\right),$$

where

$$\chi(s) = 2^s \pi^{s-1} \Gamma(1-s) \sin \frac{\pi s}{2}.$$

Therefore, first we can evaluate $\zeta(s + i\tau)$, and then, using Theorem 3.2, we can obtain the desired information on a given analytic function $f(s)$.

An application of universality in physics is given in [K.M. Bitar, N.N. Khuri, H.C. Ren (1991)].

The author thanks Professor S. Rogosin for inviting me to the school-seminar and for suggesting that write this paper.

References

[G.D. Birkhoff (1929)] G.D. Birkhoff, Démonstration d'un théorème élémentaire sur les fonctions entières, *Comptes Rendus Acad. Sci. Paris* **189** (1929), 473–475. 100

[K.M. Bitar, N.N. Khuri, H.C. Ren (1991)] K.M. Bitar, N.N. Khuri, H.C. Ren, Path integrals and Voronin's theorem on the universality of the Riemann zeta-function, *Ann. Phys.* **211**(1) (1991), 151–175. 112

[J.B. Conrey (1989)] J.B. Conrey, More than two fifths of the zeros of the Riemann zeta-functions are on the critical line, *J. Reine Angew. Math.* **399** (1989), 1–26. 98

[R. Garunkštis (2003)] R. Garunkštis, The effective universality theorem for the Riemann zeta-function, in: *Special Activity in Analytic Number Theory and Diophantine Equations, Proc. Workshop at Plank-Institut Bonn* 2002, *D. R. Heath-Brown and B. Moroz (eds.)*, Bonner Math. Schriften **360**, 2003. 102

[R. Garunkštis, A. Laurinčikas, K. Matsumoto, J. Steuding, R. Steuding (2010)] R. Garunkštis, A. Laurinčikas, K. Matsumoto, J. Steuding, R. Steuding, Effective uniform approximation by the Riemann zeta-function, *Publ. Mat.* **54** (2010), 209–219. 102

[A. Good (1981)] A. Good, On the distribution of the values of Riemann's zeta-function, *Acta Arith.* **38** (1981), 347–388. 102

[J. Hadamard (1896)] J. Hadamard, Sur les zéros de la fonction $\zeta(s)$ de Riemann, *Comptes Rendus Acad. Sci. Paries* **122** (1896), 1470–1473. 97

[G.H. Hardy (1914)] G.H. Hardy, Sur les zéros de la fonction de Riemann, *Comptes Rendus Acad. Sci. Paries* **158** (1914), 1012–1014. 98

[O. Hölder (1887)] O. Hölder, Über die Eigenschaft der Gammafunktion keinen algebraischen Differentialgleichungen zu genügen, *Math. Ann.* **28** (1887), 1–13. 111

[M.N. Huxley (2005)] M.N. Huxley, Exponential sums and the Riemann zeta-function, V, *Proc. London Math. Soc.* **90** (2005), 1–41. 99

[A. Ivič(1985)] A. Ivič, *The Riemann Zeta-function, the Theory of the Riemann Zeta-function with Applications*, Wiley, New York, 1985. 112

[A. Javtokas, A. Laurinčikas (2006)] A. Javtokas, A. Laurinčikas, Universality of the periodic Hurwitz zeta-function, *Integral Transforms Spec. Funct.* **17** (10) (2006), 711–722. 106

[J. Kaczorowski (2009)] J. Kaczorowski, Some remarks on the universality of periodic L-functions, in: *New Directions in Value Distribution Theory of Zeta and L-Functions, Würzburg Conf.* 2008, Shaker Verlag, Aachen, 2009. 105

[A. Laurinčikas (1996)] A. Laurinčikas, *Limit Theorems for the Riemann Zeta-Function*, Kluwer, Dordrecht, 1996. 99, 101

[A. Laurinčikas (2010)] A. Laurinčikas, Joint universality of zeta-functions with periodic coefficients, *Izv. Math.*74:3 (2010) 515–539. 109

[A. Laurinčikas, R. Garunkštis (2002)] A. Laurinčikas, R. Garunkštis, *The Lerch Zeta-Function*, Kluwer, Dordrecht, 2002. 104

[A. Laurinčikas, K. Matsumoto (2001)] A. Laurinčikas, K. Matsumoto, The universality of zeta-functions attached to certain cusp forms, *Acta Arith.* **98**(2001), 345–359. 107

[A. Laurinčikas, D. Šiaučiūnas (2006)] A. Laurinčikas, D. Šiaučiūnas, Remark on the universality of the periodic zeta-function, *Math. Notes*, **80** (3–4) (2006), 252–538. 105

[N. Levinson (1974)] N. Levinson, More than one third of the zeros of Riemann's zeta-function are on $\sigma = 1/2$, *Adv. Math.* **13** (1974), 383–436. 98

[J. van de Lune, H.J.J. te Riele, D.T. Winter (1986)] J. van de Lune, H.J.J. te Riele, D.T. Winter, On the zeros of the Riemann zeta-function in the critical strip, IV, *Math. Comp.* **46** (1986), 667–681. 98

[H. von Mangoldt (1895)] H. von Mangoldt, Zu Riemann's Abhandlung über die Anzahl der Primzahlen unter einer gegebenen Grösse, *J. Reine Angew. Math.* **114** (1895), 255–305. 98

[J. Marcinkiewicz (1935)] J. Marcinkiewicz, Sur les nombres dérivés, Fund. Math. **24** (1935), 305–308. 100

[S.N. Mergelyan (1951)] S.N. Mergelyan, On the representation of functions by series of polynomials on closed sets, *Doklady Akad. Nauk SSSR (N.S.)* **78** (1951), 405–408 (in Russian). 95

[S.N. Mergelyan (1952)] S.N. Mergelyan, Uniform approximations to functions of a complex variable, *Usp. Mat. Nauk.* **7** (1952), 31–122 (in Russian). 95

[The Millennium Prize problems (2006)] *The Millennium Prize problems*, J. Carlson, A. Joffe and A. Wiles (eds.), Clay Math. Inst., Cambridge, Massachusetts, 2006. 98

[B. Riemann (1859)] B. Riemann, Über die Anzahl der Primzahlen unterhalb einer gegebenen Grösse, *Monatsber. Preuss. Akad. Wiss. Berlin* (1859), 671–680. 97

[A. Selberg (1942)] A. Selberg, On the zeros of Riemann's zeta-function, *Skr. Norske Vid. Akad. Oslo* **10** (1942), 1–59. 98

[J. Steuding (2003)] J. Steuding, Upper bounds for the density of universality, *Acta Arith.* **107** (2003), 195–202. 102

[J. Steuding (2007)] J. Steuding, *Value-Distribution of L-Functions, Lectures Notes Math.* **1877**, Springer, Berlin, 2007. 101

[C.J. de la Vallée-Poussin (1896)] C.J. de la Vallée-Poussin, Recherches analytiques sur la théorie des nombres premiers, I–III, *Ann. Soc. Bruxelles* 20 (1896), 183–256, 281–362, 363–397. 97

[S.M. Voronin (1975)] S.V. Voronin, Theorem on the "universality" of the Riemann zeta-function, *Izv. Akad. Nauk SSSR, ser. matem.* **39** (1975), 475–486 (in Russian) ≡ Math. USSR Izv. **9** (1975), 443–455. 100

[S.M. Voronin (1975)] S.M. Voronin, On the functional independence of Dirichlet L-functions, *Acta Aritm.* **27** (1975), 493–503 (in Russian). 107, 108

[S.M. Voronin (1973)] S.M. Voronin, On the differential independence of ζ-functions, *Dokl. AN SSSR*, **209**(6) (1973), 1264–1266, (in Russian). 111

[J.L. Walsh (1960)] J.L. Walsh, *Interpolation and Approximation by Rational Functions in the Complex Domain*, Amer. Math. Soc. Coll. Publications, Vol. XX, 1960. 95

Index

Antanas Laurinčikas
Faculty of Mathematics and Informatics
Vilnius University
Naugarduko 24
2600 Vilnius, Lithuania
e-mail: `antanas.laurincikas@maf.vu.lt`

Advances in Applied Analysis
Trends in Mathematics, 115–145

Anomalous Diffusion: Models, Their Analysis, and Interpretation

Yury Luchko

Abstract. In this chapter, modeling of anomalous diffusion processes in terms of differential equations of an arbitrary (not necessarily integer) order is discussed. We start with micro-modeling and first deduce a probabilistic interpretation of normal and anomalous diffusion from basic random walk models. The fractional differential equations are then derived asymptotically in the Fourier-Laplace domain from random walk models and generalized master equations, in the same way as the standard diffusion equation is obtained from a Brownian motion model. The obtained equations and their generalizations are analyzed both with the help of the Laplace-Fourier transforms (the Cauchy problems) and the spectral method (initial-boundary-value problems). In particular, the maximum principle, well known for elliptic and parabolic type PDEs, is extended to initial-boundary-value problems for the generalized diffusion equation of fractional order.

Mathematics Subject Classification (2010). 26A33, 33E12, 35A05, 35B30, 35B45, 35B50, 35K99, 45K05, 60J60, 60J65.

Keywords. Anomalous diffusion, random walk models, fractional diffusion equation, initial-boundary-value problems, Fourier integral transform, Laplace integral transform, maximum principle, uniqueness theorem, spectral method, Mittag-Leffler function.

1. Introduction

Anomalous transport processes represent a very natural way for description of structural and dynamical properties of the so-called complex systems that are characterized by a large diversity of elementary particles participating in transport processes, by strong interactions between them, and by an anomalous evolution of the whole system in time (for a detailed discussion of the complex systems and anomalous transport processes see, e.g., [Luchko et al.(2010)] or [Metzler and Klafter(2000a)]). In this chapter, the focus will be on the anomalous diffusion processes; we do not consider here the effects of convection or the degradation components of general transport processes.

What is the critical attribute to distinguish between the normal and the anomalous diffusion? There exist several possibilities to decide between them. In this chapter, we define the anomalous diffusion processes as those that no longer follow the Gaussian statistics for long time intervals. Especially, a deviation from the linear time dependence of the mean squared displacement of a particle participating in an anomalous diffusion process can be observed, i.e., the relation

$$\overline{x^2(t)} \approx C_1 t, \tag{1.1}$$

no longer holds. Let us note here that the relation (1.1) can be interpreted as the main characteristic of Brownian motion. Furthermore, it is a direct consequence of the central limit theorem and of the Markovian nature of the underlying stochastic process. Instead of (1.1), the anomalous diffusion shows a non-linear growth of the mean squared displacement in time. In this chapter, we deal with the power-low pattern for the mean squared displacement in time:

$$\overline{x^2(t)} \approx C_\alpha t^\alpha. \tag{1.2}$$

It can be shown that, for anomalous diffusion, the central limit theorem no longer holds and has to be replaced by a more general Levy-Gnedenko generalized central limit theorem. Another problem connected with the anomalous diffusion described by the relation (1.2) is that not all moments of the underlying elementary transport events exist; the last property is closely connected with the non-Markovian time evolution of the whole system on a macro level. For the anomalous diffusion that is described by the relation (1.2) the following cases are distinguished:

- subdiffusion ($0 < \alpha < 1$),
- normal diffusion ($\alpha = 1$),
- superdiffusion ($1 < \alpha < 2$).

It is well known that both normal and anomalous diffusion can be described either on the micro level through an appropriate stochastic formulation in terms of random walk processes or on the macro level through deterministic diffusion equations. In this chapter, we start to compile a model for anomalous diffusion on the micro level in the framework of a continuous time random walk model and then proceed with a deterministic model on the macro level. It turns out that – under some suitable restrictions – the deterministic models of the anomalous diffusion that fulfils the relation (1.2) can be formulated in terms of partial differential equations of a fractional (non-integer) order α.

During the last few decades, partial differential equations of fractional order begun to play an important role in the modeling of anomalous phenomena and in the theory of complex systems (see, e.g., [Chechkin et al.(2005), Dubbeldam et al.(2007), Freed et al.(2002), Gorenflo and Mainardi(1998), Hilfer(2000), Kilbas et al.(2006), Luchko and Punzi(2011), Mainardi(1996), Mainardi and Tomirotti(1997), Metzler and Klafter(2000a), Podlubny(1999)]) and references therein). In this connection, the so-called time-fractional diffusion equation that is obtained from the diffusion equation by replacing the first-order time derivative by a fractional derivative of order α with $0 < \alpha < 1$ has to be especially mentioned. In the paper [Schneider and Wyss(1989)], the Green function for the time-fractional diffusion equation was shown to be a probability density with a mean square displacement proportional to t^α. As a consequence, the time-fractional diffusion equation appeared to be a suitable mathematical model for the sub-diffusion processes and thus became important and useful for different applications.

Probabilistic interpretation of anomalous diffusion processes, different random walk models, and their connection each to the other and to the fractional differential equations were considered, e.g., in [Beghin and Orsingher(2009), Gorenflo and Mainardi(2008), Gorenflo and Mainardi(2009), Jacob(2005), Mainardi et al.(2004), Meerschaert et al.(2011), Scalas et al.(2004)] to mention only a few out of many recent relevant publications.

The mathematical theory of partial differential equations of fractional order in general, and of the time-fractional diffusion equation in particular, is still far away from being at least nearly as complete as that of the PDEs. In the literature, mainly the initial-value problems for these equations have been considered until now (see, e.g., [Eidelman and Kochubei(2004), Kilbas et al.(2006), Kochubei(1989), Mainardi et al.(2001), Podlubny(1999), Voroshilov and Kilbas(2006)]). As to the boundary-value or initial-boundary-value problems, they were mainly investigated for equations with constant coefficients and in the one-dimensional case (see, e.g., [Bazhlekova(1998), Meerschaert et al.(2009), Metzler and Klafter(2000b), Zhang(2006)]).

In the recent papers [Luchko(2009b), Luchko(2010)], the case of the generalized time-fractional diffusion equation with variable coefficients and over an open bounded n-dimensional domain has been considered. This equation is obtained from the diffusion equation by replacing the first-order time derivative by a fractional derivative of order α $(0 < \alpha \leq 1)$ and the second-order spatial derivative by a more general linear second-order differential operator with variable coefficients. In the paper [Zacher(2008)], some more general linear and quasi-linear evolutionary partial integro-differential equations of second order were investigated. In particular, the global boundedness of appropriately defined weak solutions and a maximum principle for the weak solutions of such equations were established by employing a different technique compared to the one used in the papers [Luchko(2009b), Luchko(2010)].

In this chapter, a general time-fractional diffusion equation with variable coefficients is considered. Let us mention that some complex systems can show more complicated behavior compared to one that can be described by the time-fractional diffusion equation. In some cases the multi-term time-fractional diffusion equations (see, e.g., [Daftardar-Gejji and Bhalekar(2008), Luchko(2011)]) or even the fractional differential equations of the distributed order (see, e.g., [Chechkin et al.(2003), Luchko(2009a)]) with a special weight function could be a more appropriate alternative to model these complex systems.

The rest of the chapter is organized as follows. In the second section, continuous time random walk models are introduced and analyzed. It is shown that under certain restrictions these models can be written down in the form of a deterministic partial differential equation of fractional order. In the third section, the Cauchy problem for the space-time fractional diffusion equation that is obtained from the standard diffusion equation by replacing the second-order space derivative with a Riesz-Feller derivative of order $\alpha \in (0,2]$ and skewness θ ($|\theta| \le \min\{\alpha, 2-\alpha\}$), and the first-order time derivative with a Caputo derivative of order $\beta \in (0,2]$ are considered. The fundamental solution (Green function) for the Cauchy problem is investigated with respect to its scaling and similarity properties, starting from its Fourier-Laplace representation. We review the particular cases of space-fractional diffusion $0 < \alpha \le 2$, $\beta = 1$, time-fractional diffusion $\alpha = 2$, $0 < \beta \le 2$, and neutral-fractional diffusion $0 < \alpha = \beta \le 2$, for which the fundamental solution can be interpreted as a spatial probability density function evolving in time. Then, by using the Mellin integral transform, a general representation of the Green functions in terms of Mellin-Barnes integrals in the complex plane is provided. This allows us to extend the probability interpretation to the ranges $(0 < \alpha \le 2) \cap (0 < \beta \le 1)$ and $(1 < \beta \le \alpha \le 2)$. Furthermore, from this representation the explicit formulae (convergent series and asymptotic expansions) are derived. In the fourth section, the initial-boundary-value problems for generalized time-fractional diffusion equation are analyzed. First, the maximum principle, well known for partial differential equations of elliptic and parabolic type, is extended for the case of the generalized time-fractional differential equation. In the proof of the maximum principle, an appropriate extremum principle for the Caputo fractional derivative plays a very important role. Then the maximum principle is applied to show that the initial-boundary-value problem under consideration possesses at most one solution. This solution – if it exists – depends continuously on the data given in the problem. To show the existence of the solution, a notion of the generalized solution is first introduced. The generalized solution is constructed with the method of variables separation. Under certain conditions, it is shown that the generalized solution can be interpreted as a solution and thus its existence is proved. Finally, in the last section some open questions and directions for further research are summarized.

2. Continuous time random walk models

In the framework of the well-known random walk model for Brownian motion, the random walker jumps at each time step $t = 0, \Delta t, 2\Delta t, \ldots$ in a randomly selected direction, thereby covering the distance Δx, the lattice constant.

Denoting by $u(x,t)\,\Delta x$ the probability that the random walker is located between x and $x + \Delta x$ at the time t, the master equation

$$u(x, t+\Delta t) = \frac{1}{2}u(x+\Delta x, t) + \frac{1}{2}u(x-\Delta x, t) \tag{2.1}$$

can be easily derived. For the one-dimensional Brownian motion, the Taylor expansions

$$u(x, t+\Delta t) = u(x,t) + \Delta t \frac{\partial u}{\partial t} + \mathcal{O}((\Delta t)^2),$$

$$u(x \pm \Delta x, t) = u(x,t) \pm \Delta x \frac{\partial u}{\partial x} + \frac{(\Delta x)^2}{2}\frac{\partial^2 u}{\partial x^2} + \mathcal{O}((\Delta x)^3)$$

lead to

$$\frac{\partial u}{\partial t} = \frac{(\Delta x)^2}{2\Delta t}\frac{\partial^2 u}{\partial x^2} + \Delta x\, \mathcal{O}\left(\frac{(\Delta x)^2}{\Delta t}\right) + \mathcal{O}(\Delta t). \tag{2.2}$$

In the continuum limit $\Delta t \to 0$ and $\Delta x \to 0$, this equation becomes the diffusion equation

$$\frac{\partial u}{\partial t} = C_1 \frac{\partial^2 u}{\partial x^2} \tag{2.3}$$

under the condition that the diffusion coefficient

$$C_1 = \lim_{\Delta x \to 0,\ \Delta t \to 0} \frac{(\Delta x)^2}{2\Delta t}$$

is finite. Of course, the same procedure leads to the two- or three-dimensional diffusion equations for the two- or three-dimensional Brownian motion, respectively:

$$\frac{\partial u}{\partial t} = C_1 \Delta u,\ \Delta := \sum_{i=1}^{n} \frac{\partial^2}{\partial x_i^2},\ n = 2, 3. \tag{2.4}$$

When the random walker is located at the starting point $x = 0,\ x \in I\!R^n,\ n = 1, 2, 3$ at the time $t = 0$, then an initial condition

$$u(x,0) = \prod_{i=1}^{n} \delta(x_i), \qquad \int_{-\infty}^{+\infty} \delta(x)\, dx = 1,\ n = 1, 2, 3 \tag{2.5}$$

with the Dirac δ-function has to be added to the model. The solution of the diffusion equation (2.4) with the initial condition (2.5) can be found with the Laplace and Fourier integral transforms method. In the one-dimensional case we get

$$u(x,t) = \frac{1}{\sqrt{4\pi C_1 t}} \exp\left(-\frac{x^2}{4C_1 t}\right). \tag{2.6}$$

The pdf function $u(x,t)$ is a Gaussian distribution at any time point $t > 0$ with the middle value $\mu = 0$ and with the deviation $\sigma = \sqrt{2C_1 t}$, which means that the mean squared displacement of a particle participating in the anomalous diffusion process is given by $\sigma^2(t) = 2C_1 t$, that is in accordance with the equation (1.1).

In contrast to the random walk model for Brownian motion, the continuous time random walk model (CTRW model) is based on the idea that the length of a given jump, as well as the waiting time elapsing between two successive jumps, are ruled by a joint probability density function (pdf) $\psi(x,t)$ which will be referred to as the jump pdf. From $\psi(x,t)$, the jump length pdf

$$\lambda(x) = \int_0^{\infty} \psi(x,t)\, dt \tag{2.7}$$

and the waiting time pdf

$$w(t) = \int_{-\infty}^{\infty} \psi(x,t)\,dx \tag{2.8}$$

can be deduced.

The main characteristics of the CTRW processes are the characteristic waiting time

$$T = \int_{0}^{\infty} w(t)\,t\,dt \tag{2.9}$$

and the jump length variance

$$\Sigma^2 = \int_{-\infty}^{\infty} \lambda(x)\,x^2\,dx. \tag{2.10}$$

They can be finite or infinite and this makes the difference between the CTRW processes. In general, the following different cases are distinguished:

- Both T and Σ^2 are finite: Brownian motion (diffusion equation as a deterministic model)
- T diverges, Σ^2 is finite: Sub-diffusion (time-fractional diffusion equation as a deterministic model)
- T is finite, Σ^2 diverges: Levy flights (space-fractional diffusion equation as a deterministic model)
- Both T and Σ^2 are infinite: Levy flights (time-space-fractional diffusion equation as a deterministic model)

It is known that the CTRW model can be described by the master equations in the form of integral equations of convolution type (see, e.g., [Metzler and Klafter(2000a)]). Below we give a short summary of how to establish these equations.

Denote by $\eta(x,t)$ the pdf of the event that a particle arrived at position x at time t. This pdf satisfies the equation

$$\eta(x,t) = \int_{-\infty}^{+\infty} dx' \int_{0}^{t} \eta(x',t')\psi(x-x',t-t')\,dt' \;+\; \delta(x)\delta(t). \tag{2.11}$$

The pdf $u(x,t)$ of being at position x at time t is given by

$$u(x,t) = \int_{0}^{t} \eta(x,t')\Psi(t-t')\,dt', \tag{2.12}$$

where

$$\Psi(t) = 1 - \int_{0}^{t} w(t')\,dt', \tag{2.13}$$

is assigned to the probability of no jump event during the time interval $(0,t)$.

The integral equations (2.11)–(2.13) determine the one-point probability density function that is an important part of a mathematical model but of course not enough to fully characterize the underlaying stochastic process (see, e.g., [Germano et al.(2009)] for more details). Now we transform these equations into the frequency domain by applying the Fourier and the Laplace transforms to the equations (2.11)–(2.13). By using the well-known product property of the Fourier and the Laplace transforms, when applying them to the Fourier or Laplace convolutions respectively, we can deduce from the equations (2.11)–(2.13) the Fourier-Laplace transform of the jump pdf $u(x,t)$:

$$\widehat{\widetilde{u}}(\kappa, s) = \frac{1-\widetilde{w}(s)}{s} \frac{\widehat{u_0}(\kappa)}{1-\widehat{\widetilde{\psi}}(\kappa, s)}, \tag{2.14}$$

where $\widehat{u_0}(\kappa)$ denotes the Fourier transform of the initial condition $u_0(x)$. It is worth noting that a purely probabilistic proof of this equation is given in [Germano et al.(2009)].

We remind the reader that the Fourier transform is defined by

$$\widehat{f}(\kappa) = \mathcal{F}\{f(x);\kappa\} = \int_{-\infty}^{+\infty} e^{+i\kappa x}\, f(x)\, dx\,, \ \ \kappa \in R,$$

and the Laplace transform by

$$\widetilde{f}(s) = \mathcal{L}\{f(t);s\} = \int_0^{\infty} e^{-st}\, f(t)\, dt.$$

It can be shown (see, e.g., [Metzler and Klafter(2000a)]) that if both the characteristic waiting time and the jump length variance are finite for a CTRW, then its long-time limit behavior corresponds to the Brownian motion.

Now let us discuss the CTRW, where the characteristic waiting time diverges, but the jump length variance remains finite. To this end, a particular long-tailed waiting time pdf with the asymptotic behavior

$$w(t) \approx A_\alpha (\tau/t)^{1+\alpha}, \ t \to +\infty, \ 0 < \alpha < 1$$

is considered. Its asymptotics in the Laplace domain can be easily determined by the so-called Tauberian theorem and is as follows:

$$\widetilde{w}(s) \approx 1 - (s\tau)^\alpha, \ s \to 0.$$

It is important to mention that the specific form of $w(t)$ is of minor importance. In particular, the so-called Mittag-Leffler waiting time density

$$w(t) = -\frac{d}{dt} E_\alpha(-t^\alpha), \ \ E_\alpha(x) := \sum_{k=0}^{\infty} \frac{x^k}{\Gamma(\alpha\, k+1)}$$

can be taken without loss of generality. The Laplace transform of the function w can be evaluated in explicit form

$$\widetilde{w}(s) = \frac{1}{1+s^\alpha}$$

and has the desired asymptotics.

Together with the Gaussian jump length pdf

$$\lambda(x) = (4\pi\sigma^2)^{-1/2} \exp(-x^2/(4\sigma^2)), \ \ \Sigma^2 = 2\sigma^2$$

with the Fourier transform in the form

$$\widehat{\lambda}(\kappa) \approx 1 - \sigma^2\kappa^2, \ \kappa \to 0,$$

the asymptotics of the Fourier-Laplace transform of the jump pdf $u(x,t)$ becomes

$$\widehat{\widetilde{u}}(\kappa, s) \approx \frac{\widehat{u_0}(\kappa)/s}{1 + C_\alpha s^{-\alpha}\kappa^2}, \ s \to 0, \ \kappa \to 0. \tag{2.15}$$

Using the Tauberian theorems for the Laplace and Fourier transforms, the last equation can be transformed to a time-fractional partial differential equation for large t and $|x|$.

After multiplication with the denominator of the right-hand side, the equation (2.15) becomes

$$(1 + C_\alpha s^{-\alpha}\kappa^2)\, \widehat{\widetilde{u}}(\kappa, s) \approx \widehat{u_0}(\kappa)/s, \ s \to 0, \ \kappa \to 0. \tag{2.16}$$

Making use of the differentiation theorem

$$\mathcal{F}\{f''(x); \kappa\} = -\kappa^2 \mathcal{F}\{f(x); \kappa\}$$

for the Fourier transform and employing the integration rule

$$\mathcal{L}\{(I^\alpha f)(t); s\} = s^{-\alpha}\, \widetilde{f}(s),$$

for the Riemann-Liouville fractional integrals defined by

$$(I^\alpha f)(t) := \frac{1}{\Gamma(\alpha)} \int_0^t f(\tau)(t-\tau)^{\alpha-1}\, d\tau, \ \ \alpha > 0, \ \ (I^0 f)(t) = f(t),$$

the equation (2.16) can be rewritten in the form of the fractional integral equation

$$u(x,t) - u_0(x) = C_\alpha (I^\alpha \frac{\partial^2}{\partial x^2} u(x,\tau))(t) \tag{2.17}$$

for large t and $|x|$.

By application of the fractional differential operator D_t^α to (2.17), the CTRW model can be written for large t and $|x|$ in the form of the initial-value problem

$$u(x,0) = u_0(x)$$

for the so-called time-fractional diffusion equation

$$(D_t^\alpha u)(t) = C_\alpha \frac{\partial^2 u}{\partial x^2}. \tag{2.18}$$

In what follows, the fractional derivative D_t^α ($m-1 < \alpha \le m, \ m \in N$) will be defined in the Caputo sense (see, e.g., [Podlubny(1999)]):

$$(D_t^\alpha f)(t) := (I^{m-\alpha} f^{(m)})(t). \tag{2.19}$$

For the theory of fractional integrals and derivatives the reader is referred, e.g., to [Kilbas et al.(2006)] or [Podlubny(1999)].

It is worth mentioning that the integro-differential nature of the fractional differential operator in the model (2.18) ensures the non-Markovian nature of the sub-diffusive process. Indeed, calculating the Laplace transform of the mean squared displacement via the relation

$$\overline{\widetilde{x}^2(s)} = \lim_{\kappa \to 0} -\frac{d^2}{d\kappa^2} u(\kappa, s)$$

and using the Laplace inversion, the formula

$$\overline{x^2(t)} = \frac{2C_\alpha}{\Gamma(1+\alpha)} t^\alpha$$

for the mean squared displacement in time is obtained.

Now let the characteristic waiting time be finite, but the jump length variance be infinite.

As an example, consider the Levy distribution for the jump length with the Fourier transform

$$\widehat{\lambda}(\kappa) = \exp(-\sigma^\mu |\kappa|^\mu) \approx 1 - \sigma^\mu |\kappa|^\mu, \; 1 < \mu < 2, \; |\kappa| \to 0.$$

Then

$$\lambda(x) \approx A_\mu \sigma^{-\mu} |x|^{-1-\mu}, \quad x \to \infty.$$

Again, the specific form of $\lambda(x)$ is of minor importance.

Together with the Poissonian waiting time distribution

$$w(t) = \tau^{-1} \exp(-t/\tau)$$

with the Laplace transform of the form

$$\widetilde{w}(s) \approx 1 - s\tau + O(s^2), \; s \to 0,$$

the asymptotics of the Fourier-Laplace transform of the pdf u given by (2.14) can be written in the form

$$\widehat{\widetilde{u}}(\kappa, s) = \frac{1}{s + K_\mu |\kappa|^\mu}, \quad s \to 0, \; |\kappa| \to 0. \tag{2.20}$$

By inverting the Laplace and Fourier transforms in the equation (2.20), a space-fractional diffusion equation

$$\frac{\partial u}{\partial t} = K_\mu D_x^\mu u(x,t),$$

is obtained for large t and $|x|$, where D_x^μ is the Riesz operator defined by (see, e.g., [Kilbas et al.(2006)] or [Metzler and Klafter(2000a)])

$$(D_x^\mu u)(x) = \frac{\Gamma(1+\mu)}{\pi} \sin(\mu\pi/2) \int_0^\infty \frac{u(x+\zeta) - 2u(x) + u(x-\zeta)}{\zeta^{1+\mu}} \, d\zeta.$$

From this equation, the Fourier transform of the pdf u can be determined in the form

$$\widehat{u}(\kappa, t) = \exp\left(-K_\mu t |\kappa|^\mu\right), \; |\kappa| \to 0, \; t \to \infty.$$

The asymptotics of the pdf u is then given by

$$u(x,t) \approx \frac{K_\mu t}{|x|^{1+\mu}}, \; 1 < \mu < 2, \; t \to \infty, |x| \to \infty.$$

As expected, it follows from the last formula, that the mean squared displacement of the pdf u diverges.

Following the same way, we can deduce the equation for the pdf u in the case of infinite T and Σ^2 in the form of the time-space fractional diffusion equation

$$(D_t^\beta u)(t) = S_{\beta,\mu} D_x^\mu u(x,t) \tag{2.21}$$

with the Caputo fractional derivative D_t^β in time and the Riesz fractional derivative D_x^μ in space.

Further fractional models that generalize the well-known conventional models are the fractional diffusion-advection equation (anomalous diffusion with an additional velocity field) and the fractional Fokker-Plank equation (anomalous diffusion in the presence of an external field). Of course, like in the conventional case, the multi-dimensional generalizations, fractional equations with non-constant coefficients and nonlinear fractional differential equations appear in the corresponding models and should be investigated.

3. Initial-value-problems for the space-time-fractional diffusion equation

Motivated by the models introduced in the previous section, we deduce in this section the fundamental solution (the Green function) for the one-dimensional space-time-fractional diffusion equation in terms of the Mellin-Barnes integral, consider its particular cases, and give an interpretation of the Green function as a probability density function. The representation follows the results presented in [Mainardi et al.(2001)].

The space-time fractional diffusion equation is given by

$$D_t^\beta \, u(x,t) \; = \; {}_xD_\theta^\alpha \, u(x,t) \, , \quad x \in R \, , \quad t \in I\!R^+ \, , \tag{3.1}$$

where the α , θ , β are real parameters restricted as follows:

$$0 < \alpha \le 2 \, , \quad |\theta| \le \min\{\alpha, 2-\alpha\} \, , \quad 0 < \beta \le 2 \, ,$$

D_t^β is the Caputo fractional derivative of the order β $(m-1 < \beta \le m, \; m \in I\!N)$ and by ${}_xD_\theta^\alpha$ the so-called *Riesz-Feller* space-fractional derivative of order α and skewness θ is denoted. For sufficiently well-behaved functions, this derivative is defined as a pseudo-differential operator in the form

$$\mathcal{F}\{ {}_xD_\theta^\alpha \, f(x); \kappa\} = -\psi_\alpha^\theta(\kappa) \, \widehat{f}(\kappa) \, , \tag{3.2}$$

with the symbol

$$\psi_\alpha^\theta(\kappa) = |\kappa|^\alpha \, \mathrm{e}^{i(\operatorname{sign}\kappa)\theta\pi/2} \, , \; 0 < \alpha \le 2 \, , \quad |\theta| \le \min\{\alpha, 2-\alpha\}.$$

Whereas the heuristic discussion in the previous section was presented only for the case of the symmetric diffusion, the equation (3.1) is fully asymmetric. It generalises the time-space fractional diffusion equation (2.21) for the case when the jumps follow an asymmetric Lévy distribution. For $\theta = 0$, the Riesz-Feller space-fractional derivative is a symmetric operator with respect to x, which can be interpreted as

$$ {}_xD_0^\alpha = -\left(-\frac{d^2}{dx^2}\right)^{\alpha/2}, $$

as can be formally deduced by writing $-|\kappa|^\alpha = -(\kappa^2)^{\alpha/2}$.

For $0 < \alpha < 2$ and $|\theta| \leq \min\{\alpha, 2-\alpha\}$ the Riesz-Feller derivative can be represented in the form

$$ {}_xD_\theta^\alpha\, f(x) = \frac{\Gamma(1+\alpha)}{\pi} \times \left\{ \sin\left[(\alpha+\theta)\pi/2\right] \int_0^\infty \frac{f(x+\xi)-f(x)}{\xi^{1+\alpha}}\, d\xi \right. $$
$$ \left. + \sin\left[(\alpha-\theta)\pi/2\right] \int_0^\infty \frac{f(x-\xi)-f(x)}{\xi^{1+\alpha}}\, d\xi \right\}. $$

For the space-time-fractional diffusion equation (3.1) we consider the Cauchy problem with the initial condition

$$ u(x,0) = \varphi(x)\,,\ x \in I\!R\,, \quad u(\pm\infty, t) = 0\,,\ t > 0\,, \tag{3.3} $$

where $\varphi(x) \in L^c(I\!R)$ is a sufficiently well-behaved function. If $1 < \beta \leq 2$, the condition $u_t(x,0) = 0$, where $u_t(x,t) = \frac{\partial}{\partial t}u(x,t)$ is added to the initial conditions.

By the Green function (or fundamental solution) of the Cauchy problem we mean the (generalized) function $G_{\alpha,\beta}^\theta(x,t)$ corresponding to $\varphi(x) = \delta(x)$ (the Dirac delta function). It allows us to represent the solution of the Cauchy problem by the integral formula

$$ u_{\alpha,\beta}^\theta(x,t) = \int_{-\infty}^{+\infty} G_{\alpha,\beta}^\theta(\xi,t)\,\varphi(x-\xi)\,d\xi\,. $$

For a sufficiently well-behaved function f, the Laplace transform of the Caputo time-fractional derivative of order β $(m-1 < \beta \leq m\,,\ m \in I\!N)$ is given by

$$ \mathcal{L}\left\{ D_t^\beta\, f(t); s \right\} = s^\beta\, \widetilde{f}(s) - \sum_{k=0}^{m-1} s^{\beta-1-k}\, f^{(k)}(0^+). \tag{3.4} $$

We apply now the Laplace and the Fourier transforms to the equation (3.1). Taking into account the formulae (3.2)–(3.4) we get

$$ -\psi_\alpha^\theta(\kappa)\, \widehat{\widetilde{G_{\alpha,\beta}^\theta}}(\kappa,s) \;=\; s^\beta\, \widehat{\widetilde{G_{\alpha,\beta}^\theta}}(\kappa,s) - s^{\beta-1}\,, $$

where

$$ \psi_\alpha^\theta(\kappa) := |\kappa|^\alpha\, e^{i(\operatorname{sign}\kappa)\theta\pi/2}. $$

We therefore obtain the formula

$$ \widehat{\widetilde{G_{\alpha,\beta}^\theta}}(\kappa,s) = \frac{s^{\beta-1}}{s^\beta + \psi_\alpha^\theta(\kappa)} \tag{3.5} $$

for the Green function in the Fourier-Laplace domain.

First we mention that by using the known scaling rules for the Fourier and Laplace transforms,

$$f(ax) \overset{\mathcal{F}}{\leftrightarrow} a^{-1}\,\widehat{f}(\kappa/a)\,, \quad a>0\,,$$
$$f(bt) \overset{\mathcal{L}}{\leftrightarrow} b^{-1}\,\widetilde{f}(s/b)\,, \quad b>0\,,$$

we can directly (i.e., without inverting the two transforms) infer the following scaling property of the Green function,

$$G^{\theta}_{\alpha,\beta}(ax\,,\,bt) = b^{-\gamma}G^{\theta}_{\alpha,\beta}(ax/b^{\gamma}\,,\,t)\,,\ \gamma=\beta/\alpha\,.$$

Consequently, introducing the similarity variable x/t^{γ}, the Green function can be written in the form

$$G^{\theta}_{\alpha,\beta}(x,t) = t^{-\gamma}\,K^{\theta}_{\alpha,\beta}(x/t^{\gamma})\,,\ \gamma=\beta/\alpha, \tag{3.6}$$

where the one-variable function $K^{\theta}_{\alpha,\beta}$ is to be determined.

Let us invert the Laplace-Fourier transform of the right-hand side of the formula (3.5) starting with the inverse Laplace transform. To this end, the following Laplace transform pair is used (see, e.g., [Erdélyi et al. (1953)])

$$E_{\beta}(ct^{\beta}) \overset{\mathcal{L}}{\leftrightarrow} \frac{s^{\beta-1}}{s^{\beta}-c}\,,\ \Re\,(s) > |c|^{1/\beta}\,,$$

with $c\in\mathbb{C}$, $0<\beta\le 2$, where E_{β} denotes the Mittag-Leffler function of order β, defined in the complex plane by the power series

$$E_{\beta}(z) := \sum_{n=0}^{\infty}\frac{z^{n}}{\Gamma(\beta n+1)}\,,\ \beta>0\,,\ z\in\mathbb{C}\,.$$

Using the last formula the Fourier transform of the Green function can be written in the form

$$\widehat{G^{\theta}_{\alpha,\beta}}(\kappa,t) = E_{\beta}\left[-\psi^{\theta}_{\alpha}(\kappa)\,t^{\beta}\right]\,,\ \kappa\in\mathbb{R}\,,\ t\ge 0\,. \tag{3.7}$$

In what follows, the technique of the Mellin integral transform and the Mellin-Barnes integrals is employed. We remind the reader of the definition of the Mellin integral transform and some of its properties useful for further discussions. The Mellin integral transform is defined by

$$\mathcal{M}\left\{f(r);s\right\} = f^{*}(s) = \int_{0}^{+\infty} f(r)\,r^{s-1}\,dr, \quad \gamma_1<\Re\,(s)<\gamma_2,$$

and the inverse Mellin transform by

$$\mathcal{M}^{-1}\left\{f^{*}(s);r\right\} = f(r) = \frac{1}{2\pi i}\int_{\gamma-i\infty}^{\gamma+i\infty} f^{*}(s)\,r^{-s}\,ds$$

where $r>0$, $\gamma=\Re\,(s)$, $\gamma_1<\gamma<\gamma_2$. Denoting by $\overset{\mathcal{M}}{\leftrightarrow}$ the juxtaposition of a function $f(r)$ with its Mellin transform $f^{*}(s)$, the main transformation rules are:

$$f(ar) \overset{\mathcal{M}}{\leftrightarrow} a^{-s}\,f^{*}(s)\,, \quad a>0\,,$$

$$r^a\, f(r) \overset{\mathcal{M}}{\leftrightarrow} f^*(s+a)\,,$$

$$f(r^p) \overset{\mathcal{M}}{\leftrightarrow} \frac{1}{|p|}\, f^*(s/p)\,, \quad p \neq 0\,,$$

$$h(r) = \int_0^\infty \frac{1}{\rho}\, f(\rho)\, g(r/\rho)\, d\rho \overset{\mathcal{M}}{\leftrightarrow} h^*(s) = f^*(s)\, g^*(s)\,.$$

The Mellin convolution formula will be used in further discussions in treating integrals of the Fourier type for $x = |x| > 0$:

$$I_c(x) = \frac{1}{\pi} \int_0^\infty f(\kappa)\, \cos(\kappa\, x)\, d\kappa\,, \tag{3.8}$$

$$I_s(x) = \frac{1}{\pi} \int_0^\infty f(\kappa)\, \sin(\kappa\, x)\, d\kappa\,. \tag{3.9}$$

The integrals $I_c(x)$ and $I_s(x)$ can be interpreted as the Mellin convolutions between $f(\kappa)$ and the functions $g_c(\kappa)$, $g_s(\kappa)$, respectively, with $r = 1/|x|$, $\rho = \kappa$, where

$$g_c(\kappa) := \frac{1}{\pi\, |x|\, \kappa} \cos\left(\frac{1}{\kappa}\right) \overset{\mathcal{M}}{\leftrightarrow} \frac{\Gamma(1-s)}{\pi\, |x|} \sin\left(\frac{\pi s}{2}\right) := g_c^*(s),\ 0 < \Re(s) < 1\,,$$

$$g_s(\kappa) := \frac{1}{\pi\, |x|\, \kappa} \sin\left(\frac{1}{\kappa}\right) \overset{\mathcal{M}}{\leftrightarrow} \frac{\Gamma(1-s)}{\pi\, |x|} \cos\left(\frac{\pi s}{2}\right) := g_s^*(s),\ 0 < \Re(s) < 2\,.$$

Finally, the convolution theorem for the Mellin transform allows us to represent the integrals $I_c(x)$ and $I_s(x)$ in the form

$$I_c(x) = \frac{1}{\pi\, x} \frac{1}{2\pi i} \int_{\gamma-i\infty}^{\gamma+i\infty} f^*(s)\, \Gamma(1-s)\, \sin\left(\frac{\pi\, s}{2}\right) x^s\, ds\,,\ x > 0\,,\ 0 < \gamma < 1\,, \tag{3.10}$$

$$I_s(x) = \frac{1}{\pi\, x} \frac{1}{2\pi i} \int_{\gamma-i\infty}^{\gamma+i\infty} f^*(s)\, \Gamma(1-s)\, \cos\left(\frac{\pi\, s}{2}\right) x^s\, ds\,,\ x > 0\,,\ 0 < \gamma < 2\,. \tag{3.11}$$

Another formula that is very essential for further discussion is a representation of the Mittag-Leffler function in form of the inverse Mellin integral transform

$$E_\beta(z) = \frac{1}{2\pi i} \int_{L_{-\infty}} \frac{\Gamma(s)\, \Gamma(1-s)}{\Gamma(1-\beta s)} (-z)^{-s}\, ds\,, \tag{3.12}$$

where the integration is over a left-hand loop $L_{-\infty}$ drawn round all the left-hand poles $s = 0, -1, -2, \ldots$ of the integrand in a positive direction.

Now we start with the determination of the one-variable function $K^\theta_{\alpha,\beta}$ from the formula (3.6) for the Green function. First we note the symmetry relation for the function $K^\theta_{\alpha,\beta}$ in the form

$$K^\theta_{\alpha,\beta}(-x) = K^{-\theta}_{\alpha,\beta}(x)\,.$$

As a consequence, we can restrict our attention to the case $x > 0$, and obtain from (3.6)-(3.7) the representation

$$K^\theta_{\alpha,\beta}(x) = \frac{1}{2\pi} \int_{-\infty}^{+\infty} \mathrm{e}^{-i\kappa x}\, E_\beta\left[-\psi^\theta_\alpha(\kappa)\right]\, d\kappa \,=\, {}_cK^\theta_{\alpha,\beta}(x) +_s K^\theta_{\alpha,\beta}(x)\,,$$

where

$$ {}_cK^{\theta}_{\alpha,\beta}(x) = \frac{1}{\pi}\int_0^{\infty} \cos(\kappa x)\,\Re\Big[E_\beta\Big(-\kappa^\alpha\,\mathrm{e}^{i\theta\pi/2}\Big)\Big]\,d\kappa\,, \tag{3.13} $$

$$ {}_sK^{\theta}_{\alpha,\beta}(x) = \frac{1}{\pi}\int_0^{\infty} \sin(\kappa x)\,\Im\Big[E_\beta\Big(-\kappa^\alpha\,\mathrm{e}^{i\theta\pi/2}\Big)\Big]\,d\kappa\,. \tag{3.14} $$

From (3.12) the Mellin transform pair

$$ E_\beta\Big(-\kappa^\alpha\,\mathrm{e}^{i\theta\pi/2}\Big) \overset{\mathcal{M}}{\leftrightarrow} \frac{1}{\alpha}\,\frac{\Gamma(\frac{s}{\alpha})\,\Gamma(1-\frac{s}{\alpha})}{\Gamma(1-\frac{\beta}{\alpha}s)}\,\exp\left(-i\frac{\theta\pi}{2}\frac{s}{\alpha}\right), $$

where $\kappa > 0,\ \ |\theta| \le 2-\beta,\ \ 0 < \Re(s) < \alpha$ can be deduced. Using now this formula, the representations (3.13)–(3.14), as well as (3.10)–(3.11), we then obtain the representation

$$ K^{\theta}_{\alpha,\beta}(x) = \frac{1}{\pi\alpha x}\,\frac{1}{2\pi i}\int_{\gamma-i\infty}^{\gamma+i\infty} \frac{\Gamma(\frac{s}{\alpha})\,\Gamma(1-\frac{s}{\alpha})\,\Gamma(1-s)}{\Gamma(1-\frac{\beta}{\alpha}s)}\,\sin\left[\frac{s}{\alpha}\frac{\pi}{2}(\alpha-\theta)\right]\,x^s\,ds $$

for the function $K^{\theta}_{\alpha,\beta}$ in terms of a Mellin-Barnes integral. By setting

$$ \rho = \frac{\alpha-\theta}{2\,\alpha}\,, $$

and using the reflection formula for the gamma function, we finally get

$$ K^{\theta}_{\alpha,\beta}(x) = \frac{1}{\alpha x}\,\frac{1}{2\pi i}\int_{\gamma-i\infty}^{\gamma+i\infty} \frac{\Gamma(\frac{s}{\alpha})\,\Gamma(1-\frac{s}{\alpha})\,\Gamma(1-s)}{\Gamma(1-\frac{\beta}{\alpha}s)\,\Gamma(\rho\,s)\,\Gamma(1-\rho\,s)}\,x^{s}\,ds\,. \tag{3.15} $$

The representation (3.15) shows that the function $K^{\theta}_{\alpha,\beta}$ is a particular case of the general Fox H-function; it can be used both for the asymptotical analysis of $K^{\theta}_{\alpha,\beta}$, for deducing its series representations, and for simplification of the function for some special values of its parameters. In what follows, we consider shortly some of the aspects mentioned above. For a detailed presentation of these results we refer the reader to [Mainardi et al.(2001)].

In particular, for the value of $K^{\theta}_{\alpha,\beta}(x)$ at the point $x = 0$ we obtain

$$ K^{\theta}_{\alpha,\beta}(0) = \frac{1}{\pi\alpha}\,\frac{\Gamma(1/\alpha)\,\Gamma(1-1/\alpha)}{\Gamma(1-\beta/\alpha)}\,\cos\left(\frac{\theta\pi}{2\alpha}\right)\ \text{if}\ 1 < \alpha \le 2\,,\ \beta \ne 1\,, $$

$$ K^{\theta}_{\alpha,\beta}(0) = \frac{1}{\pi\alpha}\,\Gamma(1/\alpha)\,\cos\left(\frac{\theta\pi}{2\alpha}\right)\ \text{if}\ 0 < \alpha \le 2\,,\ \beta = 1\,. $$

Now we consider some particular cases of the space-time fractional diffusion equation and of the corresponding Green function:

$\alpha = 2\,,\ \beta = 1$ (standard diffusion),

$0 < \alpha \le 2\,,\ \beta = 1$ (space-fractional diffusion),

$\alpha = 2\,,\ 0 < \beta \le 2\,,\ \beta \ne 1$ (time-fractional diffusion),

$0 < \alpha = \beta \le 2$ (neutral fractional diffusion).

It is well known that for the standard diffusion ($\alpha = 2\,, \beta = 1$) the Green function is the Gaussian *pdf*

$$G^0_{2,1}(x,t) = t^{-1/2}\,\frac{1}{2\sqrt{\pi}}\exp[-x^2/(4t)]\,, \quad -\infty < x < +\infty\,,\ t \geq 0\,,$$

with similarity variable $x/t^{1/2}$, that evolves in time with moments (of even order)

$$\mu_{2n}(t) := \int_{-\infty}^{+\infty} x^{2n}\, G^0_{2,1}(x,t)\, dx = \frac{(2n)!}{n!}\, t^n\,,\ t \geq 0\,.$$

The variance $\sigma^2 := \mu_2(t) = 2t$ is thus proportional to the first power of time, according to the Einstein diffusion law.

The case of the space-fractional diffusion ($0 < \alpha \leq 2\,, \beta = 1$) includes the standard diffusion for $\alpha = 2$. In this case, the Mittag-Leffler function is reduced to the exponential function:

$$\widehat{G^\theta_{\alpha,1}}(\kappa,t) = \mathrm{e}^{-t\psi^\theta_\alpha(\kappa)}\,,$$

with $\psi^\theta_\alpha(\kappa)$ defined as in the general case.

Then the Green function of the space-fractional diffusion equation can be interpreted as a Lévy strictly stable *pdf*, evolving in time, according to

$$G^\theta_{\alpha,1}(x,t) = t^{-1/\alpha}\, L^\theta_\alpha(x/t^{1/\alpha})\,,$$

$$-\infty < x < +\infty\,,\ t \geq 0\,.$$

The stable densities admit a representation in terms of elementary functions only in the following particular cases

$\alpha = 2\,,\ \theta = 0\,,$ *Gaussian distribution* :

$$\mathrm{e}^{-\kappa^2} \overset{\mathcal{F}}{\leftrightarrow} L^0_2(x) = \frac{1}{2\sqrt{\pi}}\,\mathrm{e}^{-x^2/4}\,, \quad -\infty < x < +\infty\,;$$

$\alpha = 1/2\,,\ \theta = -1/2\,,$ *Lévy-Smirnov distribution* :

$$L^{-1/2}_{1/2}(x) = \frac{x^{-3/2}}{2\sqrt{\pi}}\,\mathrm{e}^{-1/(4x)}\,, \quad x \geq 0\,;$$

$\alpha = 1\,,\ \theta = 0\,,$ *Cauchy distribution* :

$$\mathrm{e}^{-|\kappa|} \overset{\mathcal{F}}{\leftrightarrow} L^0_1(x) = \frac{1}{\pi}\,\frac{1}{x^2+1}\,, \quad -\infty < x < +\infty\,.$$

We note that the case $\alpha = 1$ can be easily treated also for $\theta \neq 0$ taking into account elementary properties of the Fourier transform:

$\alpha = 1\,,\quad 0 < |\theta| < 1$:

$$L^\theta_1(x) = \frac{1}{\pi}\,\frac{\cos(\theta\pi/2)}{[x+\sin(\theta\pi/2)]^2 + [\cos(\theta\pi/2)]^2}\,, \qquad -\infty < x < +\infty\,;$$

$\alpha = 1\,,\quad \theta = \pm 1$:

$$L^{\pm 1}_1(x) = \delta(x \pm 1)\,, \quad -\infty < x < +\infty\,.$$

For $0 < \alpha < 2$ the stable *pdf*'s exhibit fat tails in such a way that their absolute moment of order ν is finite only if $-1 < \nu < \alpha$. In fact one can show that for non-Gaussian stable densities the asymptotic decay of the tails is

$$L_\alpha^\theta(x) = O\left(|x|^{-(\alpha+1)}\right), \quad x \to \pm\infty .$$

Consequently, the Gaussian distribution is the unique stable distribution with finite variance. Furthermore, when $0 < \alpha \leq 1$, the first absolute moment is infinite so we should use the median instead of the non-existent expected value. The asymptotic representations of the stable distributions are given by the following formulae:

$0 < \alpha < 1, \quad -\alpha < \theta \leq \alpha$:

$$L_\alpha^\theta(x) \sim \frac{1}{\pi x} \sum_{n=1}^{\infty} (-x)^n \frac{\Gamma(1+n/\alpha)}{n!} \sin\left[\frac{n\pi}{2\alpha}(\theta - \alpha)\right], \quad x \to 0^+ ,$$

$0 < \alpha < 1, \quad \theta = -\alpha$:

$$L_\alpha^{-\alpha}(x) \sim A_1 \, x^{-a_1} \, \mathrm{e}^{-b_1 \, x^{c_1}}, \quad x \to 0^+, \quad A_1 = \left\{[2\pi(1-\alpha)]^{-1} \, \alpha^{1/(1-\alpha)}\right\}^{1/2},$$

$$a_1 = \frac{2-\alpha}{2(1-\alpha)} \quad b_1 = (1-\alpha)\,\alpha^{\alpha/(1-\alpha)}, \quad c_1 = \frac{\alpha}{1-\alpha};$$

$1 < \alpha < 2, \quad \alpha - 2 < \theta \leq 2 - \alpha$:

$$L_\alpha^\theta(x) \sim \frac{1}{\pi x} \sum_{n=1}^{\infty} (-x^{-\alpha})^n \frac{\Gamma(1+n\alpha)}{n!} \sin\left[\frac{n\pi}{2}(\theta - \alpha)\right], \quad x \to \infty ,$$

$1 < \alpha < 2, \quad \theta = \alpha - 2$:

$$L_\alpha^{\alpha-2}(x) \sim A_2 \, x^{a_2} \, \mathrm{e}^{-b_2 x^{c_2}}, \quad x \to \infty, \quad A_2 = \left[2\pi(\alpha-1)\,\alpha^{1/(\alpha-1)}\right]^{-1/2},$$

$$a_2 = \frac{2-\alpha}{2(\alpha-1)}, \quad b_2 = (\alpha-1)\,\alpha^{\alpha/(\alpha-1)}, \quad c_2 = \frac{\alpha}{\alpha-1}.$$

Let us now consider the case of the time-fractional diffusion ($\alpha = 2\,, 0 < \beta < 2$) including standard diffusion for $\beta = 1$, for which we have

$$\widehat{G_{2,\beta}^0}(\kappa, t) = E_\beta\left(-\kappa^2 \, t^\beta\right), \quad \kappa \in I\!R, \; t \geq 0 .$$

Inverting the Fourier transform we get

$$G_{2,\beta}^0(x,t) = \frac{1}{2} t^{-\beta/2} \, M_{\beta/2}\left(|x|/t^{\beta/2}\right),$$

$$-\infty < x < +\infty, \; t \geq 0,$$

where M_ν denotes the Mainardi function (a particular case of the Wright function):

$$M_\nu(z) = \sum_{n=0}^{\infty} \frac{(-z)^n}{n!\,\Gamma[-\nu n + (1-\nu)]}.$$

It turns out that $M_\nu(z)$ is an entire function of order $\rho = 1/(1-\nu)$, which provides a generalization of the Gaussian and of the Airy function:

$$M_{1/2}(z) = \frac{1}{\sqrt{\pi}} \exp\left(-\,z^2/4\right), \quad M_{1/3}(z) = 3^{2/3}\,\mathrm{Ai}\left(z/3^{1/3}\right).$$

The Green function $G^0_{2,\beta}(x,t)$ with $0 < \beta < 2$ can be interpreted as a symmetric spatial *pdf* evolving in time with a stretched exponential decay. More precisely, we have

$$G^0_{2,\beta}(x,1) = \frac{1}{2}\, M_{\beta/2}(|x|) \sim A\, x^a\, \mathrm{e}^{-bx^c}, \quad x \to +\infty,$$

$$A = \left\{2\pi(2-\beta)\, 2^{\beta/(2-\beta)}\, \beta^{(2-2\beta)/(2-\beta)}\right\}^{-1/2},$$

$$a = \frac{2\beta-2}{2(2-\beta)}, \quad b = (2-\beta)\, 2^{-2/(2-\beta)}\, \beta^{\beta/(2-\beta)}, \quad c = \frac{2}{2-\beta}.$$

Furthermore, the moments (of even order) of $G^0_{2,\beta}(x,t)$ can be evaluated in the explicit form:

$$\mu_{2n}(t) := \int_{-\infty}^{+\infty} x^{2n}\, G^0_{2,\beta}(x,t)\, dx = \frac{\Gamma(2n+1)}{\Gamma(\beta n+1)} t^{\beta n}.$$

In particular, the variance

$$\sigma^2 := \mu_2 = 2\, t^\beta / \Gamma(\beta+1)$$

is now proportional to the βth power of time, consistent with anomalous slow diffusion for $0 < \beta < 1$ and with anomalous fast diffusion for $1 < \beta < 2$.

In the limit $\beta = 2$ we recover the fundamental solution of the D'Alembert wave equation, i.e.,

$$G^0_{2,2}(x,t) = \frac{\delta(x-t) + \delta(x+t)}{2} = \frac{G^{-1}_{1,1}(x,t) + G^{+1}_{1,1}(x,t)}{2},$$
$$-\infty < x < +\infty\,, \; t \geq 0\,.$$

Finally we consider neutral fractional diffusion ($0 < \alpha = \beta \leq 2$), which includes the Cauchy diffusion for $\alpha = \beta = 1$ ($\theta = 0$) and the limiting case of wave propagation for $\alpha = \beta = 2$.

In this case, we use the following Fourier transform pair related to the Mittag-Leffler function of our interest:

$$E_\alpha\left(-|\kappa|^\alpha\right) \overset{\mathcal{F}}{\leftrightarrow} \frac{1}{\pi}\, \frac{|x|^{\alpha-1} \sin\left(\alpha\pi/2\right)}{1 + 2|x|^\alpha \cos\left(\alpha\pi/2\right) + |x|^{2\alpha}}, \quad 0 < \alpha < 2\,, \quad x \in I\!R\,.$$

Using this last formula, the Green function can be written for $\alpha = \beta$, $0 < \alpha < 2$ and $x > 0$ in the form

$$K^\theta_{\alpha,\alpha}(x) = \frac{1}{\pi}\, \frac{x^{\alpha-1} \sin[\frac{\pi}{2}(\alpha-\theta)]}{1 + 2x^\alpha \cos[\frac{\pi}{2}(\alpha-\theta)] + x^{2\alpha}}.$$

This solution can be extended to negative x by setting

$$K^\theta_{\alpha,\alpha}(-x) = K^{-\theta}_{\alpha,\alpha}(x)\,,$$

and is evidently not negative in all of $I\!R$, so it can be interpreted as a probability density. In other words, $K^{\theta}_{\alpha,\alpha}(x)$ may be considered the fractional generalization with skewness of the well-known Cauchy density.

In the limiting case $\alpha \to 2^-$ (with $\theta = 0$) the density tends to the combination $[\delta(x-1) + \delta(x+1)]/2$, so we recover the Green function of the D'Alembert wave equation.

There is a problem in the physical and probabilistic interpretation of the equation (3.1) when the Caputo derivative is of order β and $1 < \beta < 2$. This case was considered in [Mainardi and Pagnini(2002), Mainardi and Pagnini(2003)].

For a detailed discussion of the composition rules for the Green function, algorithms for its numerical evaluation, and numerous plots of this function for different values of the parameters we refer the reader to [Mainardi et al.(2001)].

4. Initial-boundary-value problems for the generalized time-fractional diffusion equation

In this section, the time-fractional diffusion equation (2.18) deduced in the second section as a model for certain CTRW processes is considered. Because in real life applications we mostly deal with the two- or three-dimensional anomalous diffusion processes and both the properties of the medium where the diffusion takes place and the diffusion coefficient can depend on the spatial coordinates, we consider here the generalized time-fractional diffusion equation (GTFDE) that is obtained from the diffusion equation by replacing the first-order time derivative by a fractional derivative of order α $(0 < \alpha \leq 1)$ and the second-order spatial derivative by the general linear second-order differential operator. Unlike in the case of the linear space-time-fractional diffusion equation with constant coefficients that was considered in the previous section, this equation cannot be solved analytically in an explicit form. We thus need to use numerical algorithms to evaluate the solution on the finite domains and to try to determine the qualitative behavior of solutions on infinite domains. The first step of this analysis we deal with in this section is to consider the questions of existence and uniqueness of the solution of this equation with the appropriately chosen initial and boundary conditions.

The generalized time-fractional diffusion equation has the form

$$(D_t^{\alpha} u)(t) = -L(u) + F(x,t), \tag{4.1}$$

$$0 < \alpha \leq 1, (x,t) \in \Omega_T := G \times (0,T),\ G \subset I\!R^n,$$

where

$$L(u) := -\mathrm{div}(p(x)\,\mathrm{grad}\, u) + q(x)u,$$

$$p \in C^1(\bar{G}), q \in C(\bar{G}),\ p(x) > 0, q(x) \geq 0, x \in \bar{G}, \tag{4.2}$$

the fractional derivative D_t^{α} is defined in the Caputo sense and the domain G with boundary S is open and bounded in $I\!R^n$.

The operator $-L$ is in fact a linear elliptic differential operator of the second order:

$$-L(u) = \sum_{k=1}^{n} \left(p(x)\frac{\partial^2 u}{\partial x_k^2} + \frac{\partial p}{\partial x_k}\frac{\partial u}{\partial x_k} \right) - q(x)u,$$

that can be represented in the form

$$-L(u) = p(x)\triangle u + (\operatorname{grad} p, \operatorname{grad} u) - q(x)u, \tag{4.3}$$

$\triangle$ being the Laplace operator. For $\alpha = 1$, the equation (4.1) is reduced to a linear second-order parabolic PDE. The theory of this equation is well known, so that the main focus in the section is on the case $0 < \alpha < 1$. For the equation (4.1), the initial-boundary-value problem

$$u\big|_{t=0} = u_0(x),\ x \in \bar{G}, \tag{4.4}$$

$$u\big|_S = v(x,t),\ (x,t) \in S \times [0,T] \tag{4.5}$$

is considered. A solution of the problem (4.1), (4.4), (4.5) is called a function $u = u(x,t)$ defined in the domain $\bar{\Omega}_T := \bar{G} \times [0,T]$ that belongs to the space $C(\bar{\Omega}_T) \cap W_t^1((0,T]) \cap C_x^2(G)$ and satisfies both equation (4.1) and the initial and boundary conditions (4.4)–(4.5). By $W_t^1((0,T])$, the space of the functions $f \in C^1((0,T])$ such that $f' \in L((0,T))$ is denoted. If the problem (4.1), (4.4), (4.5) possesses a solution, then the functions F, u_0 and v given in the problem have to belong to the spaces $C(\Omega_T)$, $C(\bar{G})$ and $C(S \times [0,T])$, respectively. In further discussions, these inclusions are always supposed to be valid.

First, the uniqueness of the solution of the problem (4.1), (4.4), (4.5) is considered. The main component of the uniqueness proof is an appropriate maximum principle for equation (4.1). In its turn, the proof of the maximum principle uses an extremum principle for the Caputo fractional derivative. The results presented in this section are based on the author's papers [Luchko(2009b), Luchko(2010)] and the reader is welcome to consult these papers for more details.

Theorem 4.1. *Let a function $f \in W_t^1((0,T]) \cap C([0,T])$ attain its maximum over the interval $[0,T]$ at the point $\tau = t_0$, $t_0 \in (0,T]$. Then the Caputo fractional derivative of the function f is non-negative at the point t_0 for any α, $0 < \alpha < 1$:*

$$0 \le (D_t^\alpha f)(t_0),\ 0 < \alpha < 1. \tag{4.6}$$

To prove the theorem, let us first introduce an auxiliary function

$$g(\tau) := f(t_0) - f(\tau),\ \tau \in [0,T],$$

that possesses the properties

$$0 \le g(\tau), \tau \in [0,T], \tag{4.7}$$

$$(D_t^\alpha g)(t) = -(D_t^\alpha f)(t),\ t \in [0,T], \tag{4.8}$$

$$|g(\tau)| \le C_\epsilon |t_0 - \tau|,\ \tau \in [\epsilon, T],\ 0 < \epsilon < T, \tag{4.9}$$

due to the conditions on the function f and the properties of the Caputo fractional derivative (see, e.g., [Kilbas et al.(2006)] or [Podlubny(1999)]).

For any $\epsilon,\ 0 < \epsilon < t_0$ we get

$$(D_t^\alpha g)(t_0) = \frac{1}{\Gamma(1-\alpha)} \int_0^{t_0} (t_0-\tau)^{-\alpha} g'(\tau)\, d\tau$$
$$= \frac{1}{\Gamma(1-\alpha)} \int_0^{\epsilon} (t_0-\tau)^{-\alpha} g'(\tau)\, d\tau + \frac{1}{\Gamma(1-\alpha)} \int_\epsilon^{t_0} (t_0-\tau)^{-\alpha} g'(\tau)\, d\tau = I_1 + I_2.$$

Since $f \in W_t^1((0,T])$, the function g belongs to the space $W_t^1((0,T])$, too. This means, in particular, that $g' \in L((0,T))$. It follows from this last inclusion that

$$\forall \delta > 0\ \exists \epsilon > 0 \ \text{ such that } \ |I_1| \le \delta. \tag{4.10}$$

As to the second integral, I_2, the integration by parts formula and property (4.9) of the function g are used to get the representation

$$I_2 = -\frac{(t_0-\epsilon)^{-\alpha} g(\epsilon)}{\Gamma(1-\alpha)} + \frac{1}{\Gamma(-\alpha)} \int_\epsilon^{t_0} (t_0-\tau)^{-\alpha-1} g(\tau)\, d\tau.$$

In follows now from inequality (4.7), the inequalities $\Gamma(-\alpha) < 0,\ 0 < \Gamma(1-\alpha)$, for $0 < \alpha < 1$, and $\epsilon < t_0$, that

$$I_2 \le 0,$$

which together with (4.8) and (4.10) finishes the proof of the theorem.

The extremum principle for the Caputo fractional derivative is a foundation for the proof of a maximum principle for the generalized time-fractional diffusion equation (4.1).

Theorem 4.2. *Let a function $u \in C(\bar{\Omega}_T) \cap W_t^1((0,T]) \cap C_x^2(G)$ be a solution of the generalized time-fractional diffusion equation* (4.1) *in the domain Ω_T and $F(x,t) \le 0,\ (x,t) \in \Omega_T$.*

Then either $u(x,t) \le 0,\ (x,t) \in \bar{\Omega}_T$ or the function u attains its positive maximum on the part $S_G^T := (\bar{G} \times \{0\}) \cup (S \times [0,T])$ of the boundary of the domain Ω_T, i.e.,

$$u(x,t) \le \max_{(x,t)\in S_G^T} \{0,\ u(x,t)\},\ \forall (x,t) \in \bar{\Omega}_T. \tag{4.11}$$

To prove the theorem, we first suppose that the statement of the theorem does not hold true, i.e., $\exists (x_0,t_0),\ x_0 \in G,\ 0 < t_0 \le T$ with the property

$$u(x_0,t_0) > \max_{(x,t)\in S_G^T} \{0,\ u(x,t)\} = M > 0. \tag{4.12}$$

Let us define the number $\epsilon := u(x_0,t_0) - M > 0$ and introduce the auxiliary function

$$w(x,t) := u(x,t) + \frac{\epsilon}{2}\frac{T-t}{T},\ (x,t) \in \bar{\Omega}_T.$$

It follows from the definition of the function w and the theorem conditions that w possesses the following properties:

$$w(x,t) \le u(x,t) + \frac{\epsilon}{2},\ (x,t) \in \bar{\Omega}_T,$$
$$w(x_0,t_0) \ge u(x_0,t_0) = \epsilon + M \ge \epsilon + u(x,t)$$
$$\ge \epsilon + w(x,t) - \frac{\epsilon}{2} \ge \frac{\epsilon}{2} + w(x,t),\ (x,t) \in S_G^T.$$

The last property means that the function w cannot attain its maximum on the part S_G^T of the boundary of the domain Ω_T. If the maximum point of the function w over the domain $\bar{\Omega}_T$ is denoted by (x_1,t_1) then $x_1 \in G,\ 0 < t_1 \le T$ and

$$w(x_1,t_1) \ge w(x_0,t_0) \ge \epsilon + M > \epsilon. \tag{4.13}$$

Theorem 4.1 and the necessary conditions for the existence of the maximum in on open domain G lead then to the relations

$$\begin{cases} (D_t^\alpha w)(t_1) \ge 0, \\ \operatorname{grad} w\big|_{(x_1,t_1)} = 0,\ \triangle\, w\big|_{(x_1,t_1)} \le 0. \end{cases} \tag{4.14}$$

According to the definition of the function w, the function u satisfies the relation

$$u(x,t) = w(x,t) - \frac{\epsilon}{2}\frac{T-t}{T},\ (x,t) \in \bar{\Omega}_T. \tag{4.15}$$

The well-known formula $(0 < \alpha \le 1)$

$$(D_t^\alpha \tau^\beta)(t) = \frac{\Gamma(1+\beta)}{\Gamma(1-\alpha+\beta)} t^{\beta-\alpha},\ \beta > 0 \tag{4.16}$$

for the Caputo fractional derivative leads to

$$(D_t^\alpha u)(t) = (D_t^\alpha w)(t) + \frac{\epsilon}{2T}\frac{t^{1-\alpha}}{\Gamma(2-\alpha)}. \tag{4.17}$$

Using the formulae (4.2), (4.3), (4.13), (4.14), (4.15), and (4.17), we arrive at the following chain of equalities and inequalities for the point (x_1,t_1):

$$\begin{aligned}
&(D_t^\alpha u)(t_1) - \operatorname{div}(p \operatorname{grad} u) + qu - F \\
&= (D_t^\alpha w)(t_1) + \frac{\epsilon}{2T}\frac{t_1^{1-\alpha}}{\Gamma(2-\alpha)} - p\triangle v\big|_{(x_1,t_1)} \\
&\quad - (\operatorname{grad} p\big|_{x_1}, \operatorname{grad} w\big|_{(x_1,t_1)}) + q\left(w - \frac{\epsilon}{2}\frac{T-t_1}{T}\right) - F \\
&\ge \frac{\epsilon}{2T}\frac{t_1^{1-\alpha}}{\Gamma(2-\alpha)} + q\epsilon\left(1 - \frac{T-t_1}{2T}\right) > 0,
\end{aligned}$$

that contradicts the condition of the theorem saying that the function u is a solution of the equation (4.1). The obtained contradiction shows that the assumption made at the beginning of the theorem's proof is wrong, which proves the theorem.

Similarly to the case of the partial differential equations of parabolic ($\alpha = 1$) or elliptic ($\alpha = 0$) type, an appropriate minimum principle for $0 < \alpha < 1$ is valid, too.

Theorem 4.3. *Let a function $u \in C(\bar{\Omega}_T) \cap W_t^1((0,T]) \cap C_x^2(G)$ be a solution of the generalized time-fractional diffusion equation (4.1) in the domain Ω_T and $F(x,t) \geq 0,\ (x,t) \in \Omega_T$.*

Then either $u(x,t) \geq 0,\ (x,t) \in \bar{\Omega}_T$ or the function u attains its negative minimum on the part S_G^T of the boundary of the domain Ω_T, i.e.,

$$u(x,t) \geq \min_{(x,t)\in S_G^T} \{0,\ u(x,t)\},\ \forall (x,t) \in \bar{\Omega}_T. \tag{4.18}$$

The maximum and minimum principles can be applied to show that the problem (4.1), (4.4)–(4.5) possesses at most one solution and this solution – if it exists – continuously depends on the data given in the problem.

First, some a priori estimates for the solution norm are established.

Theorem 4.4. *Let u be a solution of the problem (4.1), (4.4)–(4.5) and F belong to the space $C(\bar{\Omega}_T)$ with the norm $M := \|F\|_{C(\bar{\Omega}_T)}$. Then the estimate*

$$\|u\|_{C(\bar{\Omega}_T)} \leq \max\{M_0,\ M_1\} + \frac{T^\alpha}{\Gamma(1+\alpha)} M \tag{4.19}$$

of the solution norm holds true, where

$$M_0 := \|u_0\|_{C(\bar{G})},\ M_1 := \|v\|_{C(S\times[0,T])}. \tag{4.20}$$

To prove the theorem, we first introduce an auxiliary function w:

$$w(x,t) := u(x,t) - \frac{M}{\Gamma(1+\alpha)} t^\alpha,\ (x,t) \in \bar{\Omega}_T.$$

Evidently, the function w is a solution of the problem (4.1), (4.4)–(4.5) with the functions $F_1(x,t) := F(x,t) - M - q(x)\frac{M}{\Gamma(1+\alpha)} t^\alpha,\ v_1(x,t) := v(x,t) - \frac{M}{\Gamma(1+\alpha)} t^\alpha$ instead of F and v, respectively. To get an expression for the function F_1, the formula (4.16) is used. The function F_1 satisfies the condition $F_1(x,t) \leq 0,\ (x,t) \in \bar{\Omega}_T$. Then the maximum principle applied to a solution w leads to the estimate

$$w(x,t) \leq \max\{M_0, M_1\},\ (x,t) \in \bar{\Omega}_T, \tag{4.21}$$

where the constants $M_0,\ M_1$ are defined as in (4.20). For the function u, we get

$$\begin{aligned} u(x,t) &= w(x,t) + \frac{M}{\Gamma(1+\alpha)} t^\alpha \\ &\leq \max\{M_0, M_1\} + \frac{T^\alpha}{\Gamma(1+\alpha)} M,\ (x,t) \in \bar{\Omega}_T. \end{aligned} \tag{4.22}$$

The minimum principle from Theorem 4.3 applied to the auxiliary function

$$w(x,t) := u(x,t) + \frac{M}{\Gamma(1+\alpha)} t^\alpha,\ (x,t) \in \bar{\Omega}_T$$

leads to the estimate ($(x,t) \in \bar{\Omega}_T$)

$$u(x,t) \geq -\max\{M_0, M_1\} - \frac{T^\alpha}{\Gamma(1+\alpha)} M,$$

that together with the estimate (4.22) finishes the proof of the theorem.

The result formulated in the next theorem follows from Theorem 4.4.

Theorem 4.5. *The initial-boundary-value problem* (4.4)–(4.5) *for the GTFDE* (4.1) *possesses at most one solution. This solution continuously depends on the data given in the problem in the sense that if*

$$\|F - \tilde{F}\|_{C(\bar{\Omega}_T)} \leq \epsilon, \|u_0 - \tilde{u}_0\|_{C(\bar{G})} \leq \epsilon_0, \ \|v - \tilde{v}\|_{C(S\times[0,T])} \leq \epsilon_1,$$

and u and $\tilde{u}$ are the classical solutions of the problem (4.1), (4.4)–(4.5) *with the source functions F and $\tilde{F}$, the initial conditions u_0 and $\tilde{u}_0$, and the boundary conditions v and $\tilde{v}$, respectively, then the norm estimate*

$$\|u - \tilde{u}\|_{C(\bar{\Omega}_T)} \leq \max\{\epsilon_0, \epsilon_1\} + \frac{T^\alpha}{\Gamma(1+\alpha)} \epsilon \tag{4.23}$$

for the solutions u and $\tilde{u}$ holds true.

Because the problem under consideration is a linear one, the uniqueness of the solution immediately follows from the fact that the homogeneous problem (4.1), (4.4)–(4.5), i.e., the problem with $F \equiv 0$, $u_0 \equiv 0$, and $v \equiv 0$ has only one solution, namely, $u(x,t) \equiv 0$, $(x,t) \in \bar{\Omega}_T$. The last statement is a simple consequence from the norm estimate (4.19) established in Theorem 4.4. The same estimate is used to prove the inequality (4.23). This time, it is applied to the function $u - \tilde{u}$ that is a solution of the problem (4.1), (4.4)–(4.5) with the functions $F - \tilde{F}$, $u_0 - \tilde{u}_0$, and $v - \tilde{v}$ instead of the functions F, u_0, and v, respectively.

The last theorem gives conditions for the uniqueness of the solution of the problem (4.1), (4.4)–(4.5). To tackle the problem of the existence of the solution, the notion of the generalized solution in the sense of Vladimirov (see [Vladimirov(1971)] is first introduced.

Definition 4.6. Let $F_k \in C(\bar{\Omega}_T)$, $u_{0k} \in C(\bar{G})$ and $v_k \in C(S \times [0,T])$, $k = 1, 2, \ldots$ be the sequences of functions that satisfy the following conditions:

1) there exist the functions F, u_0, and v, such that

$$\|F_k - F\|_{C(\bar{\Omega}_T)} \to 0 \quad \text{as} \quad k \to \infty, \tag{4.24}$$

$$\|u_{0k} - u_0\|_{C(\bar{G})} \to 0 \quad \text{as} \quad k \to \infty, \tag{4.25}$$

$$\|v_k - v\|_{C(S\times[0,T])} \to 0 \quad \text{as} \quad k \to \infty, \tag{4.26}$$

2) for any $k = 1, 2, \ldots$ there exists the solution u_k of the initial-boundary-value problem

$$u_k\big|_{t=0} = u_{0k}(x), \ x \in \bar{G}, \tag{4.27}$$

$$u_k\big|_S = v_k(x,t), \ (x,t) \in S \times [0,T], \tag{4.28}$$

for the generalized time-fractional diffusion equation

$$(D_t^\alpha u_k)(t) = -L(u_k) + F_k(x,t). \tag{4.29}$$

Suppose, there exists a function $u \in C(\bar{\Omega}_T)$ such that

$$\|u_k - u\|_{C(\bar{G})} \to 0 \ \text{ as } \ k \to \infty. \tag{4.30}$$

The function u is called a generalized solution of the problem (4.1), (4.4)–(4.5).

The generalized solution of the problem (4.1), (4.4)–(4.5) is a continuous function, not a generalized one. Still, the generalized solution is not required to be from the functional space $C(\bar{\Omega}_T) \cap W_t^1((0,T]) \cap C_x^2(G)$, where our solution has to belong to that space.

It follows from Definition 4.6 that if the problem (4.1), (4.4)–(4.5) possesses a solution then this solution is a generalized solution of the problem, too. In this sense, Definition 4.6 extends the notion of the solution of the problem (4.1), (4.4)–(4.5). This extension is needed to get some existence results. But of course one does not want to lose the uniqueness of the solution. Let us consider some properties of the generalized solution including its uniqueness.

If the problem (4.1), (4.4), (4.5) possesses a generalized solution, then the functions F, u_0 and v given in the problem have to belong to the spaces $C(\bar{\Omega}_T)$, $C(\bar{G})$ and $C(S \times [0,T])$, respectively. In further discussions, these inclusions are always supposed to be valid.

Let us show that the sequence u_k, $k = 1,2,\dots$ defined by the relations (4.24)–(4.29) of Definition 4.6 is always a uniformly convergent one in $\bar{\Omega}_T$, i.e., there always exists a function $u \in C(\bar{\Omega}_T)$ that satisfies the property (4.30). Indeed, applying the estimate (4.23) from Theorem 4.5 to the functions u_k and u_p that are solutions of the corresponding initial-boundary-value problems (4.27)–(4.28) for the equation (4.29) one gets the inequality

$$\begin{aligned}\|u_k - u_p\|_{C(\bar{\Omega}_T)} \le \max\big\{&\|u_{0k} - u_{0p}\|_{C(\bar{G})}, \|v_k - v_p\|_{C(S\times[0,T])}\big\} \\ &+ \frac{T^\alpha}{\Gamma(1+\alpha)} \|F_k - F_p\|_{C(\bar{\Omega}_T)},\end{aligned} \tag{4.31}$$

that, together with the relations (4.24)–(4.26), means that u_k, $k = 1,2,\dots$ is a Cauchy sequence in $C(\bar{\Omega}_T)$ that converges to a function $u \in C(\bar{\Omega}_T)$.

Moreover, the estimate (4.19) proved in Theorem 4.4 for the solution of the problem (4.1), (4.4)–(4.5) remains valid for the generalized solution, too. To show this, the inequality

$$\|u_k\|_{C(\bar{\Omega}_T)} \le \max\{M_{0k}, M_{1k}\} + \frac{T^\alpha}{\Gamma(1+\alpha)} M_k, \tag{4.32}$$

$$M_{0k} := \|u_{0k}\|_{C(\bar{G})},\ M_{1k} := \|v_k\|_{C(S\times[0,T])},\ M_k := \|F\|_{C(\bar{\Omega}_T)}$$

that is valid $\forall k = 1,2\dots$ is considered as k tends to $+\infty$.

The estimate (4.19) for the generalized solution is a foundation for the following important uniqueness theorem.

Theorem 4.7. *The problem* (4.1), (4.4)–(4.5) *possesses at most one generalized solution in the sense of Definition* 4.6. *The generalized solution – if it exists – continuously depends on the data given in the problem in the sense of the estimate* (4.23).

The proof of the theorem follows the lines of the proof of Theorem 4.5 and is omitted here.

Contrary to the situation with the solution of the problem (4.1), (4.4)–(4.5), existence of the generalized solution can be shown in the general case under some standard restrictions on the problem data and the boundary S of the domain G. In this section, the existence of the solution of the problem

$$(D_t^\alpha u)(t) = -L(u), \tag{4.33}$$

$$u\big|_{t=0} = u_0(x), \;\; x \in \bar{G}, \tag{4.34}$$

$$u\big|_S = 0, \;\; (x,t) \in S \times [0,T] \tag{4.35}$$

is considered to demonstrate the technique that can be used with the appropriate standard modifications in the general case, too. The generalized solution of the problem (4.33)–(4.35) can be constructed in an analytical form by using the Fourier method of the variables separation. Let us look for a particular solution u of the equation (4.33) in the form

$$u(x,t) = T(t)\,X(x), \quad (x,t) \in \bar{\Omega}_T, \tag{4.36}$$

that satisfies the boundary condition (4.35). Substitution of the function (4.36) into the equation (4.33) and separation of the variables lead to the equation

$$\frac{(D_t^\alpha T)(t)}{T(t)} = -\frac{L(X)}{X(x)} = -\lambda, \tag{4.37}$$

λ being a constant not depending on the variables t and x. The last equation, together with the boundary condition (4.35), is equivalent to the fractional differential equation

$$(D_t^\alpha T)(t) + \lambda T(t) = 0 \tag{4.38}$$

and the eigenvalue problem

$$L(X) = \lambda\, X, \tag{4.39}$$

$$X\big|_S = 0, \;\; x \in S \tag{4.40}$$

for the operator L. Due to the condition (4.2), the operator L is a positive definite and self-adjoint linear operator. The theory of the eigenvalue problems for such operators is well known (see, e.g., [Vladimirov(1971)]). In particular, the eigenvalue problem (4.39)–(4.40) has a countable number of positive eigenvalues $0 < \lambda_1 \le \lambda_2 \le \cdots$ with finite multiplicity and – if the boundary S of G is a smooth surface – any function $f \in \mathcal{M}_L$ can be represented through its Fourier series in the form

$$f(x) = \sum_{i=1}^{\infty} (f, X_i)\, X_i(x), \tag{4.41}$$

where $X_i \in \mathcal{M}_L$ are the eigenfunctions corresponding to the eigenvalues λ_i:

$$L(X_i) = \lambda_i X_i, \ i = 1, 2, \ldots. \tag{4.42}$$

We denote by $\mathcal{M}_L$ the space of functions f that satisfy the boundary condition (4.40) and the inclusions $f \in C^1(\bar{\Omega}_T) \cap C^2(G)$, $L(f) \in L^2(G)$.

The solution of the fractional differential equation (4.38) with $\lambda = \lambda_i$, $i = 1, 2, \ldots$ has the form (see, e.g., [Luchko and Gorenflo(1999), Luchko(1999)])

$$T_i(t) = c_i E_\alpha(-\lambda_i t^\alpha), \tag{4.43}$$

E_α being the Mittag-Leffler function defined by

$$E_\alpha(z) := \sum_{k=1}^{\infty} \frac{z^k}{\Gamma(\alpha\, k + 1)}. \tag{4.44}$$

Any of the functions

$$u_i(x,t) = c_i E_\alpha(-\lambda_i t^\alpha)\, X_i(x), \ i = 1, 2, \ldots \tag{4.45}$$

and thus the finite sums

$$u_k(x,t) = \sum_{i=1}^{k} c_i E_\alpha(-\lambda_i t^\alpha)\, X_i(x), \ k = 1, 2 \ldots \tag{4.46}$$

satisfy both equation (4.33) and the boundary condition (4.35). To construct a function that satisfies the initial condition (4.34), too, the notion of a formal solution is introduced.

Definition 4.8. A formal solution of the problem (4.33)–(4.35) is called a Fourier series in the form

$$u(x,t) = \sum_{i=1}^{\infty} (u_0, X_i) E_\alpha(-\lambda_i t^\alpha)\, X_i(x), \tag{4.47}$$

X_i, $i = 1, 2, \ldots$ being the eigenfunctions corresponding to the eigenvalues λ_i of the eigenvalue problem (4.39)–(4.40).

Under certain conditions, the formal solution (4.47) can be proved to be the generalized solution of problem (4.33)–(4.35).

Theorem 4.9. *Let the function u_0 in the initial condition* (4.34) *be from the space $\mathcal{M}_L$. Then the formal solution* (4.47) *of the problem* (4.33)–(4.35) *is its generalized solution.*

It can be easily proved that the functions u_k, $k = 1, 2, \ldots$ defined by (4.46) are solutions of problem (4.33)–(4.35) with initial conditions

$$u_{0k}(x) = \sum_{i=1}^{k} (u_0, X_i) X_i(x) \tag{4.48}$$

instead of u_0. Because the function u_0 is from the functional space $\mathcal{M}_L$, its Fourier series converges uniformly to the function u_0, so that

$$\|u_{0k} - u_0\|_{C(\bar{G})} \to 0 \text{ as } k \to \infty.$$

To prove the theorem, one only needs to show that the sequence $u_k,\ k = 1, 2, \ldots$ of the partial sums (4.46) converges uniformly on $\bar{\Omega}_T$. But this statement immediately follows from the estimate (see, e.g., [Podlubny(1999)])

$$|E_\alpha(-x)| \leq \frac{M}{1+x} \leq M,\ 0 \leq x,\ 0 < \alpha < 1 \tag{4.49}$$

for the Mittag-Leffler function and the fact that the Fourier series $\sum_{i=1}^{\infty}(u_0, X_i) \times X_i(x)$ of the function $u_0 \in \mathcal{M}_L$ uniformly converges on $\bar{\Omega}_T$.

In some cases, the generalized solution (4.47) can be shown to be a solution of the initial-boundary-value problem for the generalized time-fractional diffusion equation, too. One important example is given in the following theorem.

Theorem 4.10. *Let an open domain G be a one-dimensional interval $(0, l)$ and $u_0 \in \mathcal{M}_L$, $L(u_0) \in \mathcal{M}_L$. Then the solution of the initial-boundary-value problem*

$$u\big|_{t=0} = u_0(x),\ 0 \leq x \leq l,$$
$$u(0, t) = u(l, t) = 0,\ 0 \leq t \leq T$$

for the generalized time-fractional diffusion equation

$$(D_t^\alpha u)(t) = \frac{\partial}{\partial x}\left(p(x)\frac{\partial u}{\partial x}\right) - q(x)\, u$$

exists and is given by the formula (4.47).

The proof of the theorem follows mainly the lines of the proof of the same result for the one-dimensional parabolic PDE (the case $\alpha = 1$) presented in [Vladimirov(1971)] and is omitted here.

The theory presented in this section can be applied with some small modifications in the case of the infinite domain $\Omega = G \times (0, \infty)$, $G \subset \mathbb{R}^n$, too. Another direction of research is that of initial-boundary-value problems for the multi-term time-fractional diffusion equation and for the time-fractional diffusion equation of distributed order. Whereas the first results for the multi-term time-fractional diffusion equation and for the time-fractional diffusion equation of the distributed order have been already obtained (see [Daftardar-Gejji and Bhalekar(2008),Luchko(2009a)] and [Chechkin et al.(2003),Luchko(2011)], respectively), the initial-boundary-value problems for the space-fractional diffusion equation and for the time-space-fractional diffusion equation have practically not yet been considered in the literature. In particular, one should try to extend the maximum principle discussed in this section for these types of equations, too. The next step in research would be to employ the maximum principle to establish the uniqueness of some special solutions (the so-called maximum and minimum solutions) for the nonlinear fractional partial differential equations. And finally, a very recent direction of research in Fractional Calculus and its applications is the study of so-called fractional differential operators of variable order and the ordinary and partial differential equations with these operators. All topics mentioned above are beyond the scope of the present chapter and still a subject of active research.

Acknowledgment

The author would like to thank Professor Enrico Scalas who has carefully studied the whole manuscript and whose detailed constructive-critical remarks have helped in improving it.

References

[Bazhlekova(1998)] Bazhlekova, E.G. (1998). Duhamel-type representation of the solutions of nonlocal boundary value problems for the fractional diffusion-wave equation. *Proc. 2nd Int. Workshop "'TMSF'1996"'*. Bulgarian Academy of Sciences, Sofia, pp. 32–40. 117

[Beghin and Orsingher(2009)] Beghin, L. and E. Orsingher (2009). Fractional Poisson processes and related planar random motions. *Electronic Journal of Probability* **14**, paper no. 61. 117

[Chechkin et al.(2003)] Chechkin A.V., R. Gorenflo, I.M. Sokolov, and V.Yu. Gonchar (2003). Distributed order time fractional diffusion equation. *Fract. Calc. Appl. Anal.* Vol. **6**, pp. 259–279. 117, 141

[Chechkin et al.(2005)] Chechkin, A.V., R. Gorenflo, and I.M. Sokolov (2005). Fractional diffusion in inhomogeneous media. *J. Phys. A, Math. Gen.* **38**, 679–684. 116

[Daftardar-Gejji and Bhalekar(2008)] Daftardar-Gejji, V. and S. Bhalekar (2008). Boundary value problems for multi-term fractional differential equations. *J. Math. Anal. Appl.* **345**, pp. 754–765. 117, 141

[Dubbeldam et al.(2007)] Dubbeldam, J.L.A., A. Milchev, V.G. Rostiashvili, and T.A. Vilgis (2007). Polymer translocation through a nanopore: A showcase of anomalous diffusion. *Physical Review* E **76**, 010801 (R). 116

[Eidelman and Kochubei(2004)] Eidelman, S.D and A.N. Kochubei (2004). Cauchy problem for fractional diffusion equations. *J. Differ. Equations* **199**, pp. 211–255. 117

[Erdélyi et al. (1953)] Erdélyi, A., W. Magnus, F. Oberhettinger, and F.G. Tricomi (1953). *Higher Transcendental Functions*, Vol. 3. New York, McGraw-Hill. 126

[Freed et al.(2002)] Freed, A., K. Diethelm, and Yu. Luchko (2002). *Fractional-order viscoelasticity (FOV): Constitutive development using the fractional calculus.* NASA's Glenn Research Center, Ohio. 116

[Germano et al.(2009)] Germano, G., M. Politi, E. Scalas, and R.L. Schilling (2009). Stochastic calculus for uncoupled continuous-time random walks. *Physical Review E* **79**, 066102. 120, 121

[Gorenflo and Mainardi(1998)] Gorenflo, R. and F. Mainardi (1998). Random walk models for space-fractional diffusion processes. *Frac. Calc. and Appl. Anal.* **1**, pp. 167–191. 116

[Gorenflo and Mainardi(2008)] Gorenflo, R. and F. Mainardi (2008). Continuous time random walk, Mittag-Leffler waiting time and fractional diffusion: mathematical aspects. Chap. 4 in R. Klages, G. Radons and I.M. Sokolov (editors): *Anomalous*

Transport: Foundations and Applications, Wiley-VCH, Weinheim, Germany, pp. 93–127. 117

[Gorenflo and Mainardi(2009)] Gorenflo, R. and F. Mainardi (2009). Some recent advances in theory and simulation of fractional diffusion processes. *Journal of Computational and Applied Mathematics* **229**, pp. 400–415. 117

[Gorenflo et al.(2000)] Gorenflo, R., Yu. Luchko, and S. Umarov (2000). On the Cauchy and multi-point problems for partial pseudo-differential equations of fractional order. *Frac. Calc. and Appl. Anal.* **3**, pp. 249–277.

[Hilfer(2000)] Hilfer, R. (ed.) (2000). *Applications of Fractional Calculus in Physics.* World Scientific, Singapore. 116

[Jacob(2005)] Jacob, N. (2005). *Pseudo-Differential Operators and Markov Processes. Vol. 3: Markov Processes and Applications.* Imperial College Press, London. 117

[Kilbas et al.(2006)] Kilbas, A.A., H.M. Srivastava, and J.J. Trujillo (2006). *Theory and Applications of Fractional Differential Equation.* Elsevier, Amsterdam. 116, 117, 122, 123, 133

[Kochubei(1989)] Kochubei, A.N. (1989). A Cauchy problem for evolution equations of fractional order. *Differential Equations* **25**, pp. 967–974. 117

[Luchko and Gorenflo(1999)] Luchko, Yu. and R. Gorenflo (1999). An operational method for solving fractional differential equations with the Caputo derivatives. *Acta Mathematica Vietnamica* **24**, pp. 207–233. 140

[Luchko(1999)] Luchko, Yu. (1999). Operational method in fractional calculus. *Fract. Calc. Appl. Anal.* **2**, pp. 463–489. 140

[Luchko(2009a)] Luchko, Yu. (2009a) Boundary value problems for the generalized time-fractional diffusion equation of distributed order. *Fract. Calc. Appl. Anal.* **12**, pp. 409–422. 117, 141

[Luchko(2009b)] Luchko, Yu. (2009b). Maximum principle for the generalized time-fractional diffusion equation. *J. Math. Anal. and Appl.* **351**, pp. 218–223. 117, 133

[Luchko(2010)] Luchko, Yu. (2010). Some uniqueness and existence results for the initial-boundary-value problems for the generalized time-fractional diffusion equation, *Comp. and Math. with Appl.* **59**, pp. 1766–1772. 117, 133

[Luchko et al.(2010)] Luchko, Yu., M. Rivero, J.J. Trujillo, and M.P. Velasco (2010). Fractional models, non-locality, and complex systems. *Comp. and Math. with Appl.* **59**, pp. 1048–1056. 115

[Luchko(2011)] Luchko, Yu. (2011). Initial-boundary-value problems for the generalized multi-term time-fractional diffusion equation. *J. Math. Anal. and Appl.* **374**, pp. 538–548. 117, 141

[Luchko and Punzi(2011)] Luchko, Yu. and A. Punzi (2011). Modeling anomalous heat transport in geothermal reservoirs via fractional diffusion equations. *International Journal on Geomathematics*, **1**, pp. 257–276. 116

[Mainardi(1996)] Mainardi, F. (1996). Fractional relaxation-oscillation and fractional diffusion-wave phenomena. *Chaos, Solitons and Fractals* **7**, pp. 1461–1477. 116

[Mainardi and Tomirotti(1997)] Mainardi, F. and M. Tomirotti (1997). Seismic pulse propagation with constant Q and stable probability distributions. *Annali di Geofisica* **40**, pp. 1311–1328. 116

[Mainardi and Pagnini(2002)] Mainardi, F. and G. Pagnini (2002). Space-time fractional diffusion: exact solutions and probabilistic interpretation. In Monaco, R., M.P.

Bianchi, and S. Rionero (eds.): *Proceedings WASCOM* 2001, 11*th Conference on Waves and Stability in Continuous Media*, World Scientific, Singapore. 132

[Mainardi and Pagnini(2003)] Mainardi, F. and G. Pagnini (2003). The Wright function as solution of the time-fractional diffusion equation. *Applied Mathematics and Computation* **141**, pp. 51–62. 132

[Mainardi et al.(2001)] Mainardi, F., Yu. Luchko, and G. Pagnini (2001). The fundamental solution of the space-time fractional diffusion equation. *Fract. Calc. Appl. Anal.* **4**, pp. 153–192. 117, 124, 128, 132

[Mainardi et al.(2004)] Mainardi, F., R. Gorenflo, and E. Scalas (2004). A fractional generalization of the Poisson processes. *Vietnam Journal of Mathematics* **32**, pp. 53–64. 117

[Meerschaert et al.(2009)] Meerschaert, M.M., E. Nane, and P. Vellaisamy (2009). Fractional Cauchy problems in bounded domains. *The Annals of Probability* **37**, pp. 979–1007. 117

[Meerschaert et al.(2011)] Meerschaert, M.M., E. Nane, and P. Vellaisamy (2011). The fractional Poisson process and the inverse stable subordinator. Available online at http://arxiv.org/abs/1007.5051. 117

[Metzler and Klafter(2000a)] Metzler, R. and J. Klafter (2000a). The random walk's guide to anomalous diffusion: a fractional dynamics approach. *Physics Reports* **339**, pp. 1–77. 115, 116, 120, 121, 123

[Metzler and Klafter(2000b)] Metzler, R. and J. Klafter (2000b). Boundary value problems for fractional diffusion equations. *Phys. A* **278**, pp. 107–125. 117

[Podlubny(1999)] Podlubny, I. (1999). *Fractional Differential Equations.* Academic Press, San Diego. 116, 117, 122, 133, 141

[Pskhu(2005)] Pskhu, A.V. (2005). *Partial differential equations of fractional order.* Nauka, Moscow (in Russian).

[Scalas et al.(2004)] Scalas, E., R. Gorenflo, and F. Mainardi. Uncoupled continuous-time random walks: Solution and limiting behavior of the master equation. *Physical Review E* **69**, pp. 011107/1–8. 117

[Schneider and Wyss(1989)] Schneider, W.R. and W. Wyss (1989). Fractional diffusion and wave equations. *J. Math. Phys.* **30**, pp. 134–144. 117

[Vladimirov(1971)] Vladimirov, V.S. (1971). *Equations of Mathematical Physics.* Nauka, Moscow (in Russian). 137, 139, 141

[Voroshilov and Kilbas(2006)] Voroshilov, A.A. and A.A. Kilbas (2006). The Cauchy problem for the diffusion-wave equation with the Caputo partial derivative. *Differential Equations* **42**, pp. 638–649. 117

[Zacher(2008)] Zacher, R. (2008). Boundedness of weak solutions to evolutionary partial integro-differential equations with discontinuous coefficients. *J. Math. Anal. Appl.* **348**, pp. 137–149. 117

[Zhang(2006)] Zhang, S. (2006). Existence of solution for a boundary value problem of fractional order. *Acta Math. Sci., Ser. B* **26**, pp. 220–228. 117

Index

Yury Luchko
Department of Mathematics II
Beuth Technical University of Applied Sciences Berlin
Luxemburger Str. 10
D-13353 Berlin, Germany
e-mail: luchko@bht-berlin.de

Advances in Applied Analysis
Trends in Mathematics, 147–176

$\mathbb{R}$-linear and Riemann–Hilbert Problems for Multiply Connected Domains

Vladimir V. Mityushev

Abstract. The $\mathbb{R}$-linear problem with constant coefficients for arbitrary multiply connected domains has been solved. The method is based on reduction of the problem to a system of functional equations for a circular domain and to integral equations for a general domain. In previous works, the $\mathbb{R}$-linear problem and its partial cases such as the Riemann–Hilbert problem and the Dirichlet problem were solved under geometrical restrictions to the domains. In the present work, the solution is constructed for any circular multiply connected domain in the form of modified Poincaré series. Moreover, the modified alternating Schwarz method has been justified for an arbitrary multiply connected domain. This extends application of the alternating Schwarz method, since in the previous works geometrical restrictions were imposed on locations of the inclusions. The same concerns Grave's method which was worked out before only for simple closed algebraic boundaries or for a collection of confocal boundaries.

Mathematics Subject Classification (2010). 30E25.

Keywords. $\mathbb{R}$-linear problem; Riemann-Hilbert problem; multiply connected domain; Poincaré series, Schwarz alternating method; Grave method.

1. Introduction

Various boundary value problems are reduced to singular integral equations [Gakhov (1977)], [Muskhelishvili (1968)], [Vekua (1988)]. Only some of them can be solved in closed form. In the present paper we follow the lines of the book [Mityushev and Rogosin (2000)] and describe application of the functional equations method to the $\mathbb{R}$-linear problem which in a particular case yields the Riemann–Hilbert problem.

This problem can be considered as a generalization of the classical Dirichlet and Neumann problems for harmonic functions. It includes as a particular case the mixed boundary value problem. We know the famous Poisson formula which solves the Dirichlet problem for a disk. The exact solution of the Dirichlet problem for a circular annulus is also known due to Villat–Dini (see [Koppenfels and Stallman (1959)], p. 119 and [Akhiezer (1990)], p. 169). Formulae from Theorems 3.6 and 3.7 presented below can be considered as a generalization of the Poisson and Villat–Dini formulae to arbitrary circular multiply connected domains. In order to deduce our formulae we first reduce the boundary value problem to the $\mathbb{R}$-linear problem and solve the later one by use of functional equations. By *functional equations* we mean iterative functional equations [Kuczma et al. (1990)], [Mityushev and Rogosin (2000)] with shift into its domain. Hence, we do not use traditional integral equations and infinite systems of linear algebraic equations. The solution is given explicitly in terms of the known functions or constants and geometric parameters of the domain.

Despite the solution being given by exact formulae, its structure is not elementary. More precisely, it is represented in the form of integrals involving the Abelian functions [Baker (1996)] (Poincaré series [Mityushev (1998)] or their counterparts [Mityushev and Rogosin (2000)]). The reason why the solution in general is not presented by integrals involving elementary kernels is of a topological nature. In order to explain this, we briefly recall the scheme of the solution to the Riemann–Hilbert problem

$$\phi(t) + G(t)\overline{\phi(t)} = g(t), \ t \in \partial\mathbb{C}^+, \tag{1.1}$$

for the upper half-plane $\mathbb{C}^+$ following [Gakhov (1977)], [Muskhelishvili (1968)]. Define the function $\phi^-(z) := \overline{\phi(\overline{z})}$ analytic in the lower half-plane. Then the Riemann–Hilbert problem (1.1) becomes the $\mathbb{C}$-linear problem (Riemann problem)

$$\phi^+(t) + G(t)\phi^-(t) = g(t), \ t \in \partial\mathbb{C}^+. \tag{1.2}$$

The latter problem is solved in terms of Cauchy type integrals (see details in [Gakhov (1977)], [Muskhelishvili (1968)]).

Let us look at this scheme from another point of view [Zverovich (1971)]. Introduce a copy of the upper half-plane $\mathbb{C}^+$ with the local complex coordinate $\overline{z}$ and glue it with $\mathbb{C}^+$ along the real axis. Define the function $\phi^-(\overline{z}) := \overline{\phi(z)}$ analytic on the copy of $\mathbb{C}^+$. Then we again arrive at the $\mathbb{C}$-linear problem (1.2) but on the double of $\mathbb{C}^+$ which is conformally equivalent to the Riemann sphere $\hat{\mathbb{C}}$. The fundamental functionals of $\hat{\mathbb{C}}$ are expressed by means of meromorphic functions which produces the Cauchy type integrals. The same scheme holds for any n-connected domain D. As a result, we arrive at problem (1.2) on the Schottky double of D, the Riemann surface of genus $(n-1)$, where life is more complicated than on the plane, i.e., on the Riemann sphere of zero genus. It is not described by meromorphic functions. Therefore, if one tries to solve problem (1.2) on the double of D, one has to use meromorphic analogies of the Cauchy kernel on Riemann surfaces, i.e., the Abelian functions. In the case $n = 2$, the double of D becomes a torus in which meromorphic functions are replaced by the classical elliptic functions [Akhiezer (1990)]. [Crowdy (2009)], [Crowdy (2008a)], [Crowdy (2008b)] used the

Schottky–Klein prime function associated with the Schottky double of D to solve many different problems for multiply connected domains.

The paper is organized as follows. First, we describe the known results, discuss the Riemann–Hilbert and the Schwarz problems. In Section 2 we discuss functional equations and prove convergence of the method of successive approximations for these equations. In Section 3 the harmonic measures of the circular multiply connected domains and the Schwarz operator are constructed by a method which can be outlined as follows. At the beginning the Schwarz problem is written as an $\mathbb{R}$-linear problem. Then we reduce it to functional equations. Application of the method of successive approximations yields the solution in the form of a Poincaré series of weight 2. As a sequence we obtain the almost uniform convergence of the Poincaré series for any multiply connected domain. The $\mathbb{R}$-linear problem with constant coefficients is studied in Section 4. Application of the method of successive approximations has been justified to the $\mathbb{R}$-linear problem for arbitrary multiply connected domain.

1.1. Riemann–Hilbert problem

Let D be a multiply connected domain on the complex plane whose boundary ∂D consists of n simple closed Jordan curves. The positive orientation on ∂D leaves D to the left. This orientation is kept up to Section 3.

The scalar linear Riemann–Hilbert problem for D is stated as follows: Given Hölder continuous functions $\lambda(t) \neq 0$ and $f(t)$ on ∂D, to find a function $\phi(z)$ analytic in D, continuous in the closure of D with the boundary condition

$$\operatorname{Re} \overline{\lambda(t)}\phi(t) = f(t), \ t \in \partial D. \tag{1.3}$$

This condition can also be written in the form (1.1).

The problem (1.3) had been completely solved for simply connected domains ($n = 1$). Its solution and general theory of boundary value problems is presented in the classic books [Gakhov (1977)], [Muskhelishvili (1968)] and [Vekua (1988)]. In 1975 [Bancuri (1975)] solved the Riemann–Hilbert problem for circular annulus ($n = 2$).

First results concerning the Riemann–Hilbert problem for general multiply connected domains were obtained [Kveselava (1945)]. He reduced the problem to an integral equation. Beginning in 1952, I.N. Vekua and later Bojarski began to extensively study this problem. Their results are presented in the book [Vekua (1988)]. This Georgian attack on the problem, supported by a young Polish mathematician, were successful. Due to Kveselava, Vekua and Bojarski, we have a theory of solvability of problem (1.3) based on integral equations and estimations of its defect numbers, l_χ, the number of linearly independent solutions and p_χ, the number of linearly independent conditions of solvability on $f(t)$. Here, $\chi = wind_{\partial D}\lambda$ is the index of the problem. In particular, Bojarski obtained the exact estimation $l_\chi \leq \chi + 1$. In the special case $0 < \chi < n - 2$, Bojarski showed that solvability of the problem depends on a system of linear algebraic equations with 2χ unknowns. It was also demonstrated that the rank of this system differs from 2χ

on the set of zeros of an analytic function of few variables. Hence, almost always $l_\chi = max(0, 2\chi - n + 2)$. In [Zverovich (1971)] the theory was developed by reduction of the problem (1.3) to the $\mathbb{C}$-linear problem (1.2) on the Riemann surface and it was shown that the solution of the problem is expressed in terms of the fundamental functionals of the double of D.

Any multiply connected domain D can be conformally mapped onto a circular multiply connected domain ([Golusin (1969)], p. 235). Hence, it is sufficient to solve the problem (1.3) for a circular domain and after to write the solvability conditions and solution using a conformal mapping. The complete solution to problem (1.3) for an arbitrary circular multiply connected domain had been given in [Mityushev (1994)], [Mityushev (1998)], [Mityushev and Rogosin (2000)] by the method of functional equations.

1.2. $\mathbb{R}$-linear problem

Let D be a multiply connected domain described above. Let D_k $(k = 1, 2, \ldots, n)$ be simply connected domains complementing D to the extended complex plane. In the theory of composites, the domains D_k are called by inclusions. The $\mathbb{R}$-*linear conjugation problem* or simply $\mathbb{R}$-*linear problem* is stated as follows. Given Hölder continuous functions $a(t) \neq 0$, $b(t)$ and $f(t)$ on ∂D. To find a function $\phi(z)$ analytic in $\cup_{k=1}^{n} D_k \cup D$, continuous in $D_k \cup \partial D_k$ and in $D \cup \partial D$ with the conjugation condition

$$\phi^+(t) = a(t)\phi^-(t) + b(t)\overline{\phi^-(t)} + f(t), \; t \in \partial D. \tag{1.4}$$

Here $\phi^+(t)$ is the limit value of $\phi(z)$ when $z \in D$ tends to $t \in \partial D$, $\phi^-(t)$ is the limit value of $\phi(z)$ when $z \in D_k$ tends to $t \in \partial D$. In the case $|a(t)| \equiv |b(t)|$ the $\mathbb{R}$-linear problem is reduced to the Riemann–Hilbert problem (1.3) [Mikhailov (1963)].

In the case of the smooth boundary ∂D, the homogeneous $\mathbb{R}$-linear problem with constant coefficients

$$\phi^+(t) = a\phi^-(t) + b\overline{\phi^-(t)}, \; t \in \partial D \tag{1.5}$$

is equivalent to the *transmission problem* from the theory of harmonic functions

$$u^+(t) = u^-(t), \; \lambda^+ \frac{\partial u^+}{\partial n}(t) = \lambda^- \frac{\partial u^-}{\partial n}(t), \; t \in \partial D. \tag{1.6}$$

Here the real function $u(z)$ is harmonic in D and continuously differentiable in $D_k \cup \partial D_k$ and in $D \cup \partial D$, $\frac{\partial}{\partial n}$ is the normal derivative to ∂D. The conjugation conditions express the perfect contact between materials with different conductivities λ^+ and λ^-. The functions $\phi(z)$ and $u(z)$ are related by the equalities

$$\begin{aligned} u(z) &= \operatorname{Re}\phi(z), \; z \in D, \\ u(z) &= \frac{\lambda^- + \lambda^+}{2\lambda^+} \operatorname{Re}\phi(z), \; z \in D_k \; (k = 1, 2, \ldots, n). \end{aligned} \tag{1.7}$$

The coefficients are related by formulae (for details see [Mityushev and Rogosin (2000)], Sec. 2.12.)

$$a = 1,\ b = \frac{\lambda^- - \lambda^+}{\lambda^- + \lambda^+}. \tag{1.8}$$

Let us note that for positive λ^+ and λ^- we arrive at the elliptic case $|b| < |a|$ in accordance with Mikhajlov's terminology [Mikhailov (1963)].

The non-homogeneous problem (1.4) with real coefficients $a(t)$ and $b(t)$ can be written as a transmission problem (1.6). If $a(t)$ and $b(t)$ are complex the transmission problem takes a more complicated form [Mikhailov (1963)].

In 1932, using the theory of potentials, [Muskhelishvili (1932)] (see also [Muskhelishvili (1966)], p. 522) reduced the problem (1.6) to a Fredholm integral equation and proved that it has a unique solution in the case $\lambda^\pm > 0$, the most interesting in applications. [Vekua and Rukhadze (1933)], [Vekua and Rukhadze (1933)] constructed a solution of (1.6) in closed form for an annulus and an ellipse (see also papers by Ruhadze quoted in [Muskhelishvili (1966)]). Hence, the paper [Muskhelishvili (1932)] is the first result on solvability of the ℝ-linear problem, [Vekua and Rukhadze (1933)] and [Vekua and Rukhadze (1933)] published in 1933 are the first papers devoted to exact solution of the ℝ-linear problem for an annulus and an ellipse. A little bit later [Golusin (1935)] considered the ℝ-linear problem in the form (1.6) by use of the functional equations for analytic functions (see below Section 1.4). Therefore, the paper [Golusin (1935)] is the first paper which concerns constructive solution to the ℝ-linear problem for special circular multiply connected domains. In further works these first results were not associated to the ℝ-linear problem even by their authors.

[Markushevich (1946)] had stated the ℝ-linear problem in the form (1.4) and studied it in the case $a(t) = 0$, $b(t) = 1$, $f(t) = 0$ when (1.4) is not a Nöther problem. Later [Muskhelishvili (1968)] (p. 455 in Russian edition) did not determine whether (1.4) was his problem (1.6) discussed in 1932 in terms of harmonic functions. [Vekua (1967)] established that the vector-matrix problem (1.4) is Nötherian if $\det a(t) \neq 0$.

[Bojarski (1960)] showed that in the case $|b(t)| < |a(t)|$ with $a(t)$, $b(t)$ belonging to the Hölder class $H^{1-\varepsilon}$ with sufficiently small ε, the ℝ-linear problem (1.4) is qualitatively similar to the ℂ-linear problem

$$\phi^+(t) = a(t)\phi^-(t) + f(t),\ t \in \partial D. \tag{1.9}$$

More precisely, Bojarski proved the following theorem for simply connected domains. His proof is also valid for multiply connected domains. Let $wind_L a(t)$ denote the winding number (index) of $a(t)$ along L:

Theorem 1.1 ([Bojarski (1960)]). *Let the coefficients of the problem* (1.4) *satisfy the inequality*

$$|b(t)| < |a(t)|. \tag{1.10}$$

If $\chi = wind_{\partial D} a(t) \geq 0$, *the problem* (1.4) *is solvable and the homogeneous problem* (1.4) ($f(t) = 0$) *has* 2χ *ℝ-linearly independent solutions vanishing at infinity. If*

$\chi < 0$, *the problem* (1.4) *has a unique solution if and only if* $|2\chi|$ $\mathbb{R}$*-linearly independent conditions on* $f(t)$ *are fulfilled.*

Later [Mikhailov (1963)] (first published in [Mikhailov (1961)]) developed this result to continuous coefficients $a(t)$ and $b(t)$; $f(t) \in \mathcal{L}^p(\partial D)$. The case $|b(t)| < |a(t)|$ was called the elliptic case. It corresponds to the partial case of the real constant coefficients a and b considered by [Muskhelishvili (1932)].

[Mikhailov (1963)] reduced the problem (1.4) to an integral equation and justified the absolute convergence of the method of successive approximation for the later equation in the space $\mathcal{L}^p(L)$ under the restrictions $wind_L a(t) = 0$ and

$$(1 + S_p)|b(t)| < 2|a(t)|, \tag{1.11}$$

where S_p is the norm of the singular integral in $\mathcal{L}^p(L)$. Further discussion of the conditions (1.10) and (1.11) is in our Conclusion.

1.3. Schwarz problem

As we noted above the Riemann–Hilbert problem (1.3) is a partial case of the $\mathbb{R}$-linear problem. Later we will need this fact in the case $a = 1$, $b = -1$.

Theorem 1.2. *The problem*

$$\operatorname{Re} \phi(t) = f(t), t \in \partial D \tag{1.12}$$

is equivalent to the problem

$$\phi^+(t) = \phi^-(t) - \overline{\phi^-(t)} + f(t),\ t \in \partial D, \tag{1.13}$$

i.e., the problem (1.12) *is solvable if and only if* (1.13) *is solvable. If* (1.12) *has a solution* $\phi(z)$, *it is a solution of* (1.13) *in* D *and a solution of* (1.13) *in* D_k *can be found from the following simple problem for the simply connected domain* D_k *with respect to function* $2\operatorname{Im}\phi^-(z)$ *harmonic in* D_k,

$$2\operatorname{Im}\phi^-(t) = \operatorname{Im}\phi^+(t) - f(t), t \in \partial D. \tag{1.14}$$

The problem (1.14) *has a unique solution up to an arbitrary additive real constant.*

The proof of the theorem is evident. We call problem (1.12) the *Schwarz problem* for the domain D. Along similar lines (1.14) is called the Schwarz problem for the domain D_k. The operator solving the Schwarz problem is called the *Schwarz operator* (in appropriate functional space). The function $v(z) = 2\operatorname{Im}\phi(z)$ is harmonic in D_k. Therefore, the Schwarz problem (1.14) is equivalent to the Dirichlet problem

$$v(t) = \operatorname{Im}\phi^+(t) - f(t),\ t \in \partial D.$$

For multiply connected domains D, the Schwarz problem (1.12) is not equivalent to a Dirichlet problem for harmonic functions, since any function harmonic in D is represented as the real part of a single-valued analytic function plus logarithmic terms (see for instance (3.2)).

The problem

$$\operatorname{Re}\phi(t) = f(t) + c_k,\ t \in \partial D_k,\ k = 1, 2, \ldots, n, \tag{1.15}$$

with undetermined constants c_k is called the *modified Schwarz problem.* The problem (1.15) always has a unique solution up to an arbitrary additive complex constant [Mikhlin (1964)].

1.4. Functional equations

[Golusin (1934)]–[Golusin (1935)] reduced the Dirichlet problem for circular multiply connected domains to a system of functional equations and applied the method of successive approximations to obtain its solution under some geometrical restrictions. Such a restriction can be roughly presented in the following form: each disk $\mathbb{D}_k$ lies sufficiently far away from all other disks $\mathbb{D}_m$ $(m \neq k)$. Golusin's approach was developed in [Zmorovich (1958)], [Dunduchenko (1966)], [Aleksandrov and Sorokin (1972)]. [Aleksandrov and Sorokin (1972)] extended Golusin's method to an arbitrary multiply connected circular domain. However, the analytic form of the Schwarz operator was lost. More precisely, the Schwarz problem was reduced via functional equations to an infinite system of linear algebraic equations. Application of the method of truncation to this infinite system was justified.[1]

We also reduce the problem to functional equations which are similar to Golusin's. The main advantage of our modified functional equations is based on the possibility to solve them without any geometrical restriction by successive approximations. It is worth noting that this solution produced the Poincaré series discussed above.

The same story repeats with the alternating Schwarz method, which we call for non-overlapping domains the *generalized Schwarz method* [Golusin (1934)], [Mikhlin (1964)]. It is also known as a *decomposition method* [Smith et al. (1996)]. [Mikhlin (1964)] developed the alternating Schwarz method to the Dirichlet problem for multiply connected domains and proved its convergence under some geometrical restrictions coinciding with Golusin's restrictions for circular domains. Having modified this method we obtained a method convergent for any multiply connected domain (for details see [Mityushev (1994)], [Mityushev and Rogosin (2000)]).

1.5. Poincaré series

Let us consider mutually disjointed disks $\mathbb{D}_k := \{z \in \mathbb{C} : |z - a_k| < r_k\}$ $(k = 1, 2, \ldots, n)$ in the complex plane $\mathbb{C}$. Let $\mathbb{D}$ be the complement of the closed disks $|z - a_k| \leq r_k$ to the extended complex plane $\widehat{\mathbb{C}} = \mathbb{C} \cup \{\infty\}$, i.e., $\mathbb{D} := \widehat{\mathbb{C}} \backslash \cup_{k=1}^{n} (\mathbb{D}_k \cup \partial\mathbb{D}_k)$. It is assumed that $\mathbb{T}_k \cap \mathbb{T}_m = \emptyset$ for $k \neq m$.

The circles $\mathbb{T}_k := \{t \in \mathbb{C} : |t - a_k| = r_k\}$ leave $\mathbb{D}$ to the left. Let

$$z^*_{(k)} = \frac{r_k^2}{\overline{z - a_k}} + a_k$$

[1]Though the method of truncation can be effective in numeric computations, one can hardly accept that this method yields a closed form solution. Any way it depends on the definition of the term "closed form solution". A regular infinite system [Kantorovich and Krylov (1958)] can be considered as an equation with compact operator, i.e., it is no more than a discrete form of a Fredholm integral equation.

be an inversion with respect to the circle $\mathbb{T}_k$. It is known that if $\Phi(z)$ is analytic in the disk $|z-a_k|<r_k$ and continuous in its closure, $\overline{\Phi(z^*_{(k)})}$ is analytic in $|z-a_k|>r_k$ and continuous in $|z-a_k|\geq r_k$.

Introduce the composition of successive inversions with respect to the circles $\mathbb{T}_{k_1}, \mathbb{T}_{k_2}, \dots, \mathbb{T}_{k_p}$,

$$z^*_{(k_p k_{p-1}\dots k_1)} := \left(z^*_{(k_{p-1}\dots k_1)}\right)^*_{(k_p)}. \tag{1.16}$$

In the sequence $k_1, k_2, \dots, k_p$ no two neighboring numbers are equal. The number p is called the *level* of the mapping. When p is even, these are Möbius transformations. If p is odd, we have anti-Möbius transformations, i.e., Möbius transformations in $\overline{z}$. Thus, these mappings can be written in the form

$$\begin{aligned} \gamma_j(z) &= (e_j z + b_j)\,/\,(c_j z + d_j)\,,\ p \in 2\mathbb{Z}, \\ \gamma_j(\overline{z}) &= (e_j \overline{z} + b_j)\,/\,(c_j \overline{z} + d_j)\,,\ p \in 2\mathbb{Z}+1, \end{aligned} \tag{1.17}$$

where $e_j d_j - b_j c_j = 1$. Here $\gamma_0(z) := z$ (identical mapping with the level $p = 0$), $\gamma_1(\overline{z}) := z^*_{(1)}, \dots, \gamma_n(\overline{z}) := z^*_{(n)}$ (n simple inversions, $p = 1$), $\gamma_{n+1}(z) := z^*_{(12)}, \gamma_{n+2}(z) := z^*_{(13)}, \dots, \gamma_{n^2}(z) := z^*_{(n,n-1)}$ ($n^2 - n$ pairs of inversions, $p = 2$), $\gamma_{n^2+1}(\overline{z}) := z^*_{(121)}, \dots$ and so on. The set of the subscripts j of γ_j is ordered in such a way that the level p is increasing. The functions (1.17) generate a Schottky group $\mathcal{K}$. Thus, each element of $\mathcal{K}$ is presented in the form of a composition of inversions (1.16) or in the form of linearly ordered functions (1.17). Let $\mathcal{K}_m$ be such a subset of $\mathcal{K}\backslash\{\gamma_0\}$ that the last inversion of each element of $\mathcal{K}_m$ is different from $z^*_{(m)}$, i.e., $\mathcal{K}_m = \{z^*_{(k_p k_{p-1}\dots k_1)} : k_p \neq m\}$.

Let $H(z)$ be a rational function. This following series is called the Poincaré series:

$$\theta_{2q}(z) := \sum_{j=0}^{\infty} H(\gamma_j(z))(c_j z + d_j)^{-2q}, \quad (q \in \mathbb{Z}/2) \tag{1.18}$$

for $q = 1$ associated with the subgroup $\mathcal{K}$.

Definition 1.3. A point z is called a limit point of the group $\mathcal{K}$ if z is a point of accumulation of the sequence $\gamma_j(z)$ for some $z \in \widehat{\mathbb{C}}$. A point which is not a limit point is called an ordinary point.

In other words, if z runs over the extended complex plane, then the accumulation points of the sequence $\gamma_j(z)$ generate the limit set $\Lambda(\mathcal{K})$. It is assumed that in the formula (1.18) $z \in B := \widehat{\mathbb{C}}\backslash(B_1 \cup \Lambda(\mathcal{K}))$, B_1 is the set of poles of all $H(\gamma_j(z))$ and $\gamma_j(z)$. Ordinary points are characterized by the following property.

Lemma 1.4. *A point z us a regular point of $\mathcal{K}$ if there exist numbers $k_1, k_2, \dots, k_m$ such that $z^*_{k_m k_{m-1}\dots k_1}$ belongs to $D \cup \partial D$.*

The points z_1 and z_2 are called congruent if there exists such $\gamma_j \in \mathcal{K}$ that $\gamma_j(z_1) = z_2$. All limit points of the Schottky group $\mathcal{K}$ lie within the disks $D \cup \partial D$. In the neighborhood of a limit point ς there is an infinite number of distinct points congruent to any point of $\widehat{\mathbb{C}}$ with, at most, exception of ς itself and of one other point. The limit set $\Lambda(\mathcal{K})$ is transformed into itself by any $\gamma_j \in \mathcal{K}$; $\Lambda(\mathcal{K})$ is closed and dense itself.

[Poincaré (1916)] introduced the θ_2-series (1.18) associated to various types of the Kleinian groups. He did not study carefully the Schottky groups and just conjectured that the corresponding θ_2-series always diverges [Poincaré (1916)], [Burnside (1891)] (p. 51). [Burnside (1891)] gave examples of convergent series for the Schottky groups (named by him the first class of groups) and studied their absolute convergence under some geometrical restrictions. In his study W. Burnside followed Poincaré's proof of the convergence of the θ_4-series. On p. 52 [Burnside (1891)] wrote "I have endeavoured to show that, in the case of the first class of groups, this series is convergent, but at present I have not obtained a general proof. I shall offer two partial proofs of the convergency; one of which applies only to the case of Fuchsian groups, and for that case in general, while the other will also apply to Kleinian groups, but only when certain relations of inequality are satisfied." Further, on p. 57 [Burnside (1891)] gave a condition for absolute convergence in terms of the coefficients of the Möbius transformations. He also noted that convergence holds if the radii of the circles $|z - a_k| = r_k$ are sufficiently less than the distances between the centers $|a_k - a_m|$ when $k \neq m$.

[Myrberg (1916)] gave examples of absolutely divergent θ_2-series. Afterwards many mathematicians justified the absolute convergence of the Poincaré series under geometrical restrictions to the locations of the circles (see for references [Crowdy (2008b)] and [Mityushev and Rogosin (2000)]). Here, we present such a typical restriction expressed in terms of the separated parameter Δ introduced by Henrici,

$$\Delta = \max_{k \neq m} \frac{r_k + r_m}{|a_k - a_m|} < \frac{1}{(n-1)^{\frac{1}{4}}} \tag{1.19}$$

for an n-connected domain $\mathbb{D}$ bounded by the circles $|z - a_k| = r_k$ $(k = 1, 2, \ldots, n)$.

Necessary and sufficient conditions for absolute and uniform convergence of the series have been found in [Akaza (1966)], [Akaza and Inoue (1984)] in terms of the Hausdorff dimension of $\Lambda(\mathcal{K})$. This result is based on the study of the series $\sum_{j=1}^{\infty} |c_j|^{-2}$.

After [Myrberg (1916)] it seemed that the opposite conjectures of Poincaré and Burnside were both wrong. However, it was proved in [Mityushev (1998)] that θ_2-series converges uniformly for any multiply connected domain $\mathbb{D}$ without any geometrical restriction that corresponds to Burnside's conjecture. The uniform convergence does not directly imply the automorphy relation, i.e., invariance under the Schottky group of transformations, since it is forbidden to change the order of summation without absolute convergence. But this difficulty can be easily overcome by using functional equations. As a result, the Poincaré series satisfies the required automorphy relation and can be written in each fundamental domain with a prescribed summation depending on this domain [Mityushev

(1998)]. The study [Mityushev (1998)] is based on the solution to a Riemann–Hilbert problem. First, the Riemann–Hilbert problem is written as an $\mathbb{R}$-linear problem which is stated as a conjugation problem between functions analytic in the disks $\mathbb{D}_k = \{z \in \mathbb{C} : |z - a_k| < r_k\}$ $(k = 1, 2, \ldots, n)$ and in $\mathbb{D}$. Further, the latter problem is reduced to a system of functional equations (without integral terms) with respect to the functions analytic in $|z - a_k| < r_k$. The method of successive approximations is justified for this system in a functional space in which convergence is uniform. Straightforward calculations of the successive approximations yields a Poincaré type series (see for instance (3.40) in this paper).

2. Linear functional equations

2.1. Homogeneous equation

Let G be a domain on the extended complex plane whose boundary ∂G consists of simple closed Jordan curves. Introduce the Banach space $\mathcal{C}(\partial G)$ of functions continuous on the curves of ∂G with the norm $||f|| = \max_{1 \le k \le n} \max_{\partial G} |f(t)|$. Let us consider a closed subspace $\mathcal{C}_{\mathcal{A}}(G)$ of $\mathcal{C}(\partial G)$ consisting of the functions analytically continued into all disks G. Further, we usually take $\cup_{k=0}^{n} \mathbb{D}_k$ and sometimes $\mathbb{D}$ as the domain G (not necessarily connected). For brevity, the notation $\mathcal{C}_{\mathcal{A}}$ for $\mathcal{C}_{\mathcal{A}} \left(\cup_{k=0}^{n} \mathbb{D}_k\right)$ is used.

Hereafter, a point $w \in \mathbb{D} \backslash \{\infty\}$ is fixed.

Lemma 2.1. *Let given numbers ν_k have the form $\nu_k := \exp(-i\mu_k)$ with $\mu_k \in \mathbb{R}$. Consider the system of functional equations with respect to the functions $\phi_k(z)$ analytic in $\mathbb{D}_k$,*

$$\phi_k(z) = -\nu_k \sum_{m \ne k} \overline{\nu_m} \left[\overline{\phi_m\left(z_{(m)}^*\right)} - \overline{\phi_m\left(w_{(m)}^*\right)} \right], \quad |z - a_k| \le r_k \ (k = 1, 2, \ldots, n). \tag{2.1}$$

This system has only the trivial solution.

Proof. Let $\phi_m(z)$ $(m = 1, 2, \ldots, n)$ be a solution of (2.1). Then the right-hand part of (2.1) implies that the function $\phi_k(z)$ is analytic in $|z - a_k| \le r_k$ $(k = 1, 2, \ldots, n)$. Introduce the function

$$\psi(z) := -\sum_{m=1}^{n} \overline{\nu_m} \left[\overline{\phi_m\left(z_{(m)}^*\right)} - \overline{\phi_m\left(w_{(m)}^*\right)} \right],$$

analytic in the closure of $\mathbb{D}$. Then the functions ψ, ϕ_k satisfy the $\mathbb{R}$-linear boundary conditions

$$\nu_k \psi(t) = \phi_k(t) - \overline{\phi_k(t)} + \overline{\phi_k\left(w_{(k)}^*\right)}, \quad |t - a_k| = r_k, \ \ k = 1, \ldots, n.$$

One can write the later relations in the following form:

$$\operatorname{Re} \nu_k \psi(t) = c_k, \ |t - a_k| = r_k, \ \ k = 1, \ldots, n, \tag{2.2}$$

$$2 \operatorname{Im} \phi_k(t) = \operatorname{Im} \nu_k \psi(t) + d_k, \ |t - a_k| = r_k, \ \ k = 1, \ldots, n. \tag{2.3}$$

Here $\phi_k(w_{(k)}^*) = c_k + id_k$. One may consider equalities (2.2) as a boundary value problem with respect to the function $\psi(z)$ analytic in $\mathbb{D}$ and continuous in its closure, i.e., $\psi \in \mathcal{C}_A(\mathbb{D})$. The real constants c_k have to be determined. We prove that the problem (2.2) has only constant solutions: $\psi(z) \equiv c$, $c_k = \operatorname{Re} \nu_k c$. Denote by $\psi(\mathbb{D}) := \left\{ \varsigma \in \widehat{\mathbb{C}} : z \in \mathbb{D},\ \varsigma = \psi(z) \right\}$ the image of $\mathbb{D}$ under mapping ψ. It follows from the Boundary Correspondence Principle for conformal mapping that the boundary of $\psi(\mathbb{D})$ consists of the segments $\operatorname{Re} \nu_k \varsigma = c_k$ $(k = 1, 2, \ldots, n)$. But in this case the point $\varsigma = \infty \in \psi(\mathbb{D})$ corresponds to a point of $\mathbb{D}$. It contradicts boundedness of the function $\psi(z)$ in the closure of $\mathbb{D}$. Hence, $\phi(z) =$ constant and equalities (2.3) imply that $\phi_k(t) =$ constant [Gakhov (1977)]. Using (2.1) we have $\phi_k(z) \equiv 0$.

The lemma is proved. □

2.2. Non-homogeneous equation

Lemma 2.2. *Let $h \in \mathcal{C}_{\mathcal{A}}$, $|\nu_k| = 1$. Then the system of functional equations*

$$\phi_k(z) = -\nu_k \sum_{m \neq k} \overline{\nu_m} \left[\overline{\phi_m\left(z_{(m)}^*\right)} - \overline{\phi_m\left(w_{(m)}^*\right)} \right] + h_k(z), \qquad |z - a_k| \leq r_k \ (k = 1, 2, \ldots, n), \tag{2.4}$$

has a unique solution $\Phi \in \mathcal{C}_{\mathcal{A}}$. Here $\Phi(z) := \phi_k(z)$ in $|z - a_k| \leq r_k$, $k = 1, 2, \ldots, n$. This solution can be found by the method of successive approximations. The approximations converge in $\mathcal{C}_{\mathcal{A}}$.

Proof. Rewrite the system (2.4) on $\mathbb{T}_k$ in the form of a system of integral equations

$$\phi_k(t) = -\nu_k \sum_{m \neq k} \overline{\nu_m \frac{1}{2\pi i} \int_{\mathbb{T}_m^-} \phi_m(\tau) \left(\frac{1}{\tau - t_{(m)}^*} - \frac{1}{\tau - w_{(m)}^*} \right) d\tau} + h_k(t), \qquad |t - a_k| = r_k \ (k = 1, 2, \ldots, n). \tag{2.5}$$

The orientation on $\mathbb{T}_m^-$ leaves $\mathbb{D}_m$ to the left. The system (2.5) can be written as an equation in the space $\mathcal{C}(\cup_{k=1}^n \mathbb{T}_k)$:

$$\Phi = \mathbf{A}\Phi + h. \tag{2.6}$$

The integral operators from (2.5) are compact in $\mathcal{C}(\mathbb{T}_k)$; multiplication by $\overline{\nu_m}$ and complex conjugation are bounded operators in $\mathcal{C}$. Then $\mathbf{A}$ is a compact operator in $\mathcal{C}$. Since Φ is a solution of (2.6) in $\mathcal{C}$, hence $\Phi \in \mathcal{C}_{\mathcal{A}}$ (see Pumping principle from [Mityushev and Rogosin (2000)], Sec. 2.3). This follows from the properties of the Cauchy integral and the condition $h \in \mathcal{C}_{\mathcal{A}}$. Therefore, equation (2.6) in $\mathcal{C}$ and equation (2.4) in $\mathcal{C}_{\mathcal{A}}$ are equivalent when $h \in \mathcal{C}_{\mathcal{A}}$. It follows from Lemma 2.1 that the homogeneous equation $\Phi = \mathbf{A}\Phi$ has only a trivial solution. Then the Fredholm theorem implies that equation (2.6) or the system (2.4) has a unique solution.

Let us show the convergence of the method of successive approximations. By virtue of the Successive Approximation Theorem (see [Krasnosel'skii et al. (1969)] and [Mityushev and Rogosin (2000)], Sec. 2.3) it is sufficient to prove the inequality

$\rho(\mathbf{A}) < 1$, where $\rho(\mathbf{A})$ is the spectral radius of the operator $\mathbf{A}$. The inequality $\rho(\mathbf{A}) < 1$ is satisfied if for all complex numbers λ such that $|\lambda| \leq 1$, equation

$$\Phi = \lambda \mathbf{A} \Phi$$

has only a trivial solution. This equation can be rewritten in the form

$$\phi_k(z) = -\lambda \nu_k \sum_{m \neq k} \overline{\nu_m} \left[\overline{\phi_m\left(z^*_{(m)}\right)} - \overline{\phi_m\left(w^*_{(m)}\right)} \right], \quad |z - a_k| \leq r_k. \tag{2.7}$$

Consider the case $|\lambda| < 1$. Introduce the function, analytic in the closure of $\mathbb{D}$,

$$\psi(z) = -\lambda \sum_{m=0}^{n} \overline{\nu_m} \left(\overline{\phi_m\left(z^*_{(m)}\right)} - \overline{\phi_m\left(w^*_{(m)}\right)} \right).$$

Then $\psi(z)$ and $\phi_k(z)$ satisfy the $\mathbb{R}$-linear problem

$$\nu_k \psi(t) = \phi_k(t) - \lambda \overline{\phi_k(t)} + \gamma_k, \ |t - a_k| = r_k, \ k = 1, 2, \ldots, n,$$

where $\gamma_k := \lambda \overline{\phi_k\left(w^*_{(k)}\right)}$. It can be written in the form

$$\nu_k \psi_0(t) = \phi_k(t) - \lambda \overline{\phi_k(t)} + \gamma_k - \nu_k \psi(\infty), \ |t - a_k| = r_k, \ k = 1, 2, \ldots, n, \tag{2.8}$$

where $\psi_0(z) = \psi(z) - \psi(\infty)$. Theorem 1.1 implies that problem (2.8) has the unique solution

$$\psi_0(z) = 0, \ \phi_k(z) = \frac{\gamma_k - \nu_k \psi(\infty) + \lambda(\overline{\gamma_k - \nu_k \psi(\infty)})}{|\lambda|^2 - 1}, \quad k = 1, 2, \ldots, n.$$

Hence, $\phi_k(z) =$ constant. Then (2.7) yields $\phi_k(z) \equiv 0$.

Consider the case $|\lambda| = 1$. Then by substituting $\omega_k(z) = \phi_k(z)/\sqrt{\lambda}$ the system (2.7) is reduced to the same system with $\lambda = 1$. It follows from Lemma 2.1 that $\omega_k(z) = \phi_k(z) = 0$. Hence, $\rho(\mathbf{A}) < 1$.

This inequality proves the lemma. □

3. Schwarz operator

3.1. Harmonic measures

In the present section the number s is chosen from $1, 2, \ldots, n$ and fixed. The *harmonic measure* $\alpha_s(z)$ of the circle $\mathbb{T}_s$ with respect to $\partial\mathbb{D}$ is a function harmonic in $\mathbb{D}$, continuous in its closure, satisfying the boundary conditions

$$\alpha_s(t) = \delta_{sk}, \ |t - a_k| = r_k, \ k = 1, 2, \ldots, n, \tag{3.1}$$

where δ_{sk} is the Kronecker symbol. The functions α_s are infinitely $\mathbb{R}$-differentiable in the closure of $\mathbb{D}$ (see [Mityushev and Rogosin (2000)], Sec. 2.7.2). Using the Logarithmic Conjugation Theorem [Mityushev and Rogosin (2000)] we look for $\alpha_s(z)$ in the form

$$\alpha_s(z) = \operatorname{Re} \phi(z) + \sum_{m=1}^{n} A_m \ln|z - a_m| + A, \tag{3.2}$$

where A_m and A are real constants,

$$\sum_{m=1}^{n} A_m = 0. \tag{3.3}$$

The later condition follows from the limit in (3.2) as z tends to infinity. Using the boundary condition (3.1) and the representation (3.2) we arrive at the following boundary value problem:

$$\operatorname{Re}\phi(t) + \sum_{m=1}^{n} A_m \ln|t-a_k| + A = \delta_{sk}, \ |t-a_k| = r_k. \tag{3.4}$$

This problem is equivalent to the ℝ-linear problem (see Introduction)

$$\phi(t) = \phi_k(t) - \overline{\phi_k(t)} + f_k(t), \ |t-a_k| = r_k, \ k = 1, 2, \dots, n, \tag{3.5}$$

where the unknown functions $\phi \in \mathcal{C}_{\mathcal{A}}(\mathbb{D})$, $\phi_k \in \mathcal{C}_{\mathcal{A}}(\mathbb{D}_k)$,

$$\phi(w) = 0, \tag{3.6}$$

$$f_k(z) := \delta_{sk} - A - A_k \ln r_k - \sum_{m\neq k} A_m \ln(z-a_m), z \in \mathbb{D}_k. \tag{3.7}$$

The branch of $\ln(z-a_m)$ is fixed in such a way that the cut connecting the points $z = a_m$ and $z = \infty$ does not intersect the circles $\mathbb{T}_k$ for $k \neq m$ and does not pass through the point $z = w$. The function $f_k(z)$ satisfies the boundary condition

$$\operatorname{Re} f_k(t) := \delta_{sk} - A - \sum_{m=1}^{n} A_m \ln|t-a_m|, \ |t-a_k| = r_k$$

and belongs to $\mathcal{C}_{\mathcal{A}}(\mathbb{D}_k)$.

Remark 3.1. More precisely the functions ϕ, ϕ_k and f_k are infinitely ℂ-differentiable in the closures of the domains considered.

Let us introduce the function

$$\Phi(z) := \begin{cases} \phi_k(z) + \sum_{m\neq k}\left[\overline{\phi_m\left(z_{(m)}^*\right)} - \overline{\phi_m\left(w_{(m)}^*\right)}\right] - \overline{\phi_k\left(w_{(k)}^*\right)} + f_k(z), \\ \qquad\qquad |z-a_k| \leq r_k, \\ \phi(z) + \sum_{m=1}^{n}\left[\overline{\phi_m\left(z_{(m)}^*\right)} - \overline{\phi_m\left(w_{(m)}^*\right)}\right], \ z \in \mathbb{D}. \end{cases}$$

Calculate the jump across the circle $\mathbb{T}_k$,

$$\Delta_k := \Phi^+(t) - \Phi^-(t), \ t \in \mathbb{T}_k,$$

where $\Phi^+(t) := \lim_{z\to t\ z\in\mathbb{D}} \Phi(z)$, $\Phi^-(t) := \lim_{z\to t\ z\in\mathbb{D}_k} \Phi(z)$. Using (3.5), (3.7) we get $\Delta_k = 0$. It follows from the Analytic Continuation Principle that $\Phi(z)$ is analytic in the extended complex plane. Then Liouville's theorem implies that $\Phi(z)$ is a constant. Using (3.6) we calculate $\Phi(w) = 0$, hence $\Phi(z) \equiv 0$. The definition of $\Phi(z) \equiv 0$ in $|z-a_k| \leq r_k$ yields the following system of functional equations:

$$\begin{aligned}\phi_k(z) = &-\sum_{m\neq k}\left[\overline{\phi_m\left(z^*_{(m)}\right)} - \overline{\phi_m\left(w^*_{(m)}\right)}\right] - \delta_{sk} + A + A_k \ln r_k \\ &+ \sum_{m\neq k} A_m \ln(z-a_m) + \overline{\phi_k\left(w^*_{(k)}\right)}, \quad |z-a_k| \leq r_k,\end{aligned} \tag{3.8}$$

with respect to the functions $\phi_k(z) \in \mathcal{C}_{\mathcal{A}}(\mathbb{D}_k)$. The branches of logarithms are chosen in the same way as in (3.7).

The system of functional equations (3.8) is the main point to construct the harmonic measure α_s via the analytic function $\phi(z)$ by formula (3.2). If $\phi_k(z)$ are known, the required function $\phi(z)$ has the form

$$\phi(z) = -\sum_{m=1}^{n}\left[\overline{\phi_m\left(z^*_{(m)}\right)} - \overline{\phi_m\left(w^*_{(m)}\right)}\right], \quad z \in \mathbb{D} \cup \partial\mathbb{D}. \tag{3.9}$$

It is convenient to represent $\phi_k(z)$ in the form

$$\phi_k(z) = \varphi_k^{(0)}(z) + \sum_{m=1}^{n} A_m \varphi_k^{(m)}(z), \tag{3.10}$$

where $\varphi_k^{(0)}(z)$ satisfies

$$\begin{aligned}\varphi_k^{(0)}(z) = &-\sum_{m\neq k}\left[\overline{\varphi_m^{(0)}\left(z^*_{(m)}\right)} - \overline{\varphi_m^{(0)}\left(w^*_{(m)}\right)}\right] - \delta_{sk} + A + A_k \ln r_k \\ &+ \sum_{m\neq k} A_m \ln(w-a_m) + \overline{\phi_k\left(w^*_{(k)}\right)}, \quad |z-a_k| \leq r_k, \ k=1,\dots,n,\end{aligned} \tag{3.11}$$

$\varphi_k^{(m)}(z)$ satisfies

$$\begin{aligned}\varphi_k^{(m)}(z) = &-\sum_{k_1\neq k}\left[\overline{\varphi_{k_1}^{(m)}\left(z^*_{(k_1)}\right)} - \overline{\varphi_{k_1}^{(m)}\left(w^*_{(k_1)}\right)}\right] + \delta'_{km} \ln\frac{z-a_m}{w-a_m}, \\ &|z-a_k| \leq r_k, \ k=1,\dots,n \quad (m=1,\dots,n).\end{aligned} \tag{3.12}$$

In (3.12), n systems of functional equations are written, m is the number of the system, $\delta'_{km} = 1-\delta_{km}$, where δ_{km} is the Kronecker symbol. It is assumed that the constants A, A_k and $\overline{\phi_k\left(w^*_{(k)}\right)}$ are fixed in (3.11). The values of these constants will be found later. According to Lemma 2.2, functional equations (3.11)–(3.12) can be solved by the method of successive approximations. The method of successive approximations applied to (3.12) yields

$$\varphi_k^{(m)}(z) = \delta'_{km} \ln \frac{z-a_m}{w-a_m} - \sum_{k_1 \neq k} \delta_{k_1}^{(m)} \ln \overline{\frac{\overline{z^*_{(k_1)} - a_m}}{w^*_{(k_1)} - a_m}} + \sum_{k_1 \neq k} \sum_{k_2 \neq k_1} \delta_{k_2}^{(m)} \ln \frac{z^*_{(k_2k_1)} - a_m}{w^*_{(k_2k_1)} - a_m} - \cdots, \tag{3.13}$$

where the sum $\sum\limits_{k_j \neq k_{j-1}}$ contains the terms with $k_j = 1, 2, \ldots, n;\ k_j \neq k_{j-1}$. By virtue of Lemma 2.2 with $\nu_m = 1$ the series (3.13) converges uniformly in $|z - a_k| \leq r_k$. It follows from (3.11) that $\phi_k^{(0)}(z)$ $(k = 1, 2, \ldots, n)$ are constants, since the zeroth approximation is a constant and the operator from the right-hand side of (3.11) produces constants.

One can see from (3.9) that the constants $\varphi_k^{(0)}(z)$ do not impact on $\phi(z)$, hence using (3.10) we have

$$\phi(z) = -\sum_{m=1}^{n} A_m \sum_{k=1}^{n} \left[\overline{\varphi_k^{(m)}\left(z^*_{(k)}\right)} - \overline{\varphi_k^{(m)}\left(w^*_{(k)}\right)}\right]. \tag{3.14}$$

Substitution of (3.13) into (3.14) yields

$$\begin{aligned}\phi(z) = & -\sum_{m=1}^{n} A_m \sum_{k=1}^{n} \delta'_{km} \ln \overline{\frac{\overline{z^*_{(k)} - a_m}}{w^*_{(k)} - a_m}} + \sum_{m=1}^{n} A_m \sum_{k=1}^{n} \sum_{k_1 \neq k} \delta_{k_1}^{(m)} \ln \frac{z^*_{(k_1k)} - a_m}{w^*_{(k_1k)} - a_m} \\ & - \sum_{m=1}^{n} A_m \sum_{k=1}^{n} \sum_{k_1 \neq k} \sum_{k_2 \neq k_1} \delta_{k_2}^{(m)} \ln \overline{\frac{\overline{z^*_{(k_2k_1k)} - a_m}}{w^*_{(k_2k_1k)} - a_m}} + \cdots, \quad z \in \mathbb{D} \cup \partial\mathbb{D}.\end{aligned} \tag{3.15}$$

Using the properties of δ'_{km} one can rewrite (3.15) in the form

$$\begin{aligned}\phi(z) = & -\sum_{m=1}^{n} A_m \sum_{k \neq m} \ln \overline{\frac{\overline{z^*_{(k)} - a_m}}{w^*_{(k)} - a_m}} + \sum_{m=1}^{n} A_m \sum_{k=1}^{n} \sum_{k_1 \neq k, m} \ln \frac{z^*_{(k_1k)} - a_m}{w^*_{(k_1k)} - a_m} \\ & - \sum_{m=1}^{n} A_m \sum_{k=1}^{n} \sum_{k_1 \neq k} \sum_{k_2 \neq k_1, m} \ln \overline{\frac{\overline{z^*_{(k_2k_1k)} - a_m}}{w^*_{(k_2k_1k)} - a_m}} + \cdots, \quad z \in \mathbb{D} \cup \partial\mathbb{D}.\end{aligned} \tag{3.16}$$

In order to write (3.16) in more convenient form we use the following:

Lemma 3.2. *There holds the equality*

$$\sum_{k=1}^{n} \sum_{k_1 \neq k} \cdots \sum_{k_s \neq k_{s-1}, m} A_m R_{mk_sk_{s-1}\ldots k_1k} = \sum_{k_1 \neq m} \sum_{k_2 \neq k_1} \cdots \sum_{k_s \neq k_{s-1}} \sum_{k \neq k_s} A_m R_{mk_1k_2\ldots k_sk}. \tag{3.17}$$

Proof. It is sufficient to demonstrate that both parts of equality (3.17) contain the same terms. First, replace $k_1k_2\dots k_s$ by $k_sk_{s-1}\dots k_1$ in the right-hand part of (3.17) which becomes

$$\sum_{k_s\neq m}\sum_{k_{s-1}\neq k_s}\cdots\sum_{k_1\neq k_2}\sum_{k\neq k_1} A_m R_{mk_sk_{s-1}\dots k_1 k}. \tag{3.18}$$

The left-hand part of (3.17) can be written as the sum

$$\sum_{k=1}^{n}\sum_{k_1=1}^{n}\cdots\sum_{k_s=1}^{n}\delta'_{k_1k}\delta'_{k_2k_1}\cdots\delta'_{k_sk_{s-1}}\delta'_{k_sm}A_m R_{mk_sk_{s-1}\dots k_1 k}, \tag{3.19}$$

where $\delta'_{lk}=1-\delta_{lk}$, δ_{lk} is the Kronecker symbol. One can see that the sum (3.18) written in the similar form (3.19) contains the same product of the complimentary Kronecker symbols.

This proves the lemma. □

Applying Lemma 3.2 to (3.16) we obtain

$$\begin{aligned}\phi(z) = &-\sum_{m=1}^{n}A_m\sum_{k\neq m}\ln\overline{\frac{z^*_{(k)}-a_m}{w^*_{(k)}-a_m}}\\ &+\sum_{m=1}^{n}A_m\sum_{k_1\neq m}\sum_{k\neq k_1}\ln\frac{z^*_{(k_1k)}-a_m}{w^*_{(k_1k)}-a_m}\\ &-\sum_{m=1}^{n}A_m\sum_{k_1\neq m}\sum_{k_2\neq k_1}\sum_{k\neq k_2}\ln\overline{\frac{z^*_{(k_1k_2k)}-a_m}{w^*_{(k_1k_2k)}-a_m}}+\cdots,\quad z\in\mathbb{D}\cup\partial\mathbb{D}.\end{aligned} \tag{3.20}$$

It can be also written in the form

$$\phi(z)=\sum_{m=1}^{n}A_m\psi_m(z),$$

where

$$\begin{aligned}\psi_m(z) = &\ln\prod_{k\neq m}\overline{\frac{w^*_{(k)}-a_m}{z^*_{(k)}-a_m}}+\ln\prod_{k_1\neq m}\prod_{k\neq k_1}\frac{z^*_{(k_1k)}-a_m}{w^*_{(k_1k)}-a_m}\\ &+\ln\prod_{k_1\neq m}\prod_{k_2\neq k_1}\prod_{k\neq k_2}\overline{\frac{w^*_{(k_1k_2k)}-a_m}{z^*_{(k_1k_2k)}-a_m}}+\cdots,\quad z\in\mathbb{D}\cup\partial\mathbb{D}.\end{aligned} \tag{3.21}$$

Let us rewrite (3.21) in terms of the group $\mathcal{K}$,

$$\psi_m(z)=\ln\left[\prod_{j\in\mathcal{K}_m}^{\infty}\psi_m^{(j)}(z)\right], \tag{3.22}$$

where

$$\psi_m^{(j)}(z) = \begin{cases} \frac{\gamma_j(z)-a_m}{\gamma_j(w)-a_m}, & \text{if level of } \gamma_j \text{ is even,} \\ \overline{\frac{\gamma_j(\overline{w})-a_m}{\gamma_j(\overline{z})-a_m}}, & \text{if level of } \gamma_j \text{ is odd.} \end{cases}$$

The numeration on j in (3.22) is fixed with increasing level.

In order to determine the constants A and A_m, substitute $z = w^*_{(k)}$ in the real parts of (3.8):

$$\begin{aligned} 0 = &- \sum_{m \neq k} \mathrm{Re} \left[\overline{\varphi_m \left(\left(w^*_{(k)} \right)^*_{(m)} \right)} - \overline{\varphi_m \left(w^*_{(m)} \right)} \right] \\ &- \delta_{sk} + A + A_k \ln r_k + \sum_{m \neq k} A_m \ln \left| w^*_{(k)} - a_m \right|, \quad k = 1, 2, \ldots, n. \end{aligned} \tag{3.23}$$

The function φ_m has the form (3.13) and linearly depends on the unknown constants A_m. The equalities (3.3), (3.23) generate a system of $n+1$ linear algebraic equations with respect to $n+1$ unknowns $A, A_1, \ldots, A_n$. This system has a unique solution, since in the opposite case it contradicts the uniqueness of the solution to the Dirichlet problem.

Theorem 3.3. *The harmonic measures have the form*

$$\alpha_s(z) = \sum_{m=1}^{n} A_m \left[\mathrm{Re}\, \psi_m(z) + \ln |z - a_m| \right] + A, \tag{3.24}$$

where $\psi_m(z)$ is given in (3.22). The infinite product (3.22) converges uniformly on each compact subset of $\mathbb{D} \setminus \{\infty\}$. The real constants A and A_m are uniquely determined by the system (3.3), (3.23)).

Proof. Exact formulae for harmonic measures were deduced in a formal way. In order to justify them it is necessary to prove the change of the summation in (3.16) to obtain (3.20).

Using the designations of Section 2, we write the series (3.13) in the form

$$\Phi = \sum_{k=1}^{\infty} \mathbf{A}^k h, \tag{3.25}$$

where

$$\begin{aligned} h(z) = \delta'_{km} \ln \frac{z - a_m}{w - a_m}, \quad \mathbf{A}h(z) = - \sum_{k_1 \neq k} \delta'_{k_1 m} \ln \overline{\frac{z^*_{(k_1)} - a_m}{w^*_{(k_1)} - a_m}}, \\ \mathbf{A}^2 h(z) = \sum_{k_1 \neq k} \sum_{k_2 \neq k_1} \delta'_{k_2 m} \ln \frac{z^*_{(k_2 k_1)} - a_m}{w^*_{(k_2 k_1)} - a_m} - \cdots . \end{aligned} \tag{3.26}$$

It is possible to change the order of summation in each term $\mathbf{A}^k h$ of the successive approximations which contains inversions only of the level k. Therefore, it is possible to change the order of summation in the series (3.15) in terms having the same level. The series (3.20) is obtained from (3.16) by application of this rule.

This proves the theorem. □

Remark 3.4. The constants A and A_m depend on the choice of w.

Remark 3.5. The logarithmic terms in (3.24) can be included into infinite product (3.21). Then (3.24) becomes

$$\alpha_s(z) = \sum_{m=1}^{n} A_m \ln \prod_{j \in \mathcal{K}_m \cup \{0\}} |\psi_m^{(j)}(z)| + A_0,$$

where $A_0 := A - \sum_{m=1}^{n} A_m \ln |w - a_m|$.

3.2. Exact formula for the Schwarz operator

Following the previous section we construct the complex Green function $M(z,\zeta)$ and the Schwarz operator for the circular multiply connected domain $\mathbb{D}$. One can find general properties of the Schwarz operator in [Mityushev and Rogosin (2000)], [Mikhlin (1964)].

Let z and ζ belong to the closure of $\mathbb{D}$. The real *Green function* $G(z,\zeta) = g(z,\zeta) - \ln|z-\zeta|$ is introduced via the function $g(z,\zeta)$ harmonic in $\mathbb{D}$ satisfying the Dirichlet problem

$$g(t,\zeta) - \ln|t-\zeta| = 0, \quad |t - a_k| = r_k \ (k = 1,2,\dots,n) \tag{3.27}$$

with respect to the first variable. If $G(z,\zeta)$ is known, the solution of the Dirichlet problem

$$u(t) = f(t), \ |t - a_k| = r_k \ (k = 1,2,\dots,n) \tag{3.28}$$

has the form

$$u(z) = \frac{1}{2\pi} \sum_{k=1}^{n} \int_{\mathbb{T}_k} f(\zeta) \frac{\partial G}{\partial \nu}(z,\zeta) d\sigma, \tag{3.29}$$

where ν is the outward (in sense of orientation) normal vector at the point $\zeta \in \partial\mathbb{D}$.

The *complex Green function* $M(z,\zeta)$ is defined by the formula

$$M(z,\zeta) = G(z,\zeta) + iH(z,\zeta), \tag{3.30}$$

where the function $H(z,\zeta)$ is harmonically conjugated to $G(z,\zeta)$ on the variable z. It has the form

$$H(z,\zeta) = \int_w^z -\frac{\partial G}{\partial y} dx + \frac{\partial G}{\partial x} dy$$

with $z = x + iy$.

Introduce the *Schwarz kernel* (see [Mityushev and Rogosin (2000)], Sec. 2.7.2)

$$T(z,\zeta) = \frac{\partial M}{\partial \nu}(z,\zeta), \quad \zeta \in \mathbb{D}. \tag{3.31}$$

In accordance with (3.28)–(3.31) the function

$$F(z) = \frac{1}{2\pi} \sum_{k=1}^{n} \int_{\mathbb{T}_k} f(\zeta) T(z,\zeta) d\sigma \tag{3.32}$$

satisfies the boundary value problem

$$\operatorname{Re} F(t) = f(t),\ |t - a_k| = r_k\ (k = 1, 2, \ldots, n). \tag{3.33}$$

Here $u(z)$ from (3.29) is the real part of the analytic function $F(z)$ having in general multi-valued imaginary part in the multiply connected domain $\mathbb{D}$. If we are looking for single-valued $F(z)$ by (3.33), we arrive at the Schwarz problem in accordance with the terminology introduced in our Introduction.

We use the representation for the Green function (see [Mityushev and Rogosin (2000)], Sec. 2.7.2)

$$M(z,\zeta) = M_0(z,\zeta) + \sum_{k=1}^{n} \alpha_k(\zeta) \ln(z - a_k) - \ln(\zeta - z) + A(\zeta), \tag{3.34}$$

where α_k is a harmonic measure of $\mathbb{D}$, $A(\zeta)$ is a real function in ζ. The point w and the branches of $\ln(z - a_k)$ are fixed as in the previous section. Using (3.31), (3.34) we obtain

$$T(z,\zeta) = \frac{\partial M_0}{\partial \nu}(z,\zeta) + \sum_{m=1}^{n} \frac{\partial \alpha_m}{\partial \nu}(\zeta) \ln(z - a_m) - \frac{1}{\zeta - z}\frac{\partial \zeta}{\partial \nu} + \frac{\partial A}{\partial \nu}(\zeta). \tag{3.35}$$

The function $M_0(z,\zeta)$ is infinitely $\mathbb{C}$-differentiable in the closure of $\mathbb{D}$ in z and satisfies the boundary value problem which follows from (3.27) and (3.30),

$$\begin{gathered}\operatorname{Re}\left[M_0(t,\zeta) + \sum_{k=1}^{n} \alpha_k(\zeta) \ln(t - a_k) - \ln(\zeta - t) + A(\zeta)\right] = 0, \\ |t - a_k| = r_k,\ k = 1, 2, \ldots, n;\ M_0(w,\zeta) = 0.\end{gathered} \tag{3.36}$$

The problem (3.36) has a unique solution. It is reduced to the following system of functional equations:

$$\begin{aligned}\phi_k(z) = &- \sum_{m \neq k} \left[\overline{\phi_m\left(z^*_{(m)}\right)} - \overline{\phi_m\left(w^*_{(m)}\right)}\right] - \ln(\zeta - z) + A \\ &+ \alpha_k(\zeta) \ln r_k + \sum_{m \neq k} \alpha_m(\zeta) \ln(z - a_m) + \overline{\phi_k\left(w^*_{(k)}\right)}, \\ &|z - a_k| \le r_k,\ k = 1, \ldots, n\end{aligned} \tag{3.37}$$

where $\phi_k(z)$ belongs to $\mathcal{C}_{\mathcal{A}}(\mathbb{D}_k)$ and $\mathbb{C}$-infinitely differentiable in the closure of $\mathbb{D}_k$. Here ζ is considered as a parameter fixed in the closure of $\mathbb{D}$. The required function $M_0(z,\zeta)$ is related to the auxiliary functions $\phi_k(z)$ by equality

$$M_0(z,\zeta) = -\sum_{k=1}^{n} \left[\overline{\phi_k\left(z^*_{(k)}\right)} - \overline{\phi_k\left(w^*_{(k)}\right)}\right],\ z \in \mathbb{D} \cup \partial\mathbb{D} \setminus \{\zeta\}. \tag{3.38}$$

In order to solve (3.37) we consider two auxiliary systems of functional equations

$$\Psi_k(z) = -\sum_{m\neq k} \left[\overline{\Psi_m\left(z_m^*\right)} - \overline{\Psi_m\left(w_m^*\right)}\right] + A + \alpha_k(\zeta)\ln r_k$$
$$+ \sum_{m\neq k} \alpha_m(\zeta)\ln(z-a_m) + \overline{\phi_m\left(w_{(k)}^*\right)},$$
$$\Omega_k(z) = -\sum_{m\neq k} \left[\overline{\Omega_m\left(z_m^*\right)} - \overline{\Omega_m\left(w_m^*\right)}\right] - \ln(\zeta - z),\ |z-a_k| \leq r_k,\ k=1,\dots,n.$$

The first system coincides with the system (2.5) ($\nu_k = 1$), and thus can be solved by the method of successive approximations (cf. Lemma 2.2). Let us consider the second system. If $|\zeta - z| \leq r_s$ for some s, the right-hand part of the second system $-\ln(\zeta - z)$ does not belong to $\mathcal{C}_{\mathcal{A}}(\mathbb{D}_s)$. But by introducing a new unknown function $\Omega_s^{(0)}(z) := \Omega_s(z) - \ln(\zeta - z)$ we get an equation in the space $\mathcal{C}_{\mathcal{A}}(\mathbb{D}_s)$. Therefore, the method of successive approximations can be applied to the second system too, and

$$\overline{\Psi_k\left(z_k^*\right)} - \overline{\Psi_k\left(w_k^*\right)} = \sum_{m\neq k} \alpha_m(\zeta)\ln \overline{\frac{z_{(k)}^* - a_m}{w_{(k)}^* - a_m}} - \sum_{k_1\neq k}\sum_{m\neq k_1} \alpha_m(\zeta)\ln \frac{z_{(k_1k)}^* - a_m}{w_{(k_1k)}^* - a_m}$$
$$+ \sum_{k_1\neq k}\sum_{k_2\neq k_1}\sum_{m\neq k_2} \alpha_m(\zeta)\ln \overline{\frac{z_{(k_2k_1k)}^* - a_m}{w_{(k_2k_1k)}^* - a_m}} - \cdots, \tag{3.39}$$

$$\overline{\Omega_k\left(z_k^*\right)} - \overline{\Omega_k\left(w_k^*\right)} = \ln \overline{\frac{\zeta - w_{(k)}^*}{\zeta - z_{(k)}^*}} + \sum_{k_1\neq k} \ln \frac{\zeta - z_{(k_1k)}^*}{\zeta - w_{(k_1k)}^*} + \sum_{k_1\neq k}\sum_{k_2\neq k_1} \ln \overline{\frac{\zeta - z_{(k_2k_1k)}^*}{\zeta - w_{(k_2k_1k)}^*}}$$
$$+ \cdots,\ z \in \mathbb{D}\cup\partial\mathbb{D}. \tag{3.40}$$

The series (3.39), (3.40) converge uniformly in every compact subset of $\mathbb{D}\cup\partial\mathbb{D}\setminus\{\zeta\}$. We have $\phi_k(z) = \Psi_k(z) + \Omega_k(z)$, hence the values

$$\overline{\phi_k\left(z_k^*\right)} - \overline{\phi_k\left(w_k^*\right)} = \overline{\Psi_k\left(z_k^*\right)} - \overline{\Psi_k\left(w_k^*\right)} + \overline{\Omega_k\left(z_k^*\right)} - \overline{\Omega_k\left(w_k^*\right)} \tag{3.41}$$

are completely determined. It follows from (3.38) that

$$M_0(z,\zeta) = \sum_{m=1}^{n} \alpha_m(\zeta)\psi_m(z) - \omega(z,\zeta),$$

where the functions $\psi_m(z)$ have the form (3.21) or (3.22), $\alpha_m(\zeta)$ are given in Theorem 3.3,

$$\omega(z,\zeta) = \ln\left(\prod_{k=1}^{n} \overline{\frac{\zeta - z_{(k)}^*}{\zeta - w_{(k)}^*}}\right)\left(\prod_{k=1}^{n}\prod_{k_1\neq k} \frac{\zeta - w_{(k_1k)}^*}{\zeta - z_{(k_1k)}^*}\right)\left(\prod_{k=1}^{n}\prod_{k_1\neq k}\prod_{k_2\neq k_1} \overline{\frac{\zeta - z_{(k_2k_1k)}^*}{\zeta - w_{(k_2k_1k)}^*}}\right)\cdots. \tag{3.42}$$

This infinite product can be represented in the form

$$\omega(z,\zeta) = \ln \prod_{j=1}^{\infty} \omega_j(z,\zeta), \tag{3.43}$$

where

$$\omega_j(z,\zeta) = \begin{cases} \frac{\zeta-\gamma_j(z)}{\zeta-\gamma_j(w)}, & \text{if level of } \gamma_j \text{ is even,} \\ \overline{\frac{\zeta-\gamma_j(\overline{w})}{\zeta-\gamma_j(\overline{z})}}, & \text{if level of } \gamma_j \text{ is odd.} \end{cases}$$

In order to find $A(\zeta)$ we substitute $w^*_{(k)}$ in the real part of (3.37) and obtain

$$\begin{aligned} 0 = &-\sum_{m\neq k} \operatorname{Re}\left[\overline{\phi_m\left(\left(w^*_{(k)}\right)^*_{(m)}\right)} - \overline{\phi_m\left(w^*_{(m)}\right)}\right] - \ln\left|\zeta - w^*_{(k)}\right| + A(\zeta) \\ &+ \alpha_k(\zeta)\ln r_k + \sum_{m\neq k} \alpha_m(\zeta) \ln\left|w^*_{(k)} - a_m\right|, \quad k = 1,\ldots,n. \end{aligned} \tag{3.44}$$

The harmonic measures satisfy the equality

$$\sum_{m=1}^{n} \alpha_m(\zeta) = 1. \tag{3.45}$$

One can consider (3.44), (3.45) as a system of $n+1$ real linear algebraic equations with respect to $n+1$ real unknowns $\alpha_1(\zeta)$, $\alpha_2(\zeta), \ldots,$ $\alpha_n(\zeta)$, $A(\zeta)$. The systems (3.44), (3.45) and (3.3), (3.23) have the same homogeneous part. Therefore, the system (3.44), (3.45) has a unique solution We may at the beginning look for the complex Green function $M(z,\zeta)$ with undetermined periods $\alpha_k(\zeta)/2\pi$, find $\alpha_k(\zeta)$ from (3.44), (3.45) and after assert that $\alpha_k(\zeta)$ is a harmonic measure. In order to determine $A(\zeta)$, we fix for instance $k=n$ in (3.44) and obtain

$$\begin{aligned} A(\zeta) = &\sum_{m=1}^{n-1} \operatorname{Re}\left[\overline{\phi_m\left(\left(w^*_{(n)}\right)^*_{(m)}\right)} - \overline{\phi_m\left(w^*_{(m)}\right)}\right] + \ln\left|\zeta - w^*_{(n)}\right| \\ &- \alpha_k(\zeta)\ln r_k - \sum_{m=1}^{n-1} \alpha_m(\zeta)\ln\left|w^*_{(k)} - a_m\right|, \end{aligned} \tag{3.46}$$

where $\left[\overline{\phi_m\left(z^*_{(m)}\right)} - \overline{\phi_m\left(w^*_{(m)}\right)}\right]$ has the form (3.39), (3.40), (3.41).

The function (3.32) is single-valued in $\mathbb{D}$ if and only if

$$\sum_{k=1}^{n} \int_{\mathbb{T}_k} f(\zeta)\frac{\partial \alpha_m}{\partial \nu}(\zeta)d\sigma = 0, \; m = 1,2,\ldots,n. \tag{3.47}$$

Note that one of the relations (3.47) follows from the other ones. For instance, let (3.47) be valid for $m = 1,2,\ldots,n-1$. Then (3.47) for $m=n$ is fulfilled, since

$$\sum_{k=1}^{n} \int_{\mathbb{T}_k} f(\zeta)\frac{\partial \alpha_n}{\partial \nu}(\zeta)d\sigma = -\sum_{m=1}^{n-1}\sum_{k=1}^{n} \int_{\mathbb{T}_k} f(\zeta)\frac{\partial \alpha_m}{\partial \nu}(\zeta)d\sigma = 0.$$

Here the identity (3.45) is used. With the help of (3.35) the single- and the multi-valued components of the Schwarz operator can be separated:

$$T(z,\zeta) = T_s(z,\zeta) + T_m(z,\zeta),$$
$$T_s(z,\zeta) = \sum_{m=1}^{n} \frac{\partial \alpha_m}{\partial \nu}(\zeta)\left[\psi_m(z) + \ln(z-a_m)\right],$$
$$T_m(z,\zeta) = \frac{\partial \omega}{\partial \nu}(z,\zeta) - \frac{1}{\zeta - z}\frac{\partial \zeta}{\partial \nu} + \frac{\partial A}{\partial \nu}(\zeta).$$

We now proceed to calculate the normal derivatives in the later formulae. One can see that

$$\frac{\partial f}{\partial \nu} d\sigma = -\frac{1}{i}\left[\frac{\partial f}{\partial \zeta} + \left(\frac{r_k}{\zeta - a_k}\right)^2 \frac{\partial f}{\partial \overline{\zeta}}\right] d\tau, \; |\zeta - a_k| = r_k, \tag{3.48}$$

for any $f \in \mathcal{C}^1(\partial\mathbb{D})$. Recall that we deal with the outward normal to $\mathbb{D}$. In order to apply (3.48) to $\omega(z,\zeta)$ we find from (3.42) that

$$\begin{aligned}\frac{\partial \omega}{\partial \zeta}(z,\zeta) &= \sum_{k=1}^{n}\sum_{k_1 \neq k}\left(\frac{1}{\zeta - w^*_{(k_1 k)}} - \frac{1}{\zeta - z^*_{(k_1 k)}}\right)\\ &\quad + \sum_{k=1}^{n}\sum_{k_1 \neq k}\sum_{k_2 \neq k_1}\sum_{k_3 \neq k_2}\left(\frac{1}{\zeta - w^*_{(k_3 k_2 k_1 k)}} - \frac{1}{\zeta - z^*_{(k_3 k_2 k_1 k)}}\right) + \cdots\\ &= \sum_{j=1}^{\infty}{}'' \left(\frac{1}{\zeta - \gamma_j(w)} - \frac{1}{\zeta - \gamma_j(z)}\right),\end{aligned} \tag{3.49}$$

where the terms in the later sum are ordered due to increasing even level. We also have

$$\frac{\partial \omega}{\partial \overline{\zeta}}(z,\tau) = \sum_{j=1}^{\infty}{}' \left(\frac{1}{\overline{\zeta - \gamma_j(\overline{z})}} - \frac{1}{\overline{\zeta - \gamma_j(\overline{w})}}\right), \tag{3.50}$$

where elements γ_j have the odd level. Substituting (3.49), (3.50) into (3.35), (3.32) we arrive at the following

Theorem 3.6. *The Schwarz operator of* $\mathbb{D}$ *has the form*

$$\begin{aligned}\phi(z) &= \frac{1}{2\pi i}\sum_{k=1}^{n}\int_{\mathbb{T}_k} f(\zeta)\left\{\sum_{j=2}^{\infty}{}'' \left[\frac{1}{\zeta - \gamma_j(w)} - \frac{1}{\zeta - \gamma_j(z)}\right]\right.\\ &\quad \left. + \left(\frac{r_k}{\zeta - a_k}\right)^2 \sum_{j=1}^{\infty}{}' \left[\frac{1}{\overline{\zeta - \gamma_j(\overline{z})}} - \frac{1}{\overline{\zeta - \gamma_j(\overline{w})}}\right] - \frac{1}{\zeta - z}\right\} d\zeta\\ &\quad + \frac{1}{2\pi i}\sum_{k=1}^{n}\int_{\mathbb{T}_k} f(\zeta)\frac{\partial A}{\partial \nu}(\zeta) d\sigma + \sum_{m=1}^{n} A_m\left[\ln(z-a_m) + \psi_m(z)\right] + i\varsigma,\end{aligned} \tag{3.51}$$

where

$$A_m := \frac{1}{2\pi i}\sum_{k=1}^{n}\int_{\mathbb{T}_k} f(\zeta)\frac{\partial \alpha_m}{\partial \nu}(\zeta)d\sigma,\ m = 1, 2, \ldots, n,$$

$A(\zeta)$ has the form (3.46), *The functions $\alpha_m(\zeta)$ and $\psi_m(z)$ are derived in Theorem* 3.3, *ς is an arbitrary real constant, $\sum'$ contains γ_j of odd level, $\sum''$ – of even level. The series converges uniformly in each compact subset of $\mathbb{D} \cup \partial\mathbb{D} \backslash \{\infty\}$.*

The single-valued part of the Schwarz operator can be determined by solution of the modified Dirichlet problem (see [Mityushev and Rogosin (2000)], Sec. 2.7.2):

$$\operatorname{Re}\phi(t) = f(t) + c_k,\ t \in \mathbb{T}_k,\ k = 1, 2, \ldots, n, \tag{3.52}$$

where a given function $f \in \mathcal{C}(\partial\mathbb{D})$, c_k are undetermined real constants. If one of the constants c_k is fixed arbitrarily, the remaining ones are determined uniquely and $\phi(z)$ is determined up to an arbitrary additive purely imaginary constant (see [Mityushev and Rogosin (2000)], Sec. 2.7.2). Thus, we have

Theorem 3.7. *The single-valued part of the Schwarz operator of $\mathbb{D}$ corresponding to the modified Dirichlet problem* (3.52) *has the form*

$$\begin{aligned}\phi(z) = \frac{1}{2\pi i}\sum_{k=1}^{n}\int_{\mathbb{T}_k} (f(\zeta) + c_k)\Bigg\{\sum_{j=2}^{\infty}{}'' \left[\frac{1}{\zeta - \gamma_j(w)} - \frac{1}{\zeta - \gamma_j(z)}\right] \\ + \left(\frac{r_k}{\zeta - a_k}\right)^2 \sum_{j=1}^{\infty}{}' \left[\frac{1}{\overline{\zeta - \gamma_j(\overline{z})}} - \frac{1}{\overline{\zeta - \gamma_j(\overline{w})}}\right] - \frac{1}{\zeta - z}\Bigg\} d\zeta \\ + \frac{1}{2\pi i}\sum_{k=1}^{n}\int_{\mathbb{T}_k} f(\zeta)\frac{\partial A}{\partial \nu}(\zeta)d\sigma + i\varsigma.\end{aligned} \tag{3.53}$$

One of the real constants c_k can be fixed arbitrarily, the remaining ones are determined uniquely from the linear algebraic system

$$\sum_{k=1}^{n}\int_{\mathbb{T}_k} (f(\zeta) + c_k)\frac{\partial \alpha_m}{\partial \nu}(\zeta)d\sigma = 0,\ m = 1, 2, \ldots, n-1. \tag{3.54}$$

4. ℝ-linear problem

4.1. Integral equations

There are two different methods of integral equations associated to boundary value problems. The first method is known as the method of potentials. In complex analysis, it is equivalent to the method of singular integral equations [Gakhov (1977), Muskhelishvili (1968), Muskhelishvili (1966), Vekua (1988)]. The alternating method of Schwarz can be presented as a method of integral equations of another type [Mikhlin (1964), Mikhailov (1963)]. Let $L_k = \partial D_k$ be Lyapunov's simple closed curves. It is convenient to introduce the opposite orientation to the orientation considered in the above sections. So, it is assumed that each L_k leaves the inclusion D_k on the left. In the present section, we discuss the following $\mathbb{R}$-linear problem corresponding to the perfect contact between components of the composite with the external field $f(t)$,

$$\varphi^-(t) = \varphi_k(t) - \rho_k\overline{\varphi_k(t)} - f(t), \quad t \in L_k \ (k = 1, 2, \dots, n). \tag{4.1}$$

Here, the contrast parameter $\rho_k = \frac{\lambda_k+\lambda}{\lambda_k-\lambda}$ is introduced via the conductivity of the host λ and the conductivity of the kth inclusion λ_k. Introduce a space $\mathcal{H}(D^+)$ consisting of functions analytic in $D^+ = \cup_{k=1}^n D_k$ and Hölder continuous in the closure of D^+ endowed with the norm

$$||\omega|| = \sup_{t\in L} |\omega(t)| + \sup_{t_{1,2}\in L} \frac{|\omega(t_1)| - \omega(t_2)|}{|t_1 - t_2|^\alpha}, \tag{4.2}$$

where $0 < \alpha \leq 1$. The space $\mathcal{H}(D^+)$ is Banach, since the norm in $\mathcal{H}(D^+)$ coincides to the norm of functions Hölder continuous on L (inf on $D^+ \cup L$ in (4.2) is equal to inf on L). It follows from Harnack's principle that convergence in the space $\mathcal{H}(D^+)$ implies uniform convergence in the closure of D^+.

For fixed m introduce the operator

$$A_m f(z) = \frac{1}{2\pi i}\int_{L_m} \frac{f(t)dt}{t-z}, \ z \in D_m. \tag{4.3}$$

In accordance with Sokhotskij's formulae,

$$A_m f(\zeta) = \lim_{z\to\zeta} A_m f(z) = \frac{1}{2} f(\zeta) + \frac{1}{2\pi i}\int_{L_m} \frac{f(t)dt}{t-\zeta}, \ \zeta \in L_m. \tag{4.4}$$

Equations (4.3)–(4.4) determine the operator A_m in the space $\mathcal{H}(D_m)$.

Lemma 4.1. *The linear operator A_m is bounded in the space $\mathcal{H}(D_m)$.*

The proof is based on a definition of the bounded operator $||A_m f|| \leq C||f||$ and the fact that the norm in $\mathcal{H}(D_m)$ is equal to the norm of functions Hölder continuous on L_m. The estimation of the later norm follows from the boundness of the operator (4.4) in Hölder's space [Gakhov (1977)].

The lemma is proved.

The conjugation condition (4.1) can be written in the form

$$\varphi_k(t) - \varphi^-(t) = \rho_k\overline{\varphi_k(t)} + f(t), \quad t \in L_k \ (k = 1, 2, \dots, n). \tag{4.5}$$

A difference of functions analytic in D^+ and in D is in the left-hand part of the later relation. Then application of Sokhotskij's formulae yield

$$\varphi_k(z) = \sum_{m=1}^n \frac{\rho_m}{2\pi i}\int_{L_m} \frac{\overline{\varphi_m(t)}}{t-z} dt + f_k(z), \quad z \in D_k \ (k = 1, 2, \dots, n), \tag{4.6}$$

where the function

$$f_k(z) = \frac{\lambda}{\pi i(\lambda_k + \lambda)} \sum_{m=1}^n \int_{L_m} \frac{f(t)}{t-z}\, dt$$

is analytic in D_k and Hölder continuous in its closure.

The integral equations (4.6) can be continued to L_k as follows:

$$\varphi_k(z) = \sum_{m=1}^{n} \rho_m \left[\frac{\overline{\varphi_k(z)}}{2} + \frac{1}{2\pi i} \int_{L_m} \frac{\overline{\varphi_m(t)}}{t-z} dt \right] + f_k(z),\ z \in L_k\ (k = 1, 2, \ldots, n). \tag{4.7}$$

One can consider equations (4.6), (4.7) as an equation with linear bounded operator in the space $\mathcal{H}(D^+)$.

Equations (4.6), (4.7) correspond to the generalized method of Schwarz. Write, for instance, equation (4.6) in the form

$$\varphi_k(z) - \frac{\rho_k}{2\pi i} \int_{L_k} \frac{\overline{\varphi_k(t)}}{t-z} dt = \sum_{m \neq k} \frac{\rho_m}{2\pi i} \int_{L_m} \frac{\overline{\varphi_m(t)}}{t-z} dt + f_k(z),\ z \in D_k\ (k = 1, 2, \ldots, n). \tag{4.8}$$

At the zeroth approximation we arrive at the problem for the single inclusion D_k $(k = 1, 2, \ldots, n)$,

$$\varphi_k(z) - \frac{\rho_k}{2\pi i} \int_{L_k} \frac{\overline{\varphi_k(t)}}{t-z} dt = f_k(z),\ z \in D_k. \tag{4.9}$$

Let problem (4.9) be solved. Further, its solution is substituted into the right-hand part of (4.8). Then we arrive at the first-order problem etc. Therefore, the generalized method of Schwarz can be considered as a method of implicit iterations applied to integral equations (4.6), (4.7).

In the case of circular domains the integral term from the left-hand part of (4.8) becomes a constant:

$$\frac{\rho_k}{2\pi i} \int_{L_k} \frac{\overline{\varphi_k(t)}}{t-z} dt = \rho_k \overline{\varphi_k(0)},$$

since $\overline{\varphi_k(t)}$ is analytically continued out of the circle L_k.

Remark 4.2. An integral equation method was proposed in Chapter 4 of [Mityushev and Rogosin (2000)] for the Dirichlet problem. A convergent direct iteration method for these equations coincides with the modified method of Schwarz. However, the integral terms of this method contain Green's functions of the domains D_k which should be constructed. One can obtain similar equations by application of the operator $\mathcal{S}_k^{-1}$ to both sides of (4.8), where the operator $\mathcal{S}_k$ solves equation (4.9).

4.2. Method of successive approximations

We use the following general result.

Theorem 4.3 ([Krasnosel'skii et al. (1969)]). *Let A be a linear bounded operator in a Banach space $\mathcal{B}$. If for any element $f \in \mathcal{B}$ and for any complex number ν satisfying the inequality $|\nu| \leq 1$ equation*

$$x = \nu A x + f \tag{4.10}$$

has a unique solution, then the unique solution of the equation

$$x = Ax + f \tag{4.11}$$

can be found by the method of successive approximations. The approximations converge in $\mathcal{B}$ to the solution

$$x = \sum_{k=0}^{\infty} A^k f. \tag{4.12}$$

A weaker form of this theorem, valid for compact operators, is used in the proof of Lemma 2.2. Theorem 4.3 can be applied to equations (4.6), (4.7).

Theorem 4.4. *Let* $|\rho_k| < 1$. *Then the system of equations* (4.6), (4.7) *has a unique solution. This solution can be found by the method of successive approximations convergent in the space* $\mathcal{H}(D^+)$.

Proof. Let $|\nu| \leq 1$. Consider equations in $\mathcal{H}(D^+)$,

$$\varphi_k(z) = \nu \sum_{m=1}^{n} \frac{\rho_m}{2\pi i} \int_{L_m} \frac{\overline{\varphi_m(t)}}{t-z} dt + f_k(z), \quad z \in D_k \ (k = 1, 2, \ldots, n). \tag{4.13}$$

Equations on L_k look like (4.7).

Let $\varphi_k(z)$ be a solution of (4.13). Introduce the function

$$\varphi(t) = \varphi_k(t) - \nu \rho_k \overline{\varphi_k(t)} - f_k(t), \quad t \in L_k \ (k = 1, 2, \ldots, n). \tag{4.14}$$

Calculate the integral

$$I = \frac{1}{2\pi i} \int_L \frac{\varphi(t)}{t-z} dt = \sum_{m=1}^{n} \frac{1}{2\pi i} \int_{L_m} \frac{\varphi_m(t) - \nu \rho_m \overline{\varphi_m(t)} - f_m(t)}{t-z} dt, \ z \in D_k. \tag{4.15}$$

Taking into account (4.13), formulae

$$\frac{1}{2\pi i} \int_{L_m} \frac{\varphi_m(t)}{t-z} dt = 0, \ m \neq k, \ \frac{1}{2\pi i} \int_{L_k} \frac{\varphi_k(t)}{t-z} dt = \varphi_k(z), \ z \in D_k \tag{4.16}$$

and analogous formulae for $f_m(z)$, we obtain $I = 0$. The latter equality implies that $\varphi(t)$ is analytically continued into D. In accordance with Corollary to Theorem 1.1, the $\mathbb{R}$-linear problem (4.14) has a unique solution. This unique solution is the unique solution of the system (4.13).

Theorem 4.3 yields convergence of the method of successive approximations applied to the system (4.13).

This completes the proof of the theorem. □

5. Conclusion

Though the method of integral equations discussed in Section 4.2 is rather a numerical method, application of the residua for special shapes of the inclusions transforms the integral terms to compositions of the functions. Therefore, at least for the boundaries expressed by algebraic functions, one should arrive at the functional equations. An example concerning elliptical inclusions is presented in [Mityushev

(2009)]. This approach can be considered as a generalization of Grave's method reviewed in [Apel'tsin (2000)] to multiply connected domains.

In order to understand the place of the convergence results obtained in this paper, we return to Section 1.2. It was established in the previous works that for $|b(t)| < |a(t)|$ the problem has a unique solution. If the stronger condition (1.11) is fulfilled (always $S_p \geq 1$), this unique solution can be constructed by the absolutely convergent method of successive approximations. Absolute convergence implies geometrical restrictions on the geometry which can be roughly presented as follows. Each inclusion D_k is sufficiently far away from other inclusions D_m ($m \neq k$). Only after the results presented in Section 4 of the present paper does the situation become clear and simplified. In the case (1.10) the method of successive approximations can be also applied, but absolute convergence is replaced by uniform convergence. The same story with convergence repeats for other methods and problems. In all previous works beginning from Poincaré's investigations, i.e., the Schwarz operator, the Poincaré series, the Riemann–Hilbert problem, the modified alternating Schwarz method etc., all the relevant problems were studied by absolute convergent methods under geometrical restrictions. The main result of the present paper is based on the modification of these methods and study of the problems by uniform convergence methods. This replacement of absolute convergence by uniform convergence abandons all previous geometrical restrictions and yields solution to the problems and convergence of the methods for an arbitrary location of non-overlapping inclusions.

This complicated situation concerning absolute and uniform convergences can be illustrated by a simple example. Let the almost uniformly convergent series $\sum_{n=1}^{\infty}(n-z)^{-2}$ ($z \notin \mathbb{N}$) be integrated term by term,

$$\int_w^z \sum_{n=1}^{\infty} \frac{1}{(n-t)^2}\, dt = \sum_{n=1}^{\infty} \left(\frac{1}{n-z} - \frac{1}{n-w} \right).$$

One can see that this series can be convergent if and only if $w \neq \infty$. This unlucky infinity is sometimes taken as a fixed point in similar investigations by specialists in complex analysis (see for instance Michlin's study [Mikhlin (1964)] devoted to convergence of Schwarz's method).

For engineers it is interesting to get exact and approximate formulae for the effective conductivity tensor. One can find a description of such formulae based on the solution to the problems discussed in the present paper in the survey [Mityushev et al. (2008)].

References

[Akaza (1966)] Akaza, T. (1966). Singular sets of some Kleinian groups, *Nagoya Math. J.*, **26**, pp. 127–143. 155

[Akaza and Inoue (1984)] Akaza, T. and Inoue, K. (1984). Limit sets of geometrically finite free Kleinian groups, *Tohoku Math. J.*, **36**, pp. 1–16. 155

[Akhiezer (1990)] Akhiezer, N.I. (1990). *Elements of the Theory of Elliptic Functions.* (AMS, Providence). 148

[Aleksandrov and Sorokin (1972)] Aleksandrov, I.A. and Sorokin, A.S. (1972). The problem of Schwarz for multiply connected domains, *Sib. Math. Zh.*, **13**, pp. 971–1001 (in Russian). 153

[Apel'tsin (2000)] Apel'tsin, F.A. (2000). A generalization of D.A. Grave's method for plane boundary-value problems in harmonic potentials theory, *Comput. Math. Modeling*, **11**, pp. 1–14. 173

[Baker (1996)] Baker, H.F. (1996). *Abelian Functions*, (Cambridge University Press). 148

[Bancuri (1975)] Bancuri, R.D. (1975). On the Riemann–Hilbert problem for doubly connected domains, *Soobsch. AN GruzSSR*, **80**, pp. 549–552 (in Russian). 149

[Bojarski (1958)] Bojarski, B. (1958). On a boundary value problem of the theory of analytic functions, *Dokl. AN SSSR*, **119**, pp. 199–202 (in Russian).

[Bojarski (1960)] Bojarski, B. (1960). On generalized Hilbert boundary value problem, *Soobsch. AN GruzSSR*, **25**, pp. 385–390 (in Russian). 151

[Burnside (1891)] Burnside, W. (1891). On a Class of Automorphic Functions, *Proc. London Math. Soc.* **23**, pp. 49–88. 155

[Koppenfels and Stallman (1959)] Koppenfels, W. and Stallman, F. (1959). *Practice of conformal mapping*, (Springer Verlag). 148

[Chibrikova (1977)] Chibrikova, L.I. (1977). *Basic Boundary-Value Problems for Analytic Functions*, (Kazan Univ. Publ.) (in Russian).

[Crowdy (2009)] Crowdy, D. (2009). Explicit solution of a class of Riemann–Hilbert problems, *Annales Universitatis Paedagogicae Cracoviensis*, **8**, pp. 5–18. 148

[Crowdy (2008a)] Crowdy, D. (2008). The Schwarz problem in multiply connected domains and the Schottky–Klein prime function, *Complex Variables and Elliptic Equations*, **53**, pp. 221–236. 148

[Crowdy (2008b)] Crowdy, D. (2008) Geometric function theory: a modern view of a classical subject, *Nonlinearity*, **21**, pp. T205–T219. 148, 155

[Dunduchenko (1966)] Dunduchenko, L.E. (1966). On the Schwarz formula for an n-connected domain, *Dopovedi AN URSR*, n. 5, pp. 1386–1389 (in Ukranian). 153

[Gakhov (1977)] Gakhov, F.D. (1977). *Boundary Value Problems*, 3rd edn. (Nauka, Moscow) (in Russian); Engl. transl. of 1st edn. (Pergamon Press). 147, 148, 149, 157, 169, 170

[Golusin (1934)] Golusin, G.M. (1934). Solution of basic plane problems of mathematical physics for the case of Laplace equation and multiply connected domains bounded by circles (method of functional equations), *Math. zb.* **41**, pp. 246–276. 153

[Golusin (1934)] Golusin, G.M. (1934). Solution of spatial Dirichlet problem for Laplace equation and for domains enbounded by finite number of spheres, *Math. zb.* **41** pp. 277–283. 153

[Golusin (1935)] Golusin, G.M. (1935). Solution of plane heat conduction problem for multiply connected domains enclosed by circles in the case of isolated layer, *Math. zb.* **42**, pp. 191–198. 151, 153

[Golusin (1969)] Golusin, G.M. (1969). *Geometric Theory of Functions of Complex Variable.* (AMS, Providence). 150

[Kantorovich and Krylov (1958)] Kantorovich, L.V. and Krylov, V.I. (1958). *Approximate methods of higher analysis*, (Groningen, Noordhoff). 153

[Krasnosel'skii et al. (1969)] Krasnosel'skii, M.A., Vainikko, G.M., Zabreiko, P.P., Rutickii, Ja.B. and Stecenko, V.Ja. (1969). *Approximate Methods for Solution of Operator Equations*, (Nauka, Moscow) (in Russian); Engl. transl. (Wolters-Noordhoff Publ). 157, 171

[Kveselava (1945)] Kveselava, D.A. (1945). Riemann–Hilbert problem for multiply connected domain, *Soobsch. AN GruzSSR* **6**, pp. 581–590 (in Russian). 149

[Kuczma et al. (1990)] Kuczma, M., Chosewski, B. and Ger, R. (1990). *Iterative functional equations*, (Cambridge University Press). 148

[Markushevich (1946)] Markushevich, A.I. (1946) On a boundary value problem of analytic function theory, *Uch. zapiski MGU*, **1**, n. 100, pp. 20–30 (in Russian). 151

[Mikhailov (1961)] Mikhailov, L.G. (1961). On a boundary value problem, *DAN SSSR*, **139**, pp. 294–297 (in Russian). 152

[Mikhailov (1963)] Mikhailov, L.G. (1963). *New Class of Singular Integral Equations and its Applications to Differential Equations with Singular Coefficients.* (AN TadzhSSR, Dushanbe) (in Russian); English transl.: (Akademie Verlag). 150, 151, 152, 169

[Mikhlin (1964)] Mikhlin, S.G. (1964). *Integral Equations*, (Pergamon Press). 153, 164, 169, 173

[Mityushev (1994)] Mityushev, V.V. (1994). Solution of the Hilbert boundary value problem for a multiply connected domain, *Slupskie Prace Mat.-Przyr.* **9a** pp. 37–69. 150

[Mityushev (1994)] Mityushev, V.V. (1995). Generalized method of Schwarz and addition theorems in mechanics of materials containing cavities, *Arch. Mech.* **47**, pp. 1169–1181. 153

[Mityushev (1998)] Mityushev, V.V. (1998). Convergence of the Poincaré series for classical Schottky groups, *Proceedings Amer. Math. Soc.* **126** pp. 2399–2406. 148, 155, 156

[Mityushev (1998)] Mityushev, V.V. (1998). Hilbert boundary value problem for multiply connected domains, *Complex Variables* **35** pp. 283–295. 150

[Mityushev and Rogosin (2000)] Mityushev, V.V. and Rogosin, S.V. (2000). *Constructive methods to linear and non-linear boundary value problems of the analytic function. Theory and applications*, (Chapman & Hall / CRC, Monographs and Surveys in Pure and Applied Mathematics). 147, 148, 150, 151, 153, 155, 157, 158, 164, 165, 169, 171

[Mityushev et al. (2008)] Mityushev, V.V., Pesetskaya, E. and Rogosin S.V. (2008). Analytical Methods for Heat Conduction in Composites and Porous Media in *Cellular and Porous Materials: Thermal Properties Simulation and Prediction*, (Wiley, Andreas) Öchsner, Graeme E. Murch, Marcelo J.S. de Lemos (eds.). 173

[Mityushev (2009)] Mityushev, V. (2009). Conductivity of a two-dimensional composite containing elliptical inclusions, *Proc. R. Soc.* **A465**, pp. 2991–3010. 172

[Myrberg (1916)] Myrberg, P.J. (1916) Zur Theorie der Konvergenz der Poincaréschen Reihen, *Ann. Acad. Sci. Fennicae*, **A9**, 4, pp. 1–75. 155

[Muskhelishvili (1932)] Muskhelishvili, N.I. (1932). To the problem of torsion and bending of beams constituted from different materials, *Izv. AN SSSR*, n. 7, pp. 907–945 (in Russian). 151, 152

[Muskhelishvili (1968)] Muskhelishvili, N.I. (1968). *Singular Integral Equations*, 3th edn. (Nauka, Moscow) (in Russian); English translation of the 1st edn. (P. Noordhoff N.V., Groningen). 147, 148, 149, 151, 169

[Muskhelishvili (1966)] Muskhelishvili, N.I. (1966) *Some Basic Problems of Mathematical Elasticity Theory.*, 5th edn. (Nauka, Moscow) (in Russian); English translation of the 1st edn. (Noordhoff, Groningen). 151, 169

[Poincaré (1916)] Poincaré, H. (1916). *Oeuvres*, (Gauthier-Villars, Paris). 155

[Smith et al. (1996)] Smith, B., Björstad, P. and Gropp, W. (1996) *Domain decomposition. Parallel multilevel methods for elliptic partial differential equations.* (Cambridge University Press). 153

[Vekua and Rukhadze (1933)] Vekua, I.N. and Rukhadze, A.K. (1933). The problem of the torsion of circular cylinder reinforced by transversal circular beam. *Izv. AN SSSR*, n. 3, pp. 373–386. 151

[Vekua and Rukhadze (1933)] Vekua, I.N. and Rukhadze, A.K. (1933). Torsion and transversal bending of the beam compounded by two materials restricted by confocal ellipses, *Prikladnaya Matematika i Mechanika* (Leningrad) **1**, 2, pp. 167–178. 151

[Vekua (1988)] Vekua, I.N. (1988). *Generalized Analytic Functions*, (Nauka, Moscow) (in Russian). 147, 149, 169

[Vekua (1967)] Vekua, N.P. (1967). *Systems of Singular Integral Equations*, 2nd edn. (Noordhoff, Groningen). 151

[Zmorovich (1958)] Zmorovich, V.A. (1958). On a generalization of the Schwarz integral formula on n-connected domains, *Dopovedi URSR*, n. 5, pp. 489–492 (in Ukranian). 153

[Zverovich (1971)] Zverovich, E.I. (1971). Boundary value problems of analytic functions in Hölder classes on Riemann surfaces, *Uspekhi Mat. nauk*, **26** (1), pp. 113–179 (in Russian). 148, 150

Index

Vladimir V. Mityushev
Department of Computer Sciences and Computer Methods
Pedagogical University ul. Podcharazych 2
Krakow 30-084, Poland
e-mail: `vmityushev@gmail.com`

Advances in Applied Analysis
Trends in Mathematics, 177–223

Commutative Algebras Associated with Classic Equations of Mathematical Physics

S.A. Plaksa

Abstract. The idea of an algebraic-analytic approach to equations of mathematical physics means to find a commutative Banach algebra such that monogenic functions with values in this algebra have components satisfying to given equations with partial derivatives.

We obtain here a constructive description of monogenic functions taking values in a commutative algebra associated with a two-dimensional biharmonic equation by means of analytic functions of complex variables. For the mentioned monogenic functions we establish basic properties analogous to properties of analytic functions of complex variables: the Cauchy integral theorem and integral formula, the Morera theorem, the uniqueness theorem, and the Taylor and Laurent expansions. Similar results are obtained for monogenic functions which take values in a three-dimensional commutative algebra and satisfy the three-dimensional Laplace equation.

In infinite-dimensional commutative Banach algebras we construct explicitly monogenic functions which have components satisfying the three-dimensional Laplace equation. We establish that all spherical functions are components of the mentioned monogenic functions. A relation between these monogenic functions and harmonic vectors is described.

We establish that solutions of elliptic equations degenerating on an axis are constructed by means of components of analytic functions taking values in an infinite-dimensional commutative Banach algebra. In such a way we obtain integral expressions for axial-symmetric potentials and Stokes flow functions in an arbitrary simply connected domain symmetric with respect to an axis.

Mathematics Subject Classification (2010). Primary 30G35; Secondary 35J05, 31A30.

Keywords. Laplace equation; biharmonic equation; harmonic vector; axial-symmetric potential; Stokes flow function; harmonic commutative Banach algebra; biharmonic algebra; monogenic function; Cauchy–Riemann conditions; Cauchy integral theorem; Cauchy integral formula; Morera theorem; Taylor expansion; Laurent expansion.

1. Algebras associated with the Laplace equation

1.1. Spatial stationary potential solenoid vector field

Consider a spatial stationary vector field defined by means of the vector-function $\mathbf{V} \equiv \mathbf{V}(x,y,z)$ of the Cartesian coordinates x, y, z. The vector $\mathbf{V}$ is defined by means of three real scalar functions $v_1 := v_1(x,y,z)$, $v_2 := v_2(x,y,z)$, $v_3 := v_3(x,y,z)$ which give its coordinates in the point (x,y,z), videlicet: $\mathbf{V} = (v_1, v_2, v_3)$.

Defining a *potential solenoid* field in a simply connected domain Q of the three-dimensional real space $\mathbb{R}^3$, the vector-function $\mathbf{V}$ satisfies the system of equations

$$\operatorname{div} \mathbf{V} = 0\,, \qquad \operatorname{rot} \mathbf{V} = 0\,, \tag{1.1}$$

that we rewrite also in expanded form:

$$\begin{aligned}
&\frac{\partial v_1}{\partial x} + \frac{\partial v_2}{\partial y} + \frac{\partial v_3}{\partial z} = 0\,,\\
&\frac{\partial v_3}{\partial y} - \frac{\partial v_2}{\partial z} = 0\,,\\
&\frac{\partial v_1}{\partial z} - \frac{\partial v_3}{\partial x} = 0\,,\\
&\frac{\partial v_2}{\partial x} - \frac{\partial v_1}{\partial y} = 0\,.
\end{aligned} \tag{1.2}$$

Then there exists a scalar potential function $u(x,y,z)$ such that

$$\mathbf{V} = \operatorname{grad} u := \left(\frac{\partial u}{\partial x}, \frac{\partial u}{\partial y}, \frac{\partial u}{\partial z}\right),$$

and u satisfies the three-dimensional Laplace equation

$$\Delta_3 u(x,y,z) := \left(\frac{\partial^2}{\partial x^2} + \frac{\partial^2}{\partial y^2} + \frac{\partial^2}{\partial z^2}\right) u(x,y,z) = 0\,. \tag{1.3}$$

Doubly continuously differentiable functions satisfying equation (1.3) are called *harmonic* functions, and solutions of the system (1.1) are called *harmonic* vectors.

Every harmonic function $u(x,y,z)$ generates a harmonic vector $\mathbf{V} = \operatorname{grad} u$, and the coordinates of vector $\mathbf{V} = (v_1, v_2, v_3)$ are solutions of (1.3).

1.2. A relation between the two-dimensional Laplace equation and the algebra of complex numbers

An important achievement of mathematics is the description of plane potential fields by means of analytic functions of a complex variable.

A potential $u(x,y)$ and a flow function $v(x,y)$ of plane stationary potential solenoid field satisfy the Cauchy–Riemann conditions

$$\frac{\partial u(x,y)}{\partial x} = \frac{\partial v(x,y)}{\partial y}\,, \qquad \frac{\partial u(x,y)}{\partial y} = -\frac{\partial v(x,y)}{\partial x}\,,$$

and they form the complex potential $F(x+iy) = u(x,y)+iv(x,y)$, being an analytic function of complex variable $x+iy$. In turn, every analytic function $F(x+iy)$ satisfies the two-dimensional Laplace equation

$$\Delta_2 F := \frac{\partial^2 F}{\partial x^2} + \frac{\partial^2 F}{\partial y^2} \equiv F''(x+iy)\ (1^2+i^2) = 0$$

owing to the equality $1^2 + i^2 = 0$ for the unit 1 and the imaginary unit i of the algebra of complex numbers.

1.3. Attempts to find an algebra associated with the three-dimensional Laplace equation

Effectiveness of analytic function methods in the complex plane for researching plane potential fields has inspired mathematicians to develop analogous methods for spatial fields.

Apparently, W. Hamilton (1843) made the first attempts to construct an algebra associated with the three-dimensional Laplace equation (1.3) in the sense that components of hypercomplex functions satisfy (1.3). However, Hamilton's quaternions form a noncommutative algebra, and after constructing the quaternion algebra he made no attempt to construct any other algebra (see [1]).

1.4. Harmonic triads in commutative algebras. Harmonic algebras

Let $\mathbb{A}$ be a commutative associative Banach algebra of a rank n ($3 \le n \le \infty$) over either the field of real numbers $\mathbb{R}$ or the field of complex numbers $\mathbb{C}$. Let $\{e_1, e_2, e_3\}$ be a part of the basis of $\mathbb{A}$ and $E_3 := \{\zeta := xe_1 + ye_2 + ze_3 : x, y, z \in \mathbb{R}\}$ be the linear envelope generated by the vectors e_1, e_2, e_3.

A function $\Phi : Q_\zeta \to \mathbb{A}$ is *analytic* in a domain $Q_\zeta \subset E_3$ if in a certain neighborhood of every point $\zeta_0 \in Q_\zeta$ it can be represented in the form of the sum of convergent power series with coefficients belonging to the algebra $\mathbb{A}$:

$$\Phi(\zeta) = \sum_{k=0}^{\infty} c_k\ (\zeta - \zeta_0)^k, \qquad c_k \in \mathbb{A}. \tag{1.4}$$

It is obvious that if the basic elements e_1, e_2, e_3 satisfy the condition

$$e_1^2 + e_2^2 + e_3^2 = 0\,, \tag{1.5}$$

then every analytic function $\Phi : Q_\zeta \to \mathbb{A}$ satisfies equation (1.3), because

$$\Delta_3 \Phi(\zeta) \equiv \Phi''(\zeta)\ (e_1^2 + e_2^2 + e_3^2) = 0\,, \qquad \zeta = xe_1 + ye_2 + ze_3\,, \tag{1.6}$$

where $\Phi''(\zeta)$ can be obtained by a formal double differentiation of the series (1.4), i.e., $\Phi''(\zeta) = \sum\limits_{k=0}^{\infty} k(k-1)\, c_k\, (\zeta - \zeta_0)^{k-2}$.

We say that an algebra $\mathbb{A}$ is *harmonic* (see [2–4]) if in $\mathbb{A}$ there exists a triad of linearly independent vectors $\{e_1, e_2, e_3\}$ satisfying the equality (1.5) provided that $e_k^2 \ne 0$ for $k = 1, 2, 3$. We say also that such a triad $\{e_1, e_2, e_3\}$ is *harmonic*.

P.W. Ketchum [2] considered the C. Segre algebra of quaternions [5] in its relations with the three-dimensional Laplace equation. Indeed, in the Segre algebra of quaternions there is the unit 1, and the multiplication table for the basis $\{1, i, j, k\}$ is of the following form:

$$i^2 = j^2 = -1, \quad k^2 = 1, \quad i\,j = k, \quad i\,k = -j, \quad j\,k = -i.$$

Therefore, there are harmonic triads, in particular: $e_1 = \sqrt{2}, e_2 = i, e_3 = j$.

K.S. Kunz [6] developed a method for a formal construction of solutions (1.3) by using power series in any harmonic algebra over the field $\mathbb{C}$.

1.5. Differentiability in the sense of Gateaux. Monogenic functions

I. P. Mel'nichenko [7] noticed that doubly differentiable, in the sense of Gateaux, functions form the largest class of functions Φ satisfying identically the equality (1.6), where Φ'' is the Gateaux second derivative of the function Φ.

We say that a continuous function $\Phi : \Omega_\zeta \to \mathbb{A}$ is *monogenic* in a domain $\Omega_\zeta \subset E_3$ if Φ is differentiable in the sense of Gateaux in every point of Ω_ζ, i.e., if for every $\zeta \in \Omega_\zeta$ there exists an element $\Phi'(\zeta) \in \mathbb{A}$ such that

$$\lim_{\varepsilon \to 0+0} \left(\Phi(\zeta + \varepsilon h) - \Phi(\zeta)\right) \varepsilon^{-1} = h\Phi'(\zeta) \quad \forall h \in E_3. \tag{1.7}$$

Thus, if the basic elements e_1, e_2, e_3 satisfy the condition (1.5), then every doubly differentiable, in the sense of Gateaux, function $\Phi : \Omega_\zeta \to \mathbb{A}$ satisfies the equality (1.6) in the domain Ω_ζ. In turn, if there exists a doubly differentiable, in the sense of Gateaux, function $\Phi : \Omega_\zeta \to \mathbb{A}$ satisfying the equality (1.6) and the inequality $\Phi''(\zeta) \neq 0$ at least at one point $\zeta := xe_1 + ye_2 + ze_3 \in \Omega_\zeta$, then in this case the condition (1.5) is satisfied.

I.P. Mel'nichenko suggested an algebraic-analytic approach to equations of mathematical physics, which means to find a commutative Banach algebra such that differentiable, in the sense of Gateaux, functions with values in this algebra have components satisfying the given equation with partial derivatives (see [4, 7]). Inasmuch as monogenic functions taking values in a commutative Banach algebra form a functional algebra, note that a relation between these functions and solutions of given equations with partial derivatives is important for constructing mentioned solutions.

1.6. Three-dimensional harmonic algebras

The problem on finding a three-dimensional harmonic algebra $\mathbb{A}$ with the unit 1 was completely solved by I.P. Mel'nichenko [3, 4, 7]. In the paper [7] I.P. Mel'nichenko established that there does not exist a harmonic algebra of third rank with a unit over the field $\mathbb{R}$, but he constructed a three-dimensional harmonic algebra over the field $\mathbb{C}$.

Theorem 1.1 (I. Mel'nichenko, [7]). *The commutative associative algebra $\mathbb{A}$ is harmonic, if the multiplication table for the basis $\{e_1, e_2, e_3\}$ is of the following form:*

$$\begin{aligned} e_k\, e_1 &= e_k, \qquad k = 1,2,3; \\ e_2\, e_2 &= -\frac{1}{2}\, e_1 - \frac{i}{2}\,(\sin\omega)\, e_2 + \frac{i}{2}\,(\cos\omega)\, e_3\,, \\ e_2\, e_3 &= \frac{i}{2}\,(\cos\omega)\, e_2 + \frac{i}{2}\,(\sin\omega)\, e_3\,, \\ e_3\, e_3 &= -\frac{1}{2}\, e_1 + \frac{i}{2}\,(\sin\omega)\, e_2 - \frac{i}{2}\,(\cos\omega)\, e_3\,, \end{aligned} \tag{1.8}$$

where i is the imaginary complex unit and $\omega \in \mathbb{C}$.

Let $\{e_1, e_2, e_3\}$ be a harmonic basis in the algebra $\mathbb{A}$ with the multiplication table (1.8). Associate with a set $Q \subset \mathbb{R}^3$ the set $Q_\zeta := \{\zeta = xe_1 + ye_2 + ze_3 : (x, y, z) \in Q\}$ in E_3.

Theorem 1.2 (I. Mel'nichenko, [7]). *Let $\mathbb{A}$ be a harmonic algebra with the multiplication table* (1.8) *for a basis $\{e_1, e_2, e_3\}$. If a function $\Phi : \Omega_\zeta \to \mathbb{A}$ is monogenic in a domain $\Omega_\zeta \subset E_3$, then the components $U_k : \Omega \to \mathbb{R}$, $V_k : \Omega \to \mathbb{R}$, $k = 1, 2, 3$, of decomposition*

$$\Phi(xe_1 + ye_2 + ze_3) = \sum_{k=1}^{3} U_k(x, y, z)\; e_k + i \sum_{k=1}^{3} V_k(x, y, z)\; e_k, \qquad (x, y, z) \in \Omega\,,$$

generate harmonic vectors $\mathbf{V_1} := (U_1, -\frac{1}{2}\, U_2, -\frac{1}{2}\, U_3)$, $\mathbf{V_2} := (V_1, -\frac{1}{2}\, V_2, -\frac{1}{2}\, V_3)$.

I. Mel'nichenko [3, 4] found all three-dimensional harmonic algebras and developed a method for finding all harmonic bases in these algebras.

Note that there exist only four commutative associative algebras of third rank with unit 1 over the field $\mathbb{C}$. If one chooses nilpotent and idempotent elements generating these algebras, then the multiplication tables will be of the most simple form.

Let $\mathbb{A}_1$ be a semisimple algebra with idempotent elements in the basis $\{\mathcal{I}_1, \mathcal{I}_2, \mathcal{I}_3\}$ and the multiplication table

$$\mathcal{I}_1^2 = I_1, \quad \mathcal{I}_2^2 = \mathcal{I}_2, \quad \mathcal{I}_3^2 = \mathcal{I}_3, \quad \mathcal{I}_1\mathcal{I}_2 = \mathcal{I}_1\mathcal{I}_3 = \mathcal{I}_2\mathcal{I}_3 = 0.$$

Here $1 = \mathcal{I}_1 + \mathcal{I}_2 + \mathcal{I}_3$.

Other algebras contain radicals.

Let $\mathbb{A}_2$ be an algebra with basis $\{\mathcal{I}_1, \mathcal{I}_2, \rho\}$ and multiplication table

$$\mathcal{I}_1^2 = \mathcal{I}_1, \quad \mathcal{I}_2^2 = \mathcal{I}_2, \quad \mathcal{I}_1\mathcal{I}_2 = 0, \quad \rho^2 = 0, \;\; \mathcal{I}_1\rho = 0, \;\; \mathcal{I}_2\rho = \rho.$$

Here $1 = \mathcal{I}_1 + \mathcal{I}_2$, and ρ is a radical of the algebra.

Algebras $\mathbb{A}_3$ and $\mathbb{A}_4$ with the basis $\{1, \rho_1, \rho_2\}$ have no ideals generated by idempotents and ρ_1 and ρ_2 are radicals of these algebras.

The multiplication table in the algebra $\mathbb{A}_3$ is of the form

$$\rho_1^2 = \rho_2, \qquad \rho_2^2 = 0, \qquad \rho_1\rho_2 = 0\,.$$

The multiplication table in the algebra $\mathbb{A}_4$ is of the form

$$\rho_1^2 = \rho_2^2 = \rho_1\rho_2 = 0\,.$$

Theorem 1.3 (I. Mel'nichenko, [3]). *The algebra $\mathbb{A}_4$ is not harmonic. The algebras $\mathbb{A}_1$, $\mathbb{A}_2$, $\mathbb{A}_3$ are harmonic.*

All harmonic bases in the algebras $\mathbb{A}_1$, $\mathbb{A}_2$, $\mathbb{A}_3$ are described (see [4]). Note that in the semisimple algebra $\mathbb{A}_1$, in particular, there exists the family of harmonic bases constructed in Theorem 1.1.

In the algebra $\mathbb{A}_2$ there exists the following family of harmonic bases:

$$\begin{aligned} e_1 &= 1, \\ e_2 &= i(\sin\omega)\,\mathcal{I}_1 + i(\cos\omega)\,\mathcal{I}_2 - (\sin\omega)\,\rho, \\ e_3 &= i(\cos\omega)\,\mathcal{I}_1 - i(\sin\omega)\,\mathcal{I}_2 - (\cos\omega)\,\rho, \end{aligned}$$

and in the algebra $\mathbb{A}_3$ there exists the following family of harmonic bases:

$$\begin{aligned} e_1 &= 1, \\ e_2 &= i\sin\omega + (\cos\omega)\,\rho_1 + i\left(\cos\left(\frac{\pi}{6} - \omega\right)\right)\rho_2, \\ e_3 &= i\cos\omega - (\sin\omega)\,\rho_1 + i\left(\sin\left(\frac{\pi}{6} - \omega\right)\right)\rho_2, \end{aligned}$$

where $\omega \in \mathbb{C}$ (see [3, 4]).

2. Algebraic-analytic properties of monogenic functions in the algebra $\mathbb{A}_3$

2.1. Harmonic bases in the algebra $\mathbb{A}_3$

All harmonic bases in the algebra $\mathbb{A}_3$ are described in Theorem 1.6 [4], videlicet, the following statement is true:

Theorem 2.1. *A basis $\{e_1, e_2, e_3\}$ is harmonic if decompositions of its elements with respect to the basis $\{1, \rho_1, \rho_2\}$ are of the form*

$$\begin{aligned} e_1 &= 1, \\ e_2 &= n_1 + n_2\rho_1 + n_3\rho_2, \\ e_3 &= m_1 + m_2\rho_1 + m_3\rho_2, \end{aligned} \tag{2.1}$$

where n_k and m_k for $k = 1, 2, 3$ are complex numbers satisfying the system of equations

$$\begin{aligned} 1 + n_1^2 + m_1^2 &= 0, \\ n_1 n_2 + m_1 m_2 &= 0, \\ xn_2^2 + m_2^2 + 2(n_1 n_3 + m_1 m_3) &= 0 \end{aligned} \tag{2.2}$$

and the inequality $n_2m_3 - n_3m_2 \neq 0$*, and moreover, at least one of the numbers in each of the pairs* (n_1, n_2) *and* (m_1, m_2) *is not equal to zero. Any harmonic basis in* $\mathbb{A}_3$ *can be obtained as a result of multiplication of elements of harmonic basis* (2.1) *by an invertible element of the algebra* $\mathbb{A}_3$.

For example, if $n_1 = i$, $n_2 = i/2$, $n_3 = m_1 = 0$, $m_2 = -1$, $m_3 = -\sqrt{3}\,i/2$, then we have a harmonic basis $\{e_1^0, e_2^0, e_3^0\}$ with the following decomposition with respect to the basis $\{1, \rho_1, \rho_2\}$:

$$e_1^0 = 1, \qquad e_2^0 = i + \frac{i}{2}\rho_2, \qquad e_3^0 = -\rho_1 - \frac{\sqrt{3}}{2}\, i\rho_2. \tag{2.3}$$

The algebra $\mathbb{A}_3$ has the unique maximal ideal $\mathcal{I} := \{\lambda_1\rho_1 + \lambda_2\rho_2 : \lambda_1, \lambda_2 \in \mathbb{C}\}$ which is also the radical of $\mathbb{A}_3$.

Consider the linear functional $f : \mathbb{A}_3 \to \mathbb{C}$ such that the maximal ideal $\mathcal{I}$ is its kernel and $f(1) = 1$. It is well known [8, p. 135] that f is also a multiplicative functional, i.e., the equality $f(ab) = f(a)f(b)$ is fulfilled for all $a, b \in \mathbb{A}_3$.

2.2. Cauchy–Riemann conditions for functions taking values in the algebra $\mathbb{A}_3$

Let $\{e_1, e_2, e_3\}$ be a harmonic basis of the form (2.1).

Let Ω be a domain in $\mathbb{R}^3$ and $\zeta = x + ye_2 + ze_3$, where $(x, y, z) \in \Omega$. Consider the decomposition

$$\Phi(\zeta) = \sum_{k=1}^{3} U_k(x, y, z)\, e_k\,, \tag{2.4}$$

of a function $\Phi : \Omega_\zeta \longrightarrow \mathbb{A}_3$ with respect to the basis $\{e_1, e_2, e_3\}$, where the functions $U_k : \Omega \longrightarrow \mathbb{C}$ are differentiable in Ω, i.e.,

$$\begin{aligned} &U_k(x + \Delta x, y + \Delta y, z + \Delta z) - U_k(x, y, z) \\ &\quad = \frac{\partial U_k(x, y, z)}{\partial x}\Delta x + \frac{\partial U_k(x, y, z)}{\partial y}\Delta y + \frac{\partial U_k(x, y, z)}{\partial z}\Delta z \\ &\qquad + o\left(\sqrt{(\Delta x)^2 + (\Delta y)^2 + (\Delta z)^2}\right), \qquad (\Delta x)^2 + (\Delta y)^2 + (\Delta z)^2 \to 0\,. \end{aligned}$$

It follows from Theorem 1.3 [4] that the function Φ is monogenic in the domain Ω_ζ if and only if the following Cauchy–Riemann conditions are satisfied in Ω_ζ:

$$\frac{\partial \Phi}{\partial y} = \frac{\partial \Phi}{\partial x} e_2, \qquad \frac{\partial \Phi}{\partial z} = \frac{\partial \Phi}{\partial x} e_3. \tag{2.5}$$

2.3. A constructive description of monogenic functions taking values in the algebra $\mathbb{A}_3$

Below in Section 2, all stated results are obtained jointly with V.S. Shpakivskyi (see also [9]).

Let $\{e_1, e_2, e_3\}$ be a harmonic basis of the form (2.1) and $\zeta = x + ye_2 + ze_3$, where $x, y, z \in \mathbb{R}$.

It follows from the equality

$$(t-\zeta)^{-1} = \frac{1}{t-x-n_1y-m_1z} + \frac{n_2y+m_2z}{(t-x-n_1y-m_1z)^2}\,\rho_1 + \left(\frac{n_3y+m_3z}{(t-x-n_1y-m_1z)^2} + \frac{(n_2y+m_2z)^2}{(t-x-n_1y-m_1z)^3}\right)\rho_2$$
$$\forall t\in\mathbb{C} : t \neq x+n_1y+m_1z \qquad (2.6)$$

(see [4, p. 30]) that the element $\zeta = x+ye_2+ze_3 \in E_3$ is noninvertible in $\mathbb{A}_3$ if and only if the point (x,y,z) belongs to the following straight line in $\mathbb{R}^3$:

$$L: \quad \begin{cases} x+y\Re n_1 + z\Re m_1 = 0, \\ y\Im n_1 + z\Im m_1 = 0\,. \end{cases}$$

We say that the domain $\Omega \subset \mathbb{R}^3$ is *convex in the direction of the straight line* L if Ω contains every segment parallel to L and connecting two points (x_1,y_1,z_1), $(x_2,y_2,z_2)\in\Omega$.

To obtain a constructive description of monogenic functions given in the domain Ω_ζ and taking values in the algebra $\mathbb{A}_3$, consider an auxiliary statement.

Lemma 2.2. *Let a domain $\Omega\subset\mathbb{R}^3$ be convex in the direction of the straight line L and $\Phi:\Omega_\zeta\to\mathbb{A}_3$ be a monogenic function in the domain Ω_ζ. If $\zeta_1,\zeta_2\in\Omega_\zeta$ and $\zeta_2-\zeta_1\in L_\zeta$, then*

$$\Phi(\zeta_1)-\Phi(\zeta_2)\in\mathcal{I}. \qquad (2.7)$$

Proof. Let the segment connecting the points $(x_1,y_1,z_1),(x_2,y_2,z_2)\in\Omega$ be parallel to the straight line L.

Let us construct in Ω two surfaces Q and Σ satisfying the following conditions:

- Q and Σ have the same edge;
- the surface Q contains the point (x_1,y_1,z_1) and the surface Σ contains the point (x_2,y_2,z_2);
- restrictions of the functional f onto the sets Q_ζ and Σ_ζ are one-to-one mappings of these sets onto the same domain G of the complex plane;
- for every $\zeta_0\in Q_\zeta$ (and $\zeta_0\in\Sigma_\zeta$) the equality

  $$\lim_{\varepsilon\to0+0}\left(\Phi(\zeta_0+\varepsilon(\zeta-\zeta_0))-\Phi(\zeta_0)\right)\varepsilon^{-1} = \Phi'(\zeta_0)(\zeta-\zeta_0) \qquad (2.8)$$

  is fulfilled for all $\zeta\in Q_\zeta$ for which $\zeta_0+\varepsilon(\zeta-\zeta_0)\in Q_\zeta$ for all $\varepsilon\in(0,1)$ (or for all $\zeta\in\Sigma_\zeta$ for which $\zeta_0+\varepsilon(\zeta-\zeta_0)\in\Sigma_\zeta$ for all $\varepsilon\in(0,1)$, respectively).

As the surface Q, we can take an equilateral triangle having the center (x_1,y_1,z_1) and apexes A_1,A_2,A_3, and, in addition, the plane of this triangle is perpendicular to the straight line L.

To construct the surface Σ, first, consider a triangle with the center (x_2,y_2,z_2) and apexes A_1',A_2',A_3' such that the segments $A_1'A_2'$, $A_2'A_3'$, $A_1'A_3'$ are parallel to the segments A_1A_2, A_2A_3, A_1A_3, respectively, and, in addition, the length of $A_1'A_2'$ is less than the length of A_1A_2. Inasmuch as the domain Ω is convex in the direction of the straight line L, the prism with vertexes $A_1',A_2',A_3',A_1'',A_2'',A_3''$ is completely contained in Ω, where the points A_1'', A_2'', A_3'' are located in the plane of triangle $A_1A_2A_3$ and the edges $A_m'A_m''$ are parallel to L for $m = \overline{1,3}$.

Further, set a triangle with apexes B_1, B_2, B_3 such that the point B_m is located on the segment $A_m'A_m''$ for $m = \overline{1,3}$ and the truncated pyramid with vertexes $A_1, A_2, A_3, B_1, B_2, B_3$ and lateral edges A_mB_m, $m = \overline{1,3}$, is completely contained in the domain Ω.

At last, in the plane of triangle $A_1'A_2'A_3'$ set a triangle T with apexes C_1, C_2, C_3 such that the segments C_1C_2, C_2C_3, C_1C_3 are parallel to the segments $A_1'A_2'$, $A_2'A_3'$, $A_1'A_3'$, respectively, and, in addition, the length of C_1C_2 is less than the length of $A_1'A_2'$. It is evident that the truncated pyramid with vertexes B_1, B_2, B_3, C_1, C_2, C_3 and lateral edges B_mC_m, $m = \overline{1,3}$, is completely contained in the domain Ω.

Now, as the surface Σ, denote the surface formed by the triangle T and the lateral surfaces of mentioned truncated pyramids

$$A_1A_2A_3B_1B_2B_3 \quad \text{and} \quad B_1B_2B_3C_1C_2C_3.$$

For each $\xi \in G$ define two complex-valued functions H_1 and H_2 so that

$$H_1(\xi) := f(\Phi(\zeta)), \ \text{ where } \xi = f(\zeta) \text{ and } \zeta \in Q_\zeta,$$

$$H_2(\xi) := f(\Phi(\zeta)), \ \text{ where } \xi = f(\zeta) \text{ and } \zeta \in \Sigma_\zeta.$$

Inasmuch as f is a linear continuous multiplicative functional, from the equality (2.8) it follows that

$$\lim_{\varepsilon\to 0+0} \left(f(\Phi(\zeta_0 + \varepsilon(\zeta - \zeta_0))) - f(\Phi(\xi))\right)\varepsilon^{-1} = f(\Phi'(\zeta_0))(f(\zeta) - f(\zeta_0)).$$

Thus, there exist all directional derivatives of the functions H_1, H_2 in the point $f(\zeta_0) \in G$, and, moreover, these derivatives are equal for each of the functions H_1, H_2. Therefore, by Theorem 21 in [10], the functions H_1, H_2 are analytic in the domain G, i.e., they are holomorphic in the case where $\xi = \tau + i\eta$, and they are antiholomorphic in the case where $\xi = \tau - i\eta$, $\tau, \eta \in \mathbb{R}$.

Inasmuch as $H_1(\xi) \equiv H_2(\xi)$ on the boundary of domain G, this identity is fulfilled everywhere in G. Therefore, the equalities

$$f(\Phi(\zeta_2) - \Phi(\zeta_1)) = f(\Phi(\zeta_2)) - f(\Phi(\zeta_1)) = 0,$$

are fulfilled for $\zeta_1 := x_1 + y_1e_2 + z_1e_3$ and $\zeta_2 := x_2 + y_2e_2 + z_2e_3$. Thus, $\Phi(\zeta_2) - \Phi(\zeta_1)$ belongs to the kernel $\mathcal{I}$ of functional f. The lemma is proved. $\square$

Let $D := f(\Omega_\zeta)$ and A be the linear operator which assigns the function $F : D \to \mathbb{C}$ to every monogenic function $\Phi : \Omega_\zeta \to \mathbb{A}_3$ by the formula $F(\xi) := f(\Phi(\zeta))$, where $\zeta = xe_1 + ye_2 + ze_3$ and $\xi := f(\zeta) = x + n_1y + m_1z$. It follows from Lemma 2.2 that the value $F(\xi)$ does not depend on a choice of a point ζ, for which $f(\zeta) = \xi$.

Theorem 2.3. *If a domain $\Omega \subset \mathbb{R}^3$ is convex in the direction of the straight line L, then every monogenic function $\Phi : \Omega_\zeta \to \mathbb{A}_3$ can be expressed in the form*

$$\Phi(\zeta) = \frac{1}{2\pi i} \int\limits_{\Gamma_\zeta} (A\Phi)(t)(t-\zeta)^{-1} dt + \Phi_0(\zeta) \qquad \forall\, \zeta \in \Omega_\zeta\,, \tag{2.9}$$

where Γ_ζ is an arbitrary closed Jordan rectifiable curve in D that is homotopic to the point $f(\zeta)$ and embraces this point, and $\Phi_0 : \Omega_\zeta \to \mathcal{I}$ is a monogenic function taking values in the radical $\mathcal{I}$.

Proof. It is easy to see that the function Φ_0 from (2.9) belongs to the kernel of the operator A, i.e., $\Phi_0(\zeta) \in \mathcal{I}$ for all $\zeta \in \Omega_\zeta$. The theorem is proved. □

Note that the complex number $\xi = f(\zeta)$ is the spectrum of $\zeta \in \mathbb{A}_3$, and the integral in the equality (2.9) is the principal extension (see [8, p. 165]) of analytic function $F(\xi) = (A\Phi)(\xi)$ of the complex variable ξ into the domain Ω_ζ.

It follows from Theorem 2.3 that the algebra of monogenic in Ω_ζ functions is decomposed into the direct sum of the algebra of principal extensions of analytic functions of the complex variable and the algebra of monogenic in Ω_ζ functions taking values in the radical $\mathcal{I}$.

In Theorem 1.7 in [4] the principal extension of analytic function $F : D \to \mathbb{C}$ into the domain $\Pi_\zeta := \{\zeta \in E_3 : f(\zeta) \in D\}$ was explicitly constructed in the form

$$\frac{1}{2\pi i} \int\limits_{\Gamma_\zeta} F(t)(t-\zeta)^{-1} dt = F(x+n_1y+m_1z) + (n_2y+m_2z)F'(x+n_1y+m_1z)\rho_1$$
$$+ \left((n_3y+m_3z)F'(x+n_1y+m_1z) + \frac{(n_2y+m_2z)^2}{2} F''(x+n_1y+m_1z) \right) \rho_2$$
$$\forall \zeta = xe_1 + ye_2 + ze_3 \in \Pi_\zeta\,. \tag{2.10}$$

It is evident that the domain $\Pi \subset \mathbb{R}^3$ congruent to Π_ζ is an infinite cylinder, and its generatrix is parallel to L.

In the following theorem we describe all monogenic functions given in the domain Ω_ζ and taking values in the radical $\mathcal{I}$.

Theorem 2.4. *If a domain $\Omega \subset \mathbb{R}^3$ is convex in the direction of the straight line L, then every monogenic function $\Phi_0 : \Omega_\zeta \to \mathcal{I}$ can be expressed in the form*

$$\Phi_0(\zeta) = F_1(\xi)\,\rho_1 + (F_2(\xi) + (n_2y+m_2z)F_1'(\xi))\,\rho_2$$
$$\forall \zeta = xe_1 + ye_2 + ze_3 \in \Omega_\zeta\,, \tag{2.11}$$

where F_1, F_2 are complex-valued analytic functions in the domain D and $\xi = x + n_1y + m_1z$.

Proof. For a monogenic function Φ_0 of the form

$$\Phi_0(\zeta) = V_1(x,y,z)\rho_1 + V_2(x,y,z)\rho_2\,, \tag{2.12}$$

where $V_k : \Omega \to \mathbb{C}$ for $k = 1, 2$, the Cauchy–Riemann conditions (2.5) are satisfied with $\Phi = \Phi_0$.

Substituting the expressions (2.1), (2.12) into the equalities (2.5) and taking into account the uniqueness of decomposition of element of $\mathbb{A}_3$ with respect to the basis $\{1, \rho_1, \rho_2\}$, we get the following system for the determination of functions V_1, V_2:

$$\begin{aligned} \frac{\partial V_1}{\partial y} &= n_1 \frac{\partial V_1}{\partial x}\,, \\ \frac{\partial V_2}{\partial y} &= n_2 \frac{\partial V_1}{\partial x} + n_1 \frac{\partial V_2}{\partial x}\,, \\ \frac{\partial V_1}{\partial z} &= m_1 \frac{\partial V_1}{\partial x}\,, \\ \frac{\partial V_2}{\partial z} &= m_2 \frac{\partial V_1}{\partial x} + m_1 \frac{\partial V_2}{\partial x}\,. \end{aligned} \tag{2.13}$$

Inasmuch as

$$\xi = (x + y\Re\, n_1 + z\Re\, m_1) + i(y\Im\, n_1 + z\Im\, m_1) =: \tau + i\eta\,, \tag{2.14}$$

from the first and the third equations of the system (2.13) we get

$$\frac{\partial V_1}{\partial \eta}\Im\, n_1 = i\,\frac{\partial V_1}{\partial \tau}\Im\, n_1\,, \qquad \frac{\partial V_1}{\partial \eta}\Im\, m_1 = i\,\frac{\partial V_1}{\partial \tau}\Im\, m_1\,. \tag{2.15}$$

It follows from the first equation of the system (2.2) that, at least one of the numbers $\Im\, n_1$, $\Im\, m_1$ is not equal to zero. Therefore, from (2.15) we get the equality

$$\frac{\partial V_1}{\partial \eta} = i\,\frac{\partial V_1}{\partial \tau}\,. \tag{2.16}$$

Let us prove that $V_1(x_1, y_1, z_1) = V_1(x_2, y_2, z_2)$ for the points (x_1, y_1, z_1), $(x_2, y_2, z_2) \in \Omega$ such that the segment connecting these points is parallel to the straight line L. Consider two surfaces Q, Σ in Ω and the domain G in $\mathbb{C}$ that are defined in the proof of Lemma 2.2. For each $\xi \in G$ define two complex-valued functions H_1 and H_2 so that

$$\begin{aligned} H_1(\xi) &:= V_1(x, y, z) \ \text{ for } (x, y, z) \in Q, \\ H_2(\xi) &:= V_1(x, y, z) \ \text{ for } (x, y, z) \in \Sigma, \end{aligned}$$

where the correspondence between the points (x, y, z) and $\xi \in G$ is determined by the relation (2.14). The functions H_1, H_2 are analytic in the domain G owing to the equality (2.16) and Theorem 6 in [11]. Further, the identity $H_1(\xi) \equiv H_2(\xi)$ in G can be proved in the same way as in the proof of Lemma 2.2.

Thus, the function V_1 of the form $V_1(x, y, z) := F_1(\xi)$, where $F_1(\xi)$ is an arbitrary function analytic in D, is the general solution of the system consisting of the first and the third equations of the system (2.13).

Now, from the second and the fourth equations of the system (2.13) we get the following system for the determination of function $V_2(x,y,z)$:

$$\begin{aligned} \frac{\partial V_2}{\partial y} - n_1 \frac{\partial V_2}{\partial x} &= n_2 \frac{\partial F_1}{\partial x}\,, \\ \frac{\partial V_2}{\partial z} - m_1 \frac{\partial V_2}{\partial x} &= m_2 \frac{\partial F_1}{\partial x}\,. \end{aligned} \tag{2.17}$$

The function $(n_2y+m_2z)F_1'(\xi)$ is a particular solution of this system and, therefore, the general solution of the system (2.17) is represented in the form

$$V_2(x,y,z) = F_2(\xi) + (n_2y+m_2z)F_1'(\xi),$$

where F_2 is an arbitrary function analytic in D. The theorem is proved. □

It follows from the equalities (2.9), (2.11) that in the case where a domain $\Omega \subset \mathbb{R}^3$ is convex in the direction of the straight line L, any monogenic function $\Phi : \Omega_\zeta \to \mathbb{A}_3$ can be constructed by means of three complex analytic in D functions F, F_1, F_2 in the form:

$$\begin{aligned} \Phi(\zeta) = \frac{1}{2\pi i}\int\limits_{\Gamma_\zeta} F(t)(t-\zeta)^{-1}dt + \rho_1 F_1(x+n_1y+m_1z) \\ + \rho_2\Big(F_2(x+n_1y+m_1z) + (n_2y+m_2z)F_1'(x+n_1y+m_1z)\Big) \\ \forall \zeta = xe_1+ye_2+ze_3 \in \Omega_\zeta\,, \end{aligned} \tag{2.18}$$

and in this case the equality (2.10) is applicable. Thus, we can rewrite the equality (2.18) in the following form:

$$\begin{aligned} \Phi(\zeta) &= F(x+n_1y+m_1z) \\ &+ \Big(F_1(x+n_1y+m_1z) + (n_2y+m_2z)F'(x+n_1y+m_1z)\Big)\rho_1 \\ &+ \Big(F_2(x+n_1y+m_1z) + (n_2y+m_2z)F_1'(x+n_1y+m_1z) \\ &+ (n_3y+m_3z)F'(x+n_1y+m_1z) + \frac{(n_2y+m_2z)^2}{2}F''(x+n_1y+m_1z)\Big)\rho_2 \\ &\forall \zeta = xe_1+ye_2+ze_3 \in \Omega_\zeta\,. \end{aligned} \tag{2.19}$$

Using the equality (2.19), one can construct all monogenic functions $\Phi : \Omega_\zeta \to \mathbb{A}_3$ by means of arithmetic operations with arbitrary complex-valued analytic functions F, F_1, F_2 given in the domain $D \subset \mathbb{C}$.

It is evident that the following statement follows from the equality (2.19).

Theorem 2.5. *If a domain $\Omega \subset \mathbb{R}^3$ is convex in the direction of the straight line L, then every monogenic function $\Phi : \Omega_\zeta \to \mathbb{A}_3$ can be continued to a function monogenic in the domain Π_ζ.*

Note that the condition of convexity of Ω in the direction of the line L is essential for the truth of Lemma 2.2 and consequently for the truth of Theorems 2.3–2.5.

Example 2.6. Let us construct a domain Ω, which is not convex in the direction of the straight line L, and an example of monogenic function $\Phi : \Omega_\zeta \to \mathbb{A}_3$ for which the relation (2.7) is not fulfilled for certain $\zeta_1, \zeta_2 \in \Omega_\zeta$ such that $\zeta_2 - \zeta_1 \in L_\zeta$.

Consider a harmonic basis (2.3). In this case the straight line L coincide with the axis Oz. Consider the domain Ω_ζ which is the union of sets

$$\Omega_\zeta^{(1)} := \{x + ye_2^0 + ze_3^0 \in E_3 : |x+iy| < 2, 0 < z < 2, -\pi/4 < \arg(x+iy) < 3\pi/2\},$$
$$\Omega_\zeta^{(2)} := \{x + ye_2^0 + ze_3^0 \in E_3 : |x+iy| < 2, 2 \le z \le 4, \pi/2 < \arg(x+iy) < 3\pi/2\},$$
$$\Omega_\zeta^{(3)} := \{x + ye_2^0 + ze_3^0 \in E_3 : |x+iy| < 2, 4 < z < 6, \pi/2 < \arg(x+iy) < 9\pi/4\}.$$

It is evident that the domain $\Omega \subset \mathbb{R}^3$ congruent to Ω_ζ is not convex in the direction of the axis Oz.

In the domain $\{\xi \in \mathbb{C} : |\xi| < 2, -\pi/4 < \arg \xi < 3\pi/2\}$ of the complex plane consider a holomorphic branch $H_1(\xi)$ of analytic function $\operatorname{Ln} \xi$ for which $H_1(1) = 0$. In the domain $\{\xi \in \mathbb{C} : |\xi| < 2,\ \pi/2 < \arg \xi < 9\pi/4\}$ consider also a holomorphic branch $H_2(\xi)$ of function $\operatorname{Ln} \xi$ for which $H_2(1) = 2\pi i$.

Further, consider the principal extension Φ_1 of function H_1 into the set $\Omega_\zeta^{(1)} \cup \Omega_\zeta^{(2)}$ and the principal extension Φ_2 of function H_2 into the set $\Omega_\zeta^{(2)} \cup \Omega_\zeta^{(3)}$ constructed by using the formula (2.10):

$$\Phi_1(\zeta) = H_1(x+iy) - \frac{2z - iy}{2(x+iy)}\,\rho_1 - \left(\frac{\sqrt{3}iz}{2(x+iy)} + \frac{(2z-iy)^2}{8(x+iy)^2}\right)\rho_2\,,$$

$$\Phi_2(\zeta) = H_2(x+iy) - \frac{2z - iy}{2(x+iy)}\,\rho_1 - \left(\frac{\sqrt{3}iz}{2(x+iy)} + \frac{(2z-iy)^2}{8(x+iy)^2}\right)\rho_2\,,$$

where $\zeta = x + ye_2^0 + ze_3^0$.

Now, the function

$$\Phi(\zeta) := \begin{cases} \Phi_1(\zeta) & \text{for } \zeta \in \Omega_\zeta^{(1)} \cup \Omega_\zeta^{(2)}, \\ \Phi_2(\zeta) & \text{for } \zeta \in \Omega_\zeta^{(3)} \end{cases}$$

is monogenic in the domain Ω_ζ, because $\Phi_1(\zeta) \equiv \Phi_2(\zeta)$ everywhere in $\Omega_\zeta^{(2)}$. At the same time, for the points $\zeta_1 = 1 + e_3^0$ and $\zeta_2 = 1 + 5\,e_3^0$ we have $\zeta_2 - \zeta_1 \in L_\zeta$ but

$$\Phi(\zeta_2) - \Phi(\zeta_1) = 2\pi i - 4\,\rho_1 - (12 + 2\sqrt{3}\,i)\rho_2 \notin \mathcal{I}\,,$$

i.e., the relation (2.7) is not fulfilled.

Taking into account the equality (2.10), we can rewrite the equality (2.19) in the following integral form:

$$\Phi(\zeta) = \frac{1}{2\pi i} \int\limits_{\Gamma_\zeta} \Big(F(t) + \rho_1 F_1(t) + \rho_2 F_2(t)\Big)(t-\zeta)^{-1} dt \qquad \forall \zeta \in \Omega_\zeta\,, \tag{2.20}$$

where the curve Γ_ζ is the same as in Theorem 2.3.

The following statement is true for monogenic functions in an arbitrary domain Ω_ζ.

Theorem 2.7. *For every monogenic function $\Phi : \Omega_\zeta \to \mathbb{A}_3$ in an arbitrary domain Ω_ζ, the Gateaux nth derivatives $\Phi^{(n)}$ are monogenic functions in Ω_ζ for any n.*

Proof. Consider an arbitrary point $(x_0, y_0, z_0) \in \Omega$ and a ball $\mho \subset \Omega$ with the center in the point (x_0, y_0, z_0). Inasmuch as $\mho$ is a convex set, in the neighbourhood $\mho_\zeta$ of the point $\zeta_0 = x_0 + y_0 e_2 + z_0 e_3$ we have the equality (2.20), where the integral has the Gateaux nth derivative for any n and these derivatives are continuous functions in $\mho_\zeta$. Thus, the Gateaux nth derivative $\Phi^{(n)}$ is a monogenic function in $\mho_\zeta$ for any n. The theorem is proved. □

Using the integral expression (2.20) of monogenic function $\Phi : \Omega_\zeta \to \mathbb{A}_3$, we obtain the following expression for the Gateaux nth derivative $\Phi^{(n)}$:

$$\Phi^{(n)}(\zeta) = \frac{n!}{2\pi i} \int\limits_{\Gamma_\zeta} \Big(F(t) + \rho_1 F_1(t) + \rho_2 F_2(t)\Big)\Big((t-\zeta)^{-1}\Big)^{n+1} dt \qquad \forall \zeta \in \Omega_\zeta\,.$$

2.4. Isomorphism of algebras of monogenic functions

Let us establish an isomorphism between algebras of monogenic functions at transition from a harmonic basis to another one. By $\mathcal{M}(E_3, \Omega_\zeta)$ we denote the algebra of monogenic functions in a domain $\Omega_\zeta \subset E_3$.

Consider a harmonic basis (2.3) and $E_3^0 := \{\zeta = xe_1^0 + ye_2^0 + ze_3^0 : x, y, z \in \mathbb{R}\}$. Let Ω_ζ^0 be a domain in E_3^0. Consider also an arbitrary harmonic basis $\{\tilde{e}_1, \tilde{e}_2, \tilde{e}_3\}$ in $\mathbb{A}_3$ and $\widetilde{E}_3 := \{\widetilde{\zeta} = \tilde{x}\tilde{e}_1 + \tilde{y}\tilde{e}_2 + \tilde{z}\tilde{e}_3 : \tilde{x}, \tilde{y}, \tilde{z} \in \mathbb{R}\}$.

Let us specify a correspondence between E_3^0 and $\widetilde{E}_3$ under which a domain $\widetilde{\Omega}_{\widetilde{\zeta}} \subset \widetilde{E}_3$ corresponds to the domain Ω_ζ^0 and algebras $\mathcal{M}(E_3^0, \Omega_\zeta^0)$, $\mathcal{M}(\widetilde{E}_3, \widetilde{\Omega}_{\widetilde{\zeta}})$ are isomorphic.

It follows from Theorem 2.1 that elements of the basis $\{\tilde{e}_1, \tilde{e}_2, \tilde{e}_3\}$ can be represented in the form $\tilde{e}_1 = ae_1$, $\tilde{e}_2 = ae_2$, $\tilde{e}_3 = ae_3$, where a is an invertible element and decompositions of the basis $\{e_1, e_2, e_3\}$ with respect to the basis $\{1, \rho_1, \rho_2\}$ are of the form (2.1), where without restricting the generality we may assume that $\Im n_1 \neq 0$. Then the basis $\{e_1, e_2, e_3\}$ can be expressed in the form

$$\begin{aligned} e_1 &= e_1^0, \\ e_2 &= \alpha_1 e_1^0 + \alpha_2 e_2^0 \qquad + r_{21}\rho_1 + r_{22}\rho_2, \\ e_3 &= \beta_1 e_1^0 + \beta_2 e_2^0 + e_3^0 + r_{31}\rho_1 + r_{32}\rho_2, \end{aligned}$$

where $\alpha_1 := \Re n_1$, $\alpha_2 := \Im n_1 \neq 0$, $\beta_1 := \Re m_1$, $\beta_2 := \Im m_1$, $r_{21} := n_2$, $r_{22} := n_3 - \frac{1}{2} i \Im n_1$, $r_{31} := m_2 + 1$, $r_{32} := m_3 + \frac{\sqrt{3}}{2} i - \frac{1}{2} i \Im m_1$.

Theorem 2.8. *Let a correspondence between* $\zeta = xe_1^0 + ye_2^0 + ze_3^0 \in E_3^0$ *and* $\widetilde{\zeta} = \tilde{x}\tilde{e}_1 + \tilde{y}\tilde{e}_2 + \tilde{z}\tilde{e}_3 \in \widetilde{E}_3$ *be given by the equalities*

$$\begin{aligned} x &= \tilde{x} + \alpha_1\tilde{y} + \beta_1\tilde{z}, \\ y &= \qquad \alpha_2\tilde{y} + \beta_2\tilde{z}, \\ z &= \qquad\qquad\quad \tilde{z}. \end{aligned}$$

Then the algebras $\mathcal{M}(E_3^0, \Omega_\zeta^0)$, $\mathcal{M}(\widetilde{E}_3, \widetilde{\Omega}_{\widetilde{\zeta}})$ *are isomorphic, and the correspondence* $\mathcal{M}(E_3^0, \Omega_\zeta^0) \ni \Phi \longleftrightarrow \widetilde{\Phi} \in \mathcal{M}(\widetilde{E}_3, \widetilde{\Omega}_{\widetilde{\zeta}})$ *are established by the equality*

$$\widetilde{\Phi}(\widetilde{\zeta}) = \Phi(\zeta) + \Phi'(\zeta)\Big((r_{21}\tilde{y} + r_{31}\tilde{z})\rho_1 + (r_{22}\tilde{y} + r_{32}\tilde{z})\rho_2\Big) + \frac{1}{2}\Phi''(\zeta)(r_{21}\tilde{y} + r_{31}\tilde{z})^2\rho_2.$$

3. Integral theorems in the algebra $\mathbb{A}_3$

3.1. On integral theorems in hypercomplex analysis

In the paper [12] for functions differentiable in the sense of Lorch in an arbitrary convex domain of a commutative associative Banach algebra, some properties similar to properties of holomorphic functions of a complex variable (in particular, the integral Cauchy theorem and the integral Cauchy formula, the Taylor expansion and the Morera theorem) are established. The convexity of the domain in the mentioned results from [12] is excluded by E.K. Blum [13].

In this paper we establish similar results for monogenic functions $\Phi : \Omega_\zeta \to \mathbb{A}_3$ given only in a domain Ω_ζ of the linear envelope E_3 instead of the domain of the whole algebra $\mathbb{A}_3$. Let us note that *a priori* the differentiability of the function Φ in the sense of Gateaux is a restriction weaker than the differentiability of this function in the sense of Lorch. Moreover, note that the integral Cauchy formula established in the papers [12, 13] is not applicable to a monogenic function $\Phi : \Omega_\zeta \to \mathbb{A}_3$ because it deals with an integration along a curve on which the function Φ is not given, generally speaking.

Note that as well as in [12, 13], some hypercomplex analogues of the integral Cauchy theorem for a curvilinear integral are established in the papers [14, 15]. In the papers [14, 16–18] similar theorems are established for surface integrals.

Below in Section 3, all stated results are obtained jointly with V.S. Shpakivskyi (see also [19, 20]).

3.2. Cauchy integral theorem for a surface integral

Let $\{e_1 = 1, e_2, e_3\}$ be a harmonic basis in the algebra $\mathbb{A}_3$.

Along with monogenic functions satisfying the Cauchy–Riemann conditions (2.5), consider a function $\Psi : \Omega_\zeta \to \mathbb{A}_3$ having continuous partial derivatives of the first order in a domain Ω_ζ and satisfying the equation

$$\frac{\partial \Psi}{\partial x} + \frac{\partial \Psi}{\partial y} e_2 + \frac{\partial \Psi}{\partial z} e_3 = 0 \tag{3.1}$$

in every point of this domain.

In the scientific literature, different denominations are used for functions satisfying equations of the form (3.1). For example, in the papers [14, 15, 21] they are called regular functions, and in the papers [16, 17, 22] they are called monogenic functions. As well as in the papers [18, 23, 24], we call Ψ a *hyperholomorphic* function if it satisfies equation (3.1).

It is well known that in the quaternion analysis the classes of functions determined by means of conditions of the form (2.5) and (3.1) do not coincide (see [14, 25]).

Note that in the algebra $\mathbb{A}_3$ the set of monogenic functions is a subset of the set of hyperholomorphic functions, because every monogenic function $\Phi : \Omega_\zeta \to \mathbb{A}_3$ satisfies the equality (3.1) owing to conditions (1.5), (2.5). But, there exist hyperholomorphic functions which are not monogenic. For example, the function $\Psi(x + ye_2 + ze_3) = ze_2 - ye_3$ satisfies condition (3.1), but it does not satisfy equalities of the form (2.5).

Let Ω be a bounded closed set in $\mathbb{R}^3$. For a continuous function $\Psi : \Omega_\zeta \to \mathbb{A}_3$ of the form

$$\Psi(x + ye_2 + ze_3) = \sum_{k=1}^{3} U_k(x, y, z)e_k + i \sum_{k=1}^{3} V_k(x, y, z)e_k, \tag{3.2}$$

where $(x, y, z) \in \Omega$ and $U_k : \Omega \to \mathbb{R}$, $V_k : \Omega \to \mathbb{R}$, we define a volume integral by the equality

$$\int_{\Omega_\zeta} \Psi(\zeta) dxdydz := \sum_{k=1}^{3} e_k \int_{\Omega} U_k(x, y, z) dxdydz + i \sum_{k=1}^{3} e_k \int_{\Omega} V_k(x, y, z) dxdydz.$$

Let Σ be a quadrable surface in $\mathbb{R}^3$ with quadrable projections on the coordinate planes. For a continuous function $\Psi : \Sigma_\zeta \to \mathbb{A}_3$ of the form (3.2), where $(x, y, z) \in \Sigma$ and $U_k : \Sigma \to \mathbb{R}$, $V_k : \Sigma \to \mathbb{R}$, we define a surface integral on Σ_ζ with the differential form $\sigma_{\alpha_1,\alpha_2,\alpha_3} := \alpha_1 dydz + \alpha_2 dzdx e_2 + \alpha_3 dxdy e_3$, where $\alpha_1, \alpha_2, \alpha_3 \in \mathbb{R}$, by the equality

$$\int_{\Sigma_\zeta} \Psi(\zeta)\sigma_{\alpha_1,\alpha_2,\alpha_3} := \sum_{k=1}^{3} e_k \int_{\Sigma} \alpha_1 U_k(x, y, z) dydz + \sum_{k=1}^{3} e_2 e_k \int_{\Sigma} \alpha_2 U_k(x, y, z) dzdx$$

$$+ \sum_{k=1}^{3} e_3 e_k \int_{\Sigma} \alpha_3 U_k(x, y, z) dxdy + i \sum_{k=1}^{3} e_k \int_{\Sigma} \alpha_1 V_k(x, y, z) dydz$$

$$+ i \sum_{k=1}^{3} e_2 e_k \int_{\Sigma} \alpha_2 V_k(x, y, z) dzdx + i \sum_{k=1}^{3} e_3 e_k \int_{\Sigma} \alpha_3 V_k(x, y, z) dxdy.$$

A connected homeomorphic image of a square in $\mathbb{R}^3$ is called a *simple surface.* A surface is *locally-simple* if it is simple in a certain neighborhood of every point.

If a simply connected domain $\Omega \subset \mathbb{R}^3$ has a closed locally-simple piece-smooth boundary $\partial\Omega$ and a function $\Psi : \Omega_\zeta \to \mathbb{A}_3$ is continuous together with partial derivatives of the first order up to the boundary $\partial\Omega_\zeta$, then the following analogue of the Gauss–Ostrogradsky formula is true:

$$\int\limits_{\partial\Omega_\zeta} \Psi(\zeta)\sigma = \int\limits_{\Omega_\zeta} \left(\frac{\partial\Psi}{\partial x} + \frac{\partial\Psi}{\partial y} e_2 + \frac{\partial\Psi}{\partial z} e_3 \right) dxdydz, \tag{3.3}$$

where $\sigma := \sigma_{1,1,1} \equiv dydz + dzdxe_2 + dxdye_3$. Now, the next theorem is a result of formula (3.3) and equality (3.1).

Theorem 3.1. *Suppose that Ω is a simply connected domain with a closed locally-simple piece-smooth boundary $\partial\Omega$. Suppose also that the function $\Psi : \overline{\Omega_\zeta} \to \mathbb{A}_3$ is continuous in the closure $\overline{\Omega_\zeta}$ of domain Ω_ζ and is hyperholomorphic in Ω_ζ. Then*

$$\int\limits_{\partial\Omega_\zeta} \Psi(\zeta)\sigma = 0.$$

3.3. Cauchy integral theorem for a curvilinear integral

Let γ be a Jordan rectifiable curve in $\mathbb{R}^3$. For a continuous function $\Psi : \gamma_\zeta \to \mathbb{A}_3$ of the form (3.2), where $(x, y, z) \in \gamma$ and $U_k : \gamma \to \mathbb{R}$, $V_k : \gamma \to \mathbb{R}$, we define an integral along the curve γ_ζ by the equality

$$\begin{aligned}\int\limits_{\gamma_\zeta} \Psi(\zeta)d\zeta := {} & \sum_{k=1}^{3} e_k \int\limits_\gamma U_k(x,y,z)dx + \sum_{k=1}^{3} e_2 e_k \int\limits_\gamma U_k(x,y,z)dy \\ & + \sum_{k=1}^{3} e_3 e_k \int\limits_\gamma U_k(x,y,z)dz + i\sum_{k=1}^{3} e_k \int\limits_\gamma V_k(x,y,z)dx \\ & + i\sum_{k=1}^{3} e_2 e_k \int\limits_\gamma V_k(x,y,z)dy + i\sum_{k=1}^{3} e_3 e_k \int\limits_\gamma V_k(x,y,z)dz,\end{aligned}$$

where $d\zeta := dx + e_2 dy + e_3 dz$.

If a function $\Phi : \Omega_\zeta \to \mathbb{A}_3$ is continuous together with partial derivatives of the first order in a domain Ω_ζ, and Σ is a piece-smooth surface in Ω, and the edge γ of surface Σ is a rectifiable Jordan curve, then the following analogue of the Stokes formula is true:

$$\begin{aligned}\int\limits_{\gamma_\zeta} \Phi(\zeta)d\zeta = \int\limits_{\Sigma_\zeta} & \left(\frac{\partial\Phi}{\partial x} e_2 - \frac{\partial\Phi}{\partial y} \right) dxdy + \left(\frac{\partial\Phi}{\partial y} e_3 - \frac{\partial\Phi}{\partial z} e_2 \right) dydz \\ & + \left(\frac{\partial\Phi}{\partial z} - \frac{\partial\Phi}{\partial x} e_3 \right) dzdx.\end{aligned} \tag{3.4}$$

Now, the next theorem is a result of formula (3.4) and equalities (2.5).

Theorem 3.2. *Suppose that $\Phi : \Omega_\zeta \to \mathbb{A}_3$ is a monogenic function in a domain Ω_ζ, and Σ is a piece-wise smooth surface in Ω, and the edge γ of surface Σ is a rectifiable Jordan curve. Then*

$$\int_{\gamma_\zeta} \Phi(\zeta) d\zeta = 0. \tag{3.5}$$

Now, similarly to the proof of Theorem 3.2 in [13] we can prove the following

Theorem 3.3. *Let $\Phi : \Omega_\zeta \to \mathbb{A}_3$ be a monogenic function in a domain Ω_ζ. Then for every closed Jordan rectifiable curve γ homotopic to a point in Ω, the equality* (3.5) *is true.*

For functions taking values in the algebra $\mathbb{A}_3$, the following Morera theorem can be established in the usual way.

Theorem 3.4. *If a function $\Phi : \Omega_\zeta \to \mathbb{A}_3$ is continuous in a domain Ω_ζ and satisfies the equality*

$$\int_{\partial \triangle_\zeta} \Phi(\zeta) d\zeta = 0 \tag{3.6}$$

for every triangle $\triangle_\zeta$ such that $\overline{\triangle_\zeta} \subset \Omega_\zeta$, then the function Φ is monogenic in the domain Ω_ζ.

3.4. Cauchy integral formula

Inasmuch as according to Theorem 2.8 there exists an isomorphism between algebras of monogenic functions $\Phi(\zeta)$ of the variable $\zeta = xe_1 + ye_2 + ze_3$ determined in various harmonic bases, it is enough to study properties of monogenic functions given in domains of a linear envelope E_3 generated by one of its harmonic bases.

Thus, let us consider a basis (2.3) as the harmonic basis $\{e_1, e_2, e_3\}$.

It follows from equality (2.6) that

$$\zeta^{-1} = \frac{1}{x+iy} + \frac{z}{(x+iy)^2}\rho_1 + \left(\frac{i}{2}\frac{\sqrt{3}z - y}{(x+iy)^2} + \frac{z^2}{(x+iy)^3}\right)\rho_2 \tag{3.7}$$

for all $\zeta = x + ye_2 + ze_3 \in E_3 \setminus \{ze_3 : z \in \mathbb{R}\}$. Thus, it is obvious that the straight line $\{ze_3 : z \in \mathbb{R}\}$ is contained in the radical of the algebra $\mathbb{A}_3$.

Using equality (3.7), it is easy to calculate that

$$\int_{\tilde{\gamma}_\zeta} \tau^{-1} d\tau = 2\pi i, \tag{3.8}$$

where $\tilde{\gamma}_\zeta := \{\tau = x + ye_2 : x^2 + y^2 = R^2\}$.

Theorem 3.5. *Let Ω be a domain convex in the direction of the axis Oz and $\Phi : \Omega_\zeta \to \mathbb{A}_3$ be a monogenic function in the domain Ω_ζ. Then for every point $\zeta_0 \in \Omega_\zeta$ the following equality is true:*

$$\Phi(\zeta_0) = \frac{1}{2\pi i} \int\limits_{\gamma_\zeta} \Phi(\zeta)\,(\zeta - \zeta_0)^{-1}\, d\zeta, \tag{3.9}$$

where γ_ζ is an arbitrary closed Jordan rectifiable curve in Ω_ζ, which winds once around the straight line $\{\zeta_0 + ze_3 : z \in \mathbb{R}\}$ and is homotopic to the point ζ_0.

Proof. We represent the integral from the right-hand side of equality (3.9) as the sum of the following two integrals:

$$\int_{\gamma_\zeta} \Phi(\zeta)\,(\zeta - \zeta_0)^{-1}\, d\zeta = \int_{\gamma_\zeta} (\Phi(\zeta) - \Phi(\zeta_0))\,(\zeta - \zeta_0)^{-1}\, d\zeta$$
$$+ \Phi(\zeta_0) \int_{\gamma_\zeta} (\zeta - \zeta_0)^{-1}\, d\zeta =: I_1 + I_2.$$

Inasmuch as the domain Ω is convex in the direction of the axis Oz and the curve γ_ζ winds once around the straight line $\{\zeta_0 + ze_3 : z \in \mathbb{R}\}$, γ is homotopic to the circle $K(R) := \{(x-x_0)^2 + (y-y_0)^2 = R^2, z = z_0\}$, where $\zeta_0 = x_0 + y_0 e_2 + z_0 e_3$. Then using equality (3.8), we have $I_2 = 2\pi i \Phi(\zeta_0)$.

Let us prove that $I_1 = 0$. First, we choose on the curve γ two points A and B in which there are tangents to γ, and we choose also two points A_1, B_1 on the circle $K(\varepsilon)$ which is completely contained in the domain Ω. Let γ^1, γ^2 be connected components of the set $\gamma \setminus \{A, B\}$. By K^1 and K^2 we denote connected components of the set $K(\varepsilon) \setminus \{A_1, B_1\}$ in such a way that after a choice of smooth arcs Γ^1, Γ^2 each of the closed curves $\gamma^1 \cup \Gamma^2 \cup K^1 \cup \Gamma^1$ and $\gamma^2 \cup \Gamma^1 \cup K^2 \cup \Gamma^2$ will homotopic to a point of the domain $\Omega \setminus \{(x_0, y_0, z) : z \in \mathbb{R}\}$.

Then it follows from Theorem 3.3 that

$$\int\limits_{\gamma^1_\zeta \cup \Gamma^2_\zeta \cup K^1_\zeta \cup \Gamma^1_\zeta} (\Phi(\zeta) - \Phi(\zeta_0))\,(\zeta - \zeta_0)^{-1}\, d\zeta = 0, \tag{3.10}$$

$$\int\limits_{\gamma^2_\zeta \cup \Gamma^1_\zeta \cup K^2_\zeta \cup \Gamma^2_\zeta} (\Phi(\zeta) - \Phi(\zeta_0))\,(\zeta - \zeta_0)^{-1}\, d\zeta = 0. \tag{3.11}$$

Inasmuch as each of the curves Γ^1_ζ, Γ^2_ζ has different orientations in the equalities (3.10), (3.11), after addition of the mentioned equalities we obtain

$$\int\limits_{\gamma_\zeta} (\Phi(\zeta) - \Phi(\zeta_0))\,(\zeta - \zeta_0)^{-1}\, d\zeta = \int\limits_{K_\zeta(\varepsilon)} (\Phi(\zeta) - \Phi(\zeta_0))\,(\zeta - \zeta_0)^{-1}\, d\zeta, \tag{3.12}$$

where the curves $K_\zeta(\varepsilon)$, γ_ζ have the same orientation.

The integrand in the right-hand side of the equality (3.12) is bounded by a constant which does not depend on ε. Therefore, passing to the limit in the equality (3.12) as $\varepsilon \to 0$, we obtain $I_1 = 0$ and the theorem is proved. □

Using formula (3.9), we obtain the Taylor expansion of a monogenic function in the usual way (see, for example, [26, p. 107]). A uniqueness theorem for monogenic functions can also be proved in the same way as for holomorphic functions of a complex variable (see, for example, [26, p. 110]).

3.5. Different equivalent definitions of monogenic functions taking values in the algebra $\mathbb{A}_3$

Thus, the following theorem giving different equivalent definitions of monogenic functions is true:

Theorem 3.6. *A function $\Phi : \Omega_\zeta \to \mathbb{A}_3$ is monogenic in an arbitrary domain Ω_ζ if and only if one of the following conditions is satisfied:*

(I) *the components U_k, $k = \overline{1,3}$, of the decomposition* (2.4) *of the function Φ are differentiable in Ω and the conditions* (2.5) *are satisfied in the domain Ω_ζ;*

(II) *in every ball $\mho_\zeta \subset \Omega_\zeta$ the function Φ is expressed in the form* (2.19), *where the triad of holomorphisms in the domain $f(\mho_\zeta)$, functions F, F_1, F_2, is unique;*

(III) *the function Φ is continuous in Ω_ζ and satisfies the equality* (3.6) *for every triangle $\triangle_\zeta$ such that $\overline{\triangle_\zeta} \subset \Omega_\zeta$;*

(IV) *for every $\zeta_0 \in \Omega_\zeta$ there exists a neighborhood in which the function Φ is expressed as the sum of the power series* (1.4), *where $\mathbb{A} = \mathbb{A}_3$.*

4. Infinite-dimensional commutative Banach algebras and spatial potential fields

4.1. An infinite-dimensional harmonic algebra

Note that it is impossible to obtain all solutions of the three-dimensional Laplace equation (1.3) in the form of components of monogenic functions taking values in commutative algebras of third rank. In particular, for every mentioned algebra there exist spherical functions which are not components of specified hypercomplex monogenic functions.

Indeed, it is well known that there exists $(2n + 1)$ linearly independent homogeneous polynomials (of real variables x, y, z) of the degree n, which satisfy (1.3). At the same time, in every algebra $\mathbb{A}$ over the field $\mathbb{C}$ with harmonic basis $\{e_1, e_2, e_3\}$, the mentioned polynomials are real components of decomposition on the vectors $e_1, e_2, e_3, ie_1, ie_2, ie_3$ of functions of the form $a(xe_1+ye_2+ze_3)^n$, where $a \in \mathbb{A}$, but there are only six linearly independent real components for all $n \geq 3$.

Hence, for every harmonic algebra $\mathbb{A}$ of the third rank there exist spherical functions which are not components of monogenic functions taking values in the algebra $\mathbb{A}$.

In the papers [4, 27, 28] we considered an infinite-dimensional commutative Banach algebra $\mathbb{F}$ over the field of real numbers and established that any spherical function is a component of some monogenic function taking values in this algebra.

We described relations between monogenic functions and harmonic vectors in the space.

Consider an infinite-dimensional commutative associative Banach algebra

$$\mathbb{F} := \{g = \sum_{k=1}^{\infty} c_k e_k : c_k \in \mathbb{R}, \sum_{k=1}^{\infty} |c_k| < \infty\}$$

over the field $\mathbb{R}$ with the norm $\|g\|_{\mathbb{F}} := \sum\limits_{k=1}^{\infty} |c_k|$ and the basis $\{e_k\}_{k=1}^{\infty}$, where the multiplication table for elements of its basis is of the following form:

$$\begin{aligned}
e_n e_1 &= e_n, \qquad e_{2n+1}e_{2n} = \frac{1}{2}\, e_{4n} && \forall\, n \geq 1\,,\\
e_{2n+1}e_{2m} &= \frac{1}{2}\Big(e_{2n+2m} - (-1)^m e_{2n-2m}\Big) && \forall\, n > m \geq 1\,,\\
e_{2n+1}e_{2m} &= \frac{1}{2}\Big(e_{2n+2m} + (-1)^n e_{2m-2n}\Big) && \forall\, m > n \geq 1\,,\\
e_{2n+1}e_{2m+1} &= \frac{1}{2}\Big(e_{2n+2m+1} + (-1)^m e_{2n-2m+1}\Big) && \forall\, n \geq m \geq 1\,,\\
e_{2n}e_{2m} &= \frac{1}{2}\Big(-e_{2n+2m+1} + (-1)^m e_{2n-2m+1}\Big) && \forall\, n \geq m \geq 1\,.
\end{aligned}$$

It is evident that here e_1, e_2, e_3 form a harmonic triad of vectors.

Let Ω be a domain in $\mathbb{R}^3$ and $\zeta = x + ye_2 + ze_3$, where $(x, y, z) \in \Omega$. Consider the decomposition

$$\Phi(\zeta) = \sum_{k=1}^{\infty} U_k(x, y, z)\; e_k\,, \tag{4.1}$$

of a function $\Phi : \Omega_\zeta \to \mathbb{F}$ with respect to the basis $\{e_k\}_{k=1}^{\infty}$, where the functions $U_k : \Omega \to \mathbb{R}$ are differentiable in Ω.

In the following theorem we establish necessary and sufficient conditions for a function $\Phi : \Omega_\zeta \to \mathbb{F}$ to be monogenic in a domain $\Omega_\zeta \subset E_3$.

Theorem 4.1. *Let a function $\Phi : \Omega_\zeta \to \mathbb{F}$ be continuous in a domain Ω_ζ and the functions $U_k : \Omega \to \mathbb{R}$ from the decomposition* (4.1) *be differentiable in Ω. In order that the function Φ be monogenic in the domain Ω_ζ, it is necessary and sufficient that the conditions* (2.5) *be satisfied and the following relations be fulfilled in Ω:*

$$\sum_{k=1}^{\infty} \left| \frac{\partial U_k(x, y, z)}{\partial x} \right| < \infty, \tag{4.2}$$

$$\lim_{\varepsilon \to 0+0} \sum_{k=1}^{\infty} \left| U_k(x + \varepsilon h_1, y + \varepsilon h_2, z + \varepsilon h_3) - U_k(x, y, z) - \frac{\partial U_k(x, y, z)}{\partial x}\, \varepsilon h_1 \right.$$
$$\left. - \frac{\partial U_k(x, y, z)}{\partial y}\, \varepsilon h_2 - \frac{\partial U_k(x, y, z)}{\partial z}\, \varepsilon h_3 \right| \varepsilon^{-1} = 0 \quad \forall\, h_1, h_2, h_3 \in \mathbb{R}. \tag{4.3}$$

Proof. Necessity. If the function (4.1) is monogenic in the domain Ω_ζ, then for $h = e_1$ the equality (1.7) turns into the equality

$$\Phi'(\zeta) = \frac{\partial \Phi(\zeta)}{\partial x} \equiv \sum_{k=1}^{\infty} \frac{\partial U_k(x,y,z)}{\partial x} e_k \,,$$

and the relation (4.2) is fulfilled. Now, setting in series $h = e_2$ and $h = e_3$ in the equality (1.7), we obtain the conditions (2.5).

Let us write the conditions (2.5) in expanded form:

$$\begin{aligned}
\frac{\partial U_1(x,y,z)}{\partial y} &= -\frac{1}{2}\,\frac{\partial U_2(x,y,z)}{\partial x}\,,\\
\frac{\partial U_2(x,y,z)}{\partial y} &= \frac{\partial U_1(x,y,z)}{\partial x} + \frac{1}{2}\,\frac{\partial U_5(x,y,z)}{\partial x}\,,\\
\frac{\partial U_3(x,y,z)}{\partial y} &= -\frac{1}{2}\,\frac{\partial U_4(x,y,z)}{\partial x}\,,\\
\frac{\partial U_{2k}(x,y,z)}{\partial y} &= \frac{1}{2}\,\frac{\partial U_{2k-1}(x,y,z)}{\partial x} + \frac{1}{2}\,\frac{\partial U_{2k+3}(x,y,z)}{\partial x}\,, \qquad k = 2,3,\dots\,,\\
\frac{\partial U_{2k+1}(x,y,z)}{\partial y} &= -\frac{1}{2}\,\frac{\partial U_{2k-2}(x,y,z)}{\partial x} - \frac{1}{2}\,\frac{\partial U_{2k+2}(x,y,z)}{\partial x}\,, \quad k = 2,3,\dots\,,\\
\frac{\partial U_1(x,y,z)}{\partial z} &= -\frac{1}{2}\,\frac{\partial U_3(x,y,z)}{\partial x}\,,\\
\frac{\partial U_2(x,y,z)}{\partial z} &= -\frac{1}{2}\,\frac{\partial U_4(x,y,z)}{\partial x}\,,\\
\frac{\partial U_3(x,y,z)}{\partial z} &= \frac{\partial U_1(x,y,z)}{\partial x} - \frac{1}{2}\,\frac{\partial U_5(x,y,z)}{\partial x}\,,\\
\frac{\partial U_k(x,y,z)}{\partial z} &= \frac{1}{2}\,\frac{\partial U_{k-2}(x,y,z)}{\partial x} - \frac{1}{2}\,\frac{\partial U_{k+2}(x,y,z)}{\partial x}\,, \qquad k = 4,5,\dots\,.
\end{aligned} \tag{4.4}$$

At last, for $h_1, h_2, h_3 \in \mathbb{R}$ and $\varepsilon > 0$, writing $h := h_1 e_1 + h_2 e_2 + h_3 e_3$ and taking into account the equalities (4.4), we have

$$\begin{aligned}
&\big(\Phi(\zeta + \varepsilon h) - \Phi(\zeta)\big)\,\varepsilon^{-1} - h\,\Phi'(\zeta)\\
&= \left(\sum_{k=1}^{\infty}\Big(U_k(x+\varepsilon h_1, y+\varepsilon h_2, z+\varepsilon h_3) - U_k(x,y,z)\Big) e_k \right.\\
&\qquad \left. - \varepsilon\,(h_1 e_1 + h_2 e_2 + h_3 e_3) \sum_{k=1}^{\infty} \frac{\partial U_k(x,y,z)}{\partial x}\, e_k \right)\varepsilon^{-1}\\
&= \varepsilon^{-1} \sum_{k=1}^{\infty} \left(U_k(x+\varepsilon h_1, y+\varepsilon h_2, z+\varepsilon h_3) - U_k(x,y,z) \right.\\
&\qquad \left. - \frac{\partial U_k(x,y,z)}{\partial x}\,\varepsilon h_1 - \frac{\partial U_k(x,y,z)}{\partial y}\,\varepsilon h_2 - \frac{\partial U_k(x,y,z)}{\partial z}\,\varepsilon h_3 \right) e_k\,.
\end{aligned} \tag{4.5}$$

Therefore, inasmuch as the function (4.1) is monogenic in the domain Ω_ζ, the relation (4.3) is fulfilled in Ω.

Sufficiency. Let $\varepsilon > 0$ and $h := h_1e_1+h_2e_2+h_3e_3$, where $h_1, h_2, h_3 \in \mathbb{R}$. Then under the conditions (4.4) and (4.2) the equality (4.5) is true. Now, it follows from the equality (4.5) and the relation (4.3) that the function (4.1) has the Gateaux derivative $\Phi'(\zeta)$ for all $\zeta \in Q_\zeta$. The theorem is proved. □

Note that the conditions (4.4) are similar by nature to the Cauchy–Riemann conditions for monogenic functions of complex variables, and the relations (4.2), (4.3) are conditioned by the infinite dimensionality of the algebra $\mathbb{F}$.

It is clear that if the Gateaux derivative Φ' of monogenic function $\Phi : \Omega_\zeta \to \mathbb{F}$, in turn, is a monogenic function in the domain Ω_ζ, then all components U_k of decomposition (4.1) satisfy equation (1.3) in Ω in consequence of condition (1.5). At the same time, the following statement is true even independently of the relation between solutions of the system of equations (4.4) and monogenic functions.

Theorem 4.2. *If the functions $U_k : \Omega \to \mathbb{R}$ have continuous second-order partial derivatives in a domain Ω and satisfy the conditions* (4.4), *then they satisfy equation* (1.3) *in Q.*

To prove Theorem 4.2 it is easy to show that if the functions U_k are doubly continuously differentiable in the domain Ω, then the equalities $\Delta U_k(x,y,z) = 0$ for $k = 1, 2, \ldots$ are corollaries of the system (4.4).

Note that the algebra $\mathbb{F}$ is isomorphic to the algebra $\mathbf{F}$ of absolutely convergent trigonometric Fourier series

$$g(\tau) = a_0 + \sum_{k=1}^{\infty}\Big(a_k\, i^k \cos k\tau + b_k\, i^k \sin k\tau\Big)$$

with real coefficients a_0, a_k, b_k and the norm $\|g\|_{\mathbf{F}} := |a_0| + \sum\limits_{k=1}^{\infty}\Big(|a_k| + |b_k|\Big)$. In this case, we have the isomorphism $e_{2k-1} \longleftrightarrow i^{k-1}\cos{(k-1)\tau}$, $e_{2k} \longleftrightarrow i^k \sin k\tau$ between basic elements.

Let us write the expansion of a power function of the variable $\xi = xe_1 + ye_2 + ze_3$ in the basis $\{e_k\}_{k=1}^{\infty}$, using spherical coordinates ρ, θ, ϕ which have the following relations with x, y, z:

$$x = \rho\cos\theta, \quad y = \rho\sin\theta\sin\phi, \quad z = \rho\sin\theta\cos\phi\,. \tag{4.6}$$

In view of the isomorphism of the algebras $\mathbb{F}$ and $\mathbf{F}$, the construction of expansions of this sort is reduced to the determination of relevant Fourier coefficients. So, we have

$$\zeta^n = \rho^n\bigg(P_n(\cos\theta)\,e_1 + 2\sum_{m=1}^{n}\frac{n!}{(n+m)!}P_n^m(\cos\theta)\Big(\sin m\phi\, e_{2m} + \cos m\phi\, e_{2m+1}\Big)\bigg), \tag{4.7}$$

where n is a positive integer, P_n and P_n^m are Legendre polynomials and associated Legendre polynomials, respectively, namely:

$$P_n(t) := \frac{1}{2^n\, n!}\, \frac{d^n}{dt^n}\, (t^2 - 1)^n, \qquad P_n^m(t) := (1-t^2)^{m/2}\, \frac{d^m}{dt^m}\, P_n(t)\,. \tag{4.8}$$

We obtain in exactly the same way the following expansion of the exponential function:

$$e^{\zeta} = e^{\rho\cos\theta}\Big(J_0(\rho\sin\theta)e_1 + 2\sum_{m=1}^{\infty} J_m(\rho\sin\theta)\big(\sin\, m\phi\, e_{2m} + \cos\, m\phi\, e_{2m+1}\big)\Big),$$

where J_m are Bessel functions, namely:

$$J_m(t) := \frac{(-1)^m}{\pi} \int_0^{\pi} e^{it\cos\tau}\, \cos m\tau\, d\tau\,. \tag{4.9}$$

Thus, $2n+1$ linearly independent spherical functions of the nth power are components of the expansion (4.7) of the function ζ^n. Using the expansion (4.7) and rules of multiplication for basic elements of the algebra $\mathbb{F}$, it is easy to prove the following statement.

Theorem 4.3. *Every spherical function*

$$\rho^n\left(a_{n,0}P_n(\cos\theta) + \sum_{m=1}^{n}\Big(a_{n,m}\cos m\phi + b_{n,m}\sin m\phi\Big)P_n^m(\cos\theta)\right)$$

where $a_{n,0}, a_{n,m}, b_{n,m} \in \mathbb{R}$, *is the first component of expansion of the monogenic function*

$$\left(a_{n,0}\, e_1 + \sum_{m=1}^{n}(-1)^m\, \frac{(n+m)!}{n!}\, \Big(b_{n,m}\, e_{2m} + a_{n,m}\, e_{2m+1}\Big)\right)\xi^n$$

in the basis $\{e_k\}_{k=1}^{\infty}$, *where* $\xi := xe_1 + ye_2 + ze_3$, *and* x, y, z *have the relations* (4.6) *with the spherical coordinates* ρ, θ, ϕ.

4.2. Relation between the system (4.4) and harmonic vectors

Let us specify a relation between solutions of the system (4.4) and spatial potential fields.

Theorem 4.4. *Every solution of the system* (4.4) *in a domain* $\Omega \subset \mathbb{R}^3$ *generates a harmonic vector* $\mathbf{V} := (U_1, -\frac{1}{2}\, U_2, -\frac{1}{2}\, U_3)$ *in* Ω.

Proof. The equalities (1.2) for a vector $\mathbf{V} := (U_1, -\frac{1}{2}\, U_2, -\frac{1}{2}\, U_3)$ are corollaries of the system (4.4). Really, the system (4.4) contains the third equation and the fourth equation from (1.2). The second equation from (1.2) is a corollary of the third condition and the seventh condition of the system (4.4). Finally, the first equation from (1.2) is a corollary of the second condition and the eighth condition of the system (4.4). The theorem is proved. □

Theorem 4.5. *For any function $U_1 : \Omega \to \mathbb{R}$ harmonic in a simply connected domain $\Omega \subset \mathbb{R}^3$ there exist harmonic functions $U_k : \Omega \to \mathbb{R}$, $k = 2, 3, \ldots$, such that the conditions* (4.4) *are fulfilled in* Ω.

Proof. Let us add and subtract the second condition and the eighth condition as well as the third condition and the seventh condition of the system (4.4). Let us also add and subtract the fourth condition for $k = m$ and the ninth condition of the system (4.4) for $k = 2m+1$, where $m = 2, 3, \ldots$. Let us else add and subtract the fifth condition for $k = m$ and the ninth condition of system (4.4) for $k = 2m$, where $m = 2, 3, \ldots$. Thus, we rewrite the system (4.4) in the following equivalent form:

$$
\begin{aligned}
&\frac{\partial U_1}{\partial x} - \frac{1}{2}\frac{\partial U_2}{\partial y} - \frac{1}{2}\frac{\partial U_3}{\partial z} = 0\,,\\
&\frac{\partial U_3}{\partial y} - \frac{\partial U_2}{\partial z} = 0\,,\\
&\frac{\partial U_1}{\partial y} + \frac{1}{2}\frac{\partial U_2}{\partial x} = 0\,,\\
&\frac{\partial U_1}{\partial z} + \frac{1}{2}\frac{\partial U_3}{\partial x} = 0\,,\\
&\frac{\partial U_{2k}}{\partial x} = -\frac{\partial U_{2k-2}}{\partial z} - \frac{\partial U_{2k-1}}{\partial y}\,,\\
&\frac{\partial U_{2k+1}}{\partial x} = \frac{\partial U_{2k-2}}{\partial y} - \frac{\partial U_{2k-1}}{\partial z}\,,\\
&\frac{\partial U_{2k}}{\partial z} - \frac{\partial U_{2k+1}}{\partial y} = \frac{\partial U_{2k-2}}{\partial x}\,,\\
&\frac{\partial U_{2k}}{\partial y} + \frac{\partial U_{2k+1}}{\partial z} = \frac{\partial U_{2k-1}}{\partial x} \qquad k = 2, 3, \ldots .
\end{aligned}
\tag{4.10}
$$

First of all, note that there exists a harmonic vector $\mathbf{V_0} := (U_1, v_2^0, v_3^0)$ in the domain Ω. Moreover, for any vector $\mathbf{V} := (U_1, v_2, v_3)$ harmonic in Ω, the components v_2, v_3 are determined accurate within the real part and the imaginary part of any function $f_1(t)$ holomorphic in the domain $\{t = z + iy : (x, y, z) \in \Omega\}$ of the complex plane, i.e., the equalities

$$v_2(x, y, z) = v_2^0(x, y, z) + \Re f_1(z + iy), \;\; v_3(x, y, z) = v_3^0(x, y, z) + \Im f_1(z + iy)$$

are true for all $(x, y, z) \in \Omega$. Then, using Theorem 4.4, we find the functions U_2 and U_3, namely:

$$U_2 := -2\,v_2, \quad U_3 := -2\,v_3.$$

Now, let us show that the last four conditions of the system (4.10) allow us to determine the functions U_{2k}, U_{2k+1}, if the functions $U_2, U_3, \ldots, U_{2k-1}$ are already determined. Really, integrating the fifth equation and the sixth equation of the system (4.10), we obtain the expressions

$$U_{2k}(x,y,z) = -\int\limits_{x_0}^{x}\left(\frac{\partial U_{2k-2}(\tau,y,z)}{\partial z} + \frac{\partial U_{2k-1}(\tau,y,z)}{\partial y}\right) d\tau + \widetilde{u}_{2k}(z,y),$$

$$U_{2k+1}(x,y,z) = \int\limits_{x_0}^{x}\left(\frac{\partial U_{2k-2}(\tau,y,z)}{\partial y} - \frac{\partial U_{2k-1}(\tau,y,z)}{\partial z}\right) d\tau + \widetilde{u}_{2k+1}(z,y)$$

for (x,y,z) belonging to a certain neighborhood $\mathcal{N}\subset\Omega$ of any point $(x_0,y_0,z_0)\in\Omega$.

Substituting these expressions into the seventh equation and eighth equation of the system (4.10) and taking into account that U_{2k-2}, U_{2k-1} are harmonic functions in the domain $\mathcal{N}$, we obtain the following inhomogeneous Cauchy–Riemann system (see, for example, [29]) for finding the functions $\widetilde{u}_{2k}$, $\widetilde{u}_{2k+1}$:

$$\begin{aligned} \frac{\partial\widetilde{u}_{2k}(z,y)}{\partial z} - \frac{\partial\widetilde{u}_{2k+1}(z,y)}{\partial y} &= \left.\frac{\partial U_{2k-2}(x,y,z)}{\partial x}\right|_{x=x_0}, \\ \frac{\partial\widetilde{u}_{2k}(z,y)}{\partial y} + \frac{\partial\widetilde{u}_{2k+1}(z,y)}{\partial z} &= \left.\frac{\partial U_{2k-1}(x,y,z)}{\partial x}\right|_{x=x_0}. \end{aligned} \tag{4.11}$$

Solutions of the system (4.11) are determined accurate within the real part and the imaginary part of any function holomorphic in the domain $\{t = z+iy : (x,y,z)\in\mathcal{N}\}$ of the complex plane. Therefore, inasmuch as the domain Ω is simply connected, taking into account the uniqueness theorem for spatial harmonic functions, it is easy to continue the functions U_{2k}, U_{2k+1} defined in the neighborhood $\mathcal{N}$ into the domain Ω. The theorem is proved. □

Let us note that the functions U_{2m}, U_{2m+1} satisfying the last four conditions of the system (4.10) for $k = m \geq 2$ are determined accurate within the real part and the imaginary part of any function $f_m(t)$ holomorphic in the domain $\{t = z+iy : (x,y,z)\in\Omega\}$ of the complex plane, i.e., the equalities

$$U_{2m}(x,y,z) = U^0_{2m}(x,y,z) + \Re\, f_m(z+iy)\,,$$
$$U_{2m+1}(x,y,z) = U^0_{2m+1}(x,y,z) + \Im\, f_m(z+iy)$$

are true for all $(x,y,z)\in\Omega$, where U^0_{2m}, U^0_{2m+1} are functions forming together with the functions $U_1, U_2, \ldots, U_{2m-1}$ a particular solution of the system (4.10), in which $k = 2,3,\ldots,m$.

4.3. Monogenic functions and axial-symmetric potential fields

In the case where a spatial potential field is symmetric with respect to the axis Ox, a potential function $u(x,y,z)$ satisfying equation (1.3) is also symmetric with respect to the axis Ox, i.e., $u(x,y,z) = \varphi(x,r) = \varphi(x,-r)$, where $r := \sqrt{y^2+z^2}$, and φ is known as the *axial-symmetric potential.* Then in a meridian plane xOr there exists a function $\psi(x,r)$ known as the *Stokes flow function* such that the functions φ and ψ satisfy the following system of equations degenerating on the axis Ox:

$$r\,\frac{\partial\varphi(x,r)}{\partial x}=\frac{\partial\psi(x,r)}{\partial r}\,,\qquad r\,\frac{\partial\varphi(x,r)}{\partial r}=-\frac{\partial\psi(x,r)}{\partial x}\,. \tag{4.12}$$

Under the condition that there exist continuous second-order partial derivatives of the functions $\varphi(x,r)$ and $\psi(x,r)$, the system (4.12) implies the equation

$$r\Delta\varphi(x,r)+\frac{\partial\varphi(x,r)}{\partial r}=0 \tag{4.13}$$

for the axial-symmetric potential and the equation

$$r\Delta\psi(x,r)-\frac{\partial\psi(x,r)}{\partial r}=0 \tag{4.14}$$

for the Stokes flow function, where $\Delta:=\dfrac{\partial^2}{\partial x^2}+\dfrac{\partial^2}{\partial r^2}$.

We proved in the papers [4, 30] that in a domain convex in the direction of the axis Or the functions φ and ψ can be constructed by means of components of principal extensions of analytic functions of a complex variable into a corresponding domain of a special two-dimensional vector manifold in an infinite-dimensional commutative Banach algebra $\mathbb{H}_{\mathbb{C}}$ over the field of complex numbers.

In such a way for solutions of the system (4.12) we obtained integral expressions which were generalized for domains of general form (see [4, 31]). Using integral expressions for solutions of the system (4.12), in the papers [4, 32–35] we developed methods for solving boundary problems for axial-symmetric potentials and Stokes flow functions that have various applications in mathematical physics. In particular, the developed methods are applicable for solving a boundary problem about a streamline of the ideal incompressible fluid along an axial-symmetric body (see [4, 36]).

Now, let us consider a subalgebra

$$\mathbb{H}:=\{a=\sum_{k=1}^{\infty}a_k e_{2k-1}\ :\ a_k\in\mathbb{R},\ \sum_{k=1}^{\infty}|a_k|<\infty\}$$

of the algebra $\mathbb{F}$. In the paper [37] I. Mel'nichenko offered the algebra $\mathbb{H}$ for describing spatial axial-symmetric potential fields.

As in the papers [30, 38], consider a complexification

$$\mathbb{H}_{\mathbb{C}}:=\mathbb{H}\oplus i\mathbb{H}\equiv\{c=a+ib\ :\ a,b\in\mathbb{H}\}$$

of the algebra $\mathbb{H}$ such that the norm of element $g:=\sum\limits_{k=1}^{\infty}c_k e_{2k-1}\in\mathbb{H}_{\mathbb{C}}$ is given by means of the equality $\|g\|_{\mathbb{H}_{\mathbb{C}}}:=\sum\limits_{k=1}^{\infty}|c_k|$.

The algebra $\mathbb{H}_{\mathbb{C}}$ is isomorphic to the algebra $\mathbf{F_{cos}}$ of absolutely convergent trigonometric Fourier series

$$c(\tau) = \sum_{k=1}^{\infty} c_k\, i^{k-1} \cos(k-1)\tau$$

with complex coefficients c_k and the norm $\|c\|_{\mathbf{F_{cos}}} := \sum\limits_{k=1}^{\infty} |c_k|$. In this case, we have the isomorphism $e_{2k-1} \longleftrightarrow i^{k-1} \cos(k-1)\tau$ between basic elements.

The structure of maximal ideals of the algebra of absolutely convergent complex Fourier series is described in the monograph [39]. This description allows us to make a conclusion about the structure of maximal ideals of the algebra $\mathbb{H}_{\mathbb{C}}$ and about the fact that there do not exist inverse elements in $\mathbb{H}_{\mathbb{C}}$ for the basic elements e_k, where $k = 2, 3, \ldots$.

In particular, the set

$$\mathcal{I}_0 := \left\{ g \in \mathbb{H}_{\mathbb{C}} : \sum_{k=1}^{\infty} (-1)^k (\Re c_{2k-1} - \Im c_{2k}) = 0, \sum_{k=1}^{\infty} (-1)^k (\Re c_{2k} + \Im c_{2k-1}) = 0 \right\}$$

is a maximum ideal of the algebra $\mathbb{H}_{\mathbb{C}}$.

Let $f_{\mathcal{I}_0} : \mathbb{H}_{\mathbb{C}} \to \mathbb{C}$ be the linear functional defined by the equality

$$f_{\mathcal{I}_0}\left(\sum_{k=1}^{\infty} c_k e_{2k-1} \right) := -\sum_{k=1}^{\infty} (-1)^k (\Re\, c_{2k-1} - \Im\, c_{2k}) - i \sum_{k=1}^{\infty} (-1)^k (\Re\, c_{2k} + \Im\, c_{2k-1}).$$

It is evident that the ideal $\mathcal{I}_0$ is the kernel of the functional $f_{\mathcal{I}_0}$ and $f(e_1) = 1$, i.e., $f_{\mathcal{I}_0}$ is a continuous multiplicative functional.

Consider the Cartesian plane $\mu := \{\zeta = xe_1 + re_3 \ : \ x, r \in \mathbb{R}\}$. For a domain $E \subset \mathbb{R}^2$ we use coordinated denotations for congruent domains of the plane μ and the complex plane $\mathbb{C}$, namely: $E_\zeta := \{\zeta = xe_1 + re_3 \ : \ (x, r) \in E\} \subset \mu$ and $E_z := \{z = x + ir \ : \ (x, r) \in E\} \subset \mathbb{C}$.

We say that a continuous function $\Phi : E_\zeta \to \mathbb{H}_{\mathbb{C}}$ is *monogenic* in a domain E_ζ if Φ is differentiable in the sense of Gateaux in every point of E_ζ, i.e., for every $\zeta \in E_\zeta$ there exists an element $\Phi'(\zeta) \in \mathbb{H}_{\mathbb{C}}$ such that the equality (1.7) is fulfilled for all $h \in \mu$.

The proof of the following theorem is similar to the proof of Theorem 4.1.

Theorem 4.6. *Let a function $\Phi : E_\zeta \to \mathbb{H}_{\mathbb{C}}$ be continuous in a domain $E_\zeta \subset \mu$ and the functions $U_k : E \to \mathbb{C}$ from the decomposition*

$$\Phi(xe_1 + re_3) = \sum_{k=1}^{\infty} U_k(x, r)\ e_{2k-1}$$

be differentiable in E. In order that the function Φ be monogenic in the domain E_ζ, it is necessary and sufficient that the conditions

$$\frac{\partial \Phi(\zeta)}{\partial y} = \frac{\partial \Phi(\zeta)}{\partial x}\, e_3 \tag{4.15}$$

be satisfied and that the following relations be fulfilled in E:

$$\sum_{k=1}^{\infty} \left| \frac{\partial U_k(x,r)}{\partial x} \right| < \infty, \tag{4.16}$$

$$\lim_{\varepsilon \to 0+0} \sum_{k=1}^{\infty} \left| U_k(x+\varepsilon h_1, r+\varepsilon h_2,) - U_k(x,r) - \frac{\partial U_k(x,r)}{\partial x}\, \varepsilon h_1 \right.$$
$$\left. - \frac{\partial U_k(x,r)}{\partial r}\, \varepsilon h_2 \right| \varepsilon^{-1} = 0 \quad \forall\, h_1, h_2 \in \mathbb{R}\,. \tag{4.17}$$

Note that the conditions (4.15) rewritten in expanded form

$$\frac{\partial U_1(x,r)}{\partial r} = -\frac{1}{2}\, \frac{\partial U_2(x,r)}{\partial x},$$
$$\frac{\partial U_2(x,r)}{\partial r} = \frac{\partial U_1(x,r)}{\partial x} - \frac{1}{2}\, \frac{\partial U_3(x,r)}{\partial x},$$
$$\frac{\partial U_k(x,r)}{\partial r} = \frac{1}{2}\, \frac{\partial U_{k-1}(x,r)}{\partial x} - \frac{1}{2}\, \frac{\partial U_{k+1}(x,r)}{\partial x}, \qquad k = 3, 4, \ldots,$$

are similar by nature to the Cauchy–Riemann conditions for monogenic functions of a complex variable, and the relations (4.16), (4.17) are conditioned by the infinite dimensionality of the algebra $\mathbb{H}_{\mathbb{C}}$.

In the papers [4, 30] we established relations between monogenic functions taking values in the algebra $\mathbb{H}_{\mathbb{C}}$ and solutions of the system (4.12) in so-called proper domains.

We call E_z a *proper* domain in $\mathbb{C}$, provided that for every $z \in E_z$ with $\Im\, z \neq 0$ the domain E_z contains the segment connecting points z and $\bar{z}$. In this case E_ζ is also called a *proper* domain in μ.

Let A be the linear operator which assigns the function $F : E_z \to \mathbb{C}$ to every function $\Phi : E_\zeta \to \mathbb{H}_{\mathbb{C}}$ by the formula $F(z) := f_{\mathcal{I}_0}(\Phi(\zeta))$, where $z = x + ir$ and $\zeta = xe_1 + re_3$. It is easy to prove that if Φ is a monogenic function in E_ζ, then F is an analytic function in E_z.

Theorem 4.7 ([4, 30]). *If E_ζ is a proper domain in μ, then every monogenic function $\Phi : E_\zeta \to \mathbb{H}_{\mathbb{C}}$ can be expressed in the form*

$$\Phi(\zeta) = \frac{1}{2\pi i} \int_\gamma (A\Phi)(t)(te_1 - \zeta)^{-1}\, dt + \Phi_0(\zeta) \qquad \forall\, \zeta \in E_\zeta\,, \tag{4.18}$$

where γ is an arbitrary closed Jordan rectifiable curve in D_z that embraces the segment connecting the points $z = f_{\mathcal{I}_0}(\zeta)$ and $\bar{z}$, and $\Phi_0 : E_\zeta \to \mathcal{I}_0$ is a monogenic function taking values in the ideal $\mathcal{I}_0$.

Note that under the conditions of Theorem 4.7 the integral in the equality (4.18) is the principal extension (see [8, p. 165]) of an analytic function $F = A\Phi$ into the domain D_ζ. Thus, the algebra of monogenic functions in E_ζ can be decomposed into the direct sum of the algebra of principal extensions of complex-valued analytic functions and the algebra of monogenic functions taking values in the ideal $\mathcal{I}_0$.

In the papers [4, 30], for every function $F : E_z \to \mathbb{C}$ analytic in a proper domain E_z we constructed explicitly the decomposition of the principal extension of F into the domain E_ζ with respect to the basis $\{e_{2k-1}\}_{k=1}^{\infty}$:

$$\frac{1}{2\pi i}\int\limits_{\gamma}(te_1-\zeta)^{-1}F(t)\,dt = U_1(x,r)\;e_1 + 2\sum_{k=2}^{\infty}\;U_k(x,r)\;e_{2k-1}\,, \tag{4.19}$$

where

$$U_k(x,r) := \frac{1}{2\pi i}\int\limits_{\gamma}\frac{F(t)}{\sqrt{(t-z)(t-\bar z)}}\left(\frac{\sqrt{(t-z)(t-\bar z)}-(t-x)}{r}\right)^{k-1}dt\,, \tag{4.20}$$

$\zeta = xe_1 + re_3$ and $z = x + ir$ for $(x,r) \in E$, and the curve γ has the same properties as in Theorem 4.7, and $\sqrt{(t-z)(t-\bar z)}$ is a continuous branch of this function analytic with respect to t outside of the segment mentioned in Theorem 4.7. Note that for every $z \in E_z$ with $\Im z = 0$, we define $\sqrt{(t-z)(t-\bar z)} := t - z$.

In the following theorem we describe relations between principal extensions of analytic functions into the plane μ and solutions of the system (4.12).

Theorem 4.8 ([4, 30]). *If $F : E_z \to \mathbb{C}$ is an analytic function in a proper domain E_z, then the first and the second components of principal extension* (4.19) *of function F into the domain E_ζ generate the solutions φ and ψ of system* (4.12) *in E by the formulas*

$$\varphi(x,r) = U_1(x,r), \quad \psi(x,r) = r\,U_2(x,r)\,. \tag{4.21}$$

From the relations (4.19)–(4.21) it follows that the functions

$$\varphi(x,r) = \frac{1}{2\pi i}\int\limits_{\gamma}\frac{F(t)}{\sqrt{(t-z)(t-\bar z)}}\,dt\,, \tag{4.22}$$

$$\psi(x,r) = -\frac{1}{2\pi i}\int\limits_{\gamma}\frac{F(t)\;(t-x)}{\sqrt{(t-z)(t-\bar z)}}\,dt\,, \qquad z = x + ir\,, \tag{4.23}$$

are solutions of system (4.12) in the domain E.

In the following theorem we describe relations between components U_k of hypercomplex analytic function (4.19) and solutions of elliptic equations degenerating on the axis Ox.

Theorem 4.9 ([40, 41]). *If $F : E_z \to \mathbb{C}$ is an analytic function in a proper domain E_z, then the components U_k of principal extension* (4.19) *of function F into the domain E_ζ satisfy the equations*

$$r^2\,\Delta U_k(x,r) + r\,\frac{\partial U_k(x,r)}{\partial r} - (k-1)^2\,U_k(x,r) = 0\,, \quad k = 1,2,\ldots\,,$$

in the domain E. In addition, the function

$$\psi_k(x,r) := r^{k-1}\, U_k(x,r)$$

is a solution in E of the equation

$$r\,\Delta\psi_k(x,r) - (2k-3)\,\frac{\partial\psi_k(x,r)}{\partial r} = 0\,, \quad k = 1,2,\ldots\,.$$

In the following theorem we establish an expression of generalized axial-symmetric potential via components U_k of the hypercomplex analytic function (4.19).

Theorem 4.10 ([42]). *If $F : E_z \to \mathbb{C}$ is an analytic function in a proper domain E_z, then the function*

$$u(x,r) := \left(1 + \sum_{n=1}^{\infty} \frac{(m-1)(m-3)\ldots(m-4n+1)}{2^{4n}(2n!)}\, C_{2n}^{n}\right) U_1(x,r) \tag{4.24}$$
$$+ \sum_{k=1}^{\infty}\left(\sum_{n=k}^{\infty} \frac{(m-1)(m-3)\ldots(m-4n+1)}{2^{4n}(2n!)}\, C_{2n}^{n-k}\right) U_{4k+1}(x,r)$$
$$+ \sum_{k=0}^{\infty}\left(\sum_{n=k}^{\infty} \frac{(m-1)(m-3)\ldots(m-4n-1)}{2^{4n+2}(2n+1)!}\, C_{2n+1}^{n-k}\right) U_{4k+3}(x,r),$$

where C_n^k are the binomial coefficients, satisfies the equation

$$\Delta u(x,r) + \frac{m}{r}\,\frac{\partial u(x,r)}{\partial r} = 0 \tag{4.25}$$

on the set $\{(x,r) \in D : r \neq 0\}$ for $m \geq 1$.

Furthermore, the function (4.24) is expressed in the form

$$u(x,r) = \frac{2^{(m-3)/2}}{\pi i\, |r|^{m-1}} \int_{\gamma'} \frac{F(t)}{\sqrt{(t-z)(t-\bar{z})}}\, [(t-z)(t-\bar{z})]^{(m-1)/2}\, dt\,,$$

where γ' is an arbitrary closed Jordan rectifiable curve in E_z that contains the points z and $\bar{z}$ and embraces the set $\{x + i\eta : |\eta| < |r|\}$, and $[(t-z)(t-\bar{z})]^{(m-1)/2}$ is a continuous branch of this function, analytic with respect to t outside of the cut $\{x + i\eta : |\eta| \geq |r|\}$.

Let us write expansions with respect to the basis $\{e_{2k-1}\}_{k=1}^{\infty}$ of some elementary analytic functions of the variable $\zeta = xe_1 + re_3$ (note that in view of the isomorphism between algebras $\mathbb{H}_{\mathbb{C}}$ and $\mathbf{F}_{\mathbf{cos}}$, the construction of expansions of this sort is reduced to a determination of relevant Fourier coefficients). The expansion of a power function has the form

$$\zeta^n = (x^2 + r^2)^{n/2}\left(P_n(\cos\vartheta)\, e_1 + 2\sum_{k=1}^{n} \frac{(\operatorname{sgn} r)^k\, n!}{(n+k)!}\, P_n^k(\cos\vartheta)\, e_{2k+1}\right),$$

where n is a positive integer, $\cos\vartheta := x(x^2+r^2)^{-1/2}$,

$$\operatorname{sgn} r := \begin{cases} 1, & \text{for } r \geq 0, \\ -1, & \text{for } r < 0, \end{cases}$$

and Legendre polynomials P_n and associated Legendre polynomials P_n^m are defined to be the equalities (4.8).

For the functions e^ζ, $\sin\zeta$ and $\cos\zeta$ we have

$$e^\zeta = e^x\Bigl(J_0(r)\,e_1 + 2\sum_{k=1}^{\infty} J_k(r)\,e_{2k+1}\Bigr),$$

$$\sin\zeta = \sin x\Bigl(J_0(ir)\,e_1 + 2\sum_{k=1}^{\infty} J_{2k}(ir)\,e_{4k+1}\Bigr) - 2i\,\cos x\sum_{k=1}^{\infty} J_{2k-1}(ir)\,e_{4k-1},$$

$$\cos\zeta = \cos x\Bigl(J_0(ir)\,e_1 + 2\sum_{k=1}^{\infty} J_{2k}(ir)\,e_{4k+1}\Bigr) + 2i\,\sin x\sum_{k=1}^{\infty} J_{2k-1}(ir)\,e_{4k-1},$$

where Bessel functions J_m are defined by the equality (4.9).

For the functions ζ^{-1} and $\operatorname{Ln}\zeta$ we obtain

$$\zeta^{-1} = \begin{cases} \dfrac{1}{\sqrt{x^2+r^2}}\Bigl(e_1 + 2\sum\limits_{k=1}^{\infty}(-1)^k\Bigl(\dfrac{\sqrt{x^2+r^2}-x}{r}\Bigr)^k e_{2k+1}\Bigr) & \text{for } x>0, \\ -\dfrac{1}{\sqrt{x^2+r^2}}\Bigl(e_1 + 2\sum\limits_{k=1}^{\infty}\Bigl(\dfrac{\sqrt{x^2+r^2}+x}{r}\Bigr)^k e_{2k+1}\Bigr) & \text{for } x<0, \end{cases}$$

$$\operatorname{Ln}\zeta = \begin{cases} \Bigl(\ln\dfrac{\sqrt{x^2+r^2}+x}{2} + 2m\pi i\Bigr)e_1 \\ \qquad + 2\sum\limits_{k=1}^{\infty}\dfrac{(-1)^{k+1}}{k}\Bigl(\dfrac{\sqrt{x^2+r^2}-x}{r}\Bigr)^k e_{2k+1} & \text{for } x>0, \\ \Bigl(\ln\dfrac{\sqrt{x^2+r^2}-x}{2} + (2m+1)\pi i\Bigr)e_1 \\ \qquad + 2\sum\limits_{k=1}^{\infty}\dfrac{1}{k}\Bigl(\dfrac{\sqrt{x^2+r^2}+x}{r}\Bigr)^k e_{2k+1} & \text{for } x<0, \end{cases}$$

where m is an integer number. In this case, the functions ζ^{-1} and $\operatorname{Ln}\zeta$ are not defined for $x=0$.

4.4. Integral expressions for axial-symmetric potential and the Stokes flow function

In the papers [4, 31] we generalized integral expressions (4.22) and (4.23) for the axial-symmetric potential and the Stokes flow function, respectively, to the case of an arbitrary simply connected domain symmetric with respect to the axis Ox.

Below in Section 4, E is a bounded simply connected domain symmetric with respect to the axis Ox. For every $z \in E_z$ with $\Im z \neq 0$, we fix an arbitrary Jordan rectifiable curve $\Gamma_{z\bar z}$ in E_z which connects the points z and $\bar z$. In this case, let

$\sqrt{(t-z)(t-\bar{z})}$ be a continuous branch of this function analytic with respect to t outside of the cut along $\Gamma_{z\bar{z}}$. As earlier, for every $z \in E_z$ with $\Im z = 0$ we define $\sqrt{(t-z)(t-\bar{z})} := t - z$.

Theorem 4.11 ([31, 35]). *If F is an analytic function in the domain E_z, then the functions (4.22) and (4.23) are solutions of system (4.12) in E, where γ is an arbitrary closed Jordan rectifiable curve in E_z which embraces $\Gamma_{z\bar{z}}$. Functions (4.22) and (4.23) are also solutions of equations (4.13) and (4.14), respectively.*

It is evident that if the boundary ∂E_z is a Jordan rectifiable curve and the function F belongs to the Smirnov class E_1 (see [43], p. 205) in the domain E_z, then the formulas (4.22) and (4.23) can be transformed to the form

$$\varphi(x,r) = \frac{1}{2\pi i} \int_{\partial E_z} \frac{F(t)}{\sqrt{(t-z)(t-\bar{z})}}\, dt\,, \tag{4.26}$$

$$\psi(x,r) = -\frac{1}{2\pi i} \int_{\partial E_z} \frac{F(t)\,(t-x)}{\sqrt{(t-z)(t-\bar{z})}}\, dt\,, \qquad z = x + ir\,, \tag{4.27}$$

for all $(x,r) \in E$, where $F(t)$ is the angular boundary value of the function F which is known to exist almost everywhere on ∂E_z.

If $z \in E_z$ with $\Im z \neq 0$ and $\alpha \in \mathbb{R}$, then $\big((t-z)(t-\bar{z})\big)^{\alpha}$ is understood as a continuous branch of the function $L(t) := \big((t-z)(t-\bar{z})\big)^{\alpha}$ analytic with respect to t outside of the cut along a Jordan curve that successively connects the points z, ∞ and $\bar{z}$, and the only common points of which with the set $\Gamma_{z\bar{z}} \cup \mathbb{R}$ are the points z and $\bar{z}$. In this case $L(t) > 0$ for all $t > \max_{\tau \in \Gamma_{z\bar{z}}} \Re\, \tau$.

In the following theorem we establish an integral expression of generalized axial-symmetric potential that is a generalisation of integral expressions obtained by A.G. Mackie [44], P. Henrici [45], Yu.P. Krivenkov [46] and G.N. Polozhii [47].

Theorem 4.12 ([48]). *If $m > 0$ and F is an analytic function in the domain E_z, then the function*

$$u(x,y) = \frac{1}{2\pi i |y|^{m-1}} \int_{\Gamma_{z\bar{z}}} F(t)\big((t-z)(t-\bar{z})\big)^{m/2-1}\, dt\,, \quad z = x + iy, \tag{4.28}$$

satisfies equation (4.25) on the set $\{(x,y) \in E : y \neq 0\}$. Moreover, there exists the limit

$$\lim_{(x,y)\to(x_0,0)} u(x,y) = \frac{\mathrm{B}\left(\frac{m}{2},\frac{1}{2}\right)}{2\pi} F(x_0) \qquad \forall\, x_0 \in \mathbb{R} : (x_0,0) \in E\,,$$

where $\mathrm{B}(p,q)$ is the Euler beta function.

We proved converse theorems on integral expressions of axial symmetric potentials and Stokes flow functions in domains of a meridian plane.

Theorem 4.13 ([32, 35]). *Suppose that the axial-symmetric potential $\varphi(x,r)$ is even with respect to the variable r in the domain E. Then there exists the unique function F analytic in the domain E_z and satisfying the condition*

$$F(\bar{z}) = \overline{F(z)} \qquad \forall\, z \in D_z \tag{4.29}$$

such that the equality (4.22) *is fulfilled for all* $(x,r) \in E$.

Theorem 4.14 ([34, 35]). *Suppose that the Stokes flow function $\psi(x,r)$ is even with respect to the variable r in the domain E and satisfies the additional assumption*

$$\psi(x,0) \equiv 0 \quad \forall\,(x,0) \in E\,. \tag{4.30}$$

Then there exists a function F_0 analytic in the domain E_z such that the equality (4.23) *is fulfilled with $F = F_0$ for all $(x,r) \in E$. Moreover, any analytic function F which satisfies the condition* (4.29) *and the equality* (4.23) *for all $(x,r) \in D$ is expressed in the form $F(z) = F_0(z) + C$, where C is a real constant.*

Note that the requirement (4.30) is natural. For example, for the model of steady flow of an ideal incompressible fluid without sources and vortexes it means that the axis Ox is a line of flow.

Using integral expressions (4.26) and (4.27), we developed a method for effectively solving boundary problems for axial-symmetric potential fields (see [4, 32–36]).

5. Monogenic functions in the biharmonic algebra

5.1. Biharmonic algebra

We say that an associative commutative two-dimensional algebra $\mathbb{B}$ with the unit 1 over the field $\mathbb{C}$ is *biharmonic* if in $\mathbb{B}$ there exists a *biharmonic* basis $\{e_1, e_2\}$ satisfying the conditions

$$(e_1^2 + e_2^2)^2 = 0, \qquad e_1^2 + e_2^2 \neq 0\,. \tag{5.1}$$

V.F. Kovalev and I.P. Mel'nichenko [49] found a multiplication table for a biharmonic basis $\{e_1, e_2\}$:

$$e_1 = 1, \qquad e_2^2 = e_1 + 2ie_2. \tag{5.2}$$

In the paper [50] I.P. Mel'nichenko proved that there exists a unique biharmonic algebra $\mathbb{B}$ with a non-biharmonic basis $\{1, \rho\}$, for which $\rho^2 = 0$. Moreover, he constructed all biharmonic bases in the form:

$$e_1 = \alpha_1 + \alpha_2\,\rho\,, \qquad e_2 = \pm i\left(\alpha_1 + \left(\alpha_2 - \frac{1}{2\alpha_1}\right)\rho\right), \tag{5.3}$$

where complex numbers $\alpha_1 \neq 0$, α_2 can be chosen arbitrarily. In particular, for the basis (5.2) in the equalities (5.3) we choose $\alpha_1 = 1$, $\alpha_2 = 0$ and $+$ of the double sign:

$$e_1 = 1, \qquad e_2 = i - \frac{i}{2}\,\rho, \tag{5.4}$$

Note that every analytic function $\Phi(\zeta)$ of the variable $\zeta = xe_1 + ye_2$ satisfies the two-dimensional biharmonic equation

$$(\Delta_2)^2 U(x,y) := \left(\frac{\partial^4}{\partial x^4} + 2\,\frac{\partial^4}{\partial x^2 \partial y^2} + \frac{\partial^4}{\partial y^4}\right) U(x,y) = 0 \tag{5.5}$$

owing to the relations (5.1) and $(\Delta_2)^2\Phi = \Phi^{(4)}(\zeta)\,(e_1^2 + e_2^2)^2$.

The algebra $\mathbb{B}$ has the unique maximum ideal $\mathcal{I} := \{\lambda\rho : \lambda \in \mathbb{C}\}$ which is also the radical of $\mathbb{B}$. In what follows, $f : \mathbb{B} \to \mathbb{C}$ is the linear functional such that the maximum ideal $\mathcal{I}$ is its kernel and $f(1) = 1$.

5.2. Monogenic functions given in a biharmonic plane. Cauchy–Riemann conditions

Consider a *biharmonic plane* $\mu_{e_1,e_2} := \{\zeta = x\,e_1 + y\,e_2 : x, y \in \mathbb{R}\}$ which is a linear envelope generated by the elements e_1, e_2 of biharmonic basis (5.3). In what follows, $\zeta = x\,e_1 + y\,e_2$ and $x, y \in \mathbb{R}$.

Let G_ζ be a domain in the biharmonic plane μ_{e_1,e_2}. Inasmuch as divisors of zero don't belong to the plane μ_{e_1,e_2}, the Gateaux derivative of function $\Phi : G_\zeta \to \mathbb{B}$ coincides with the derivative

$$\Phi'(\zeta) := \lim_{h\to 0,\, h\in\mu_{e_1,e_2}} \big(\Phi(\zeta + h) - \Phi(\zeta)\big)\, h^{-1}\,.$$

Therefore, we define *monogenic* functions as functions $\Phi : G_\zeta \to \mathbb{B}$ for which the derivative $\Phi'(\zeta)$ exists in every point $\zeta \in G_\zeta$.

It is established in the paper [49] that a function $\Phi(\zeta)$ is monogenic in a domain of biharmonic plane generated by the biharmonic basis (5.4) if and only if the following Cauchy–Riemann condition is satisfied:

$$\frac{\partial \Phi(\zeta)}{\partial y} = \frac{\partial \Phi(\zeta)}{\partial x}\, e_2. \tag{5.6}$$

It can similarly be proved that a function $\Phi : G_\zeta \to \mathbb{B}$ is monogenic in a domain G_ζ of an arbitrary biharmonic plane μ_{e_1,e_2} if and only if the following equality is fulfilled:

$$\frac{\partial \Phi(\zeta)}{\partial y}\, e_1 = \frac{\partial \Phi(\zeta)}{\partial x}\, e_2 \qquad \forall\, \zeta = xe_1 + ye_2 \in G_\zeta. \tag{5.7}$$

Below in Section 5, all stated results are obtained jointly with S.V. Gryshchuk (see also [51]).

5.3. A constructive description of monogenic functions given in a biharmonic plane

Let $D := f(G_\zeta)$ and A be the linear operator which assigns the function $F : D \to \mathbb{C}$ to every function $\Phi : G_\zeta \to \mathbb{B}$ by the formula $F(\xi) := f(\Phi(\zeta))$, where $\xi := f(\zeta) = \alpha_1(x \pm iy)$.

It is evident that if Φ is a monogenic function in the domain G_ζ, then F is an analytic function in the domain D, i.e., F is either holomorphic in the case where $\xi = \alpha_1(x + iy)$ or antiholomorphic in the case where $\xi = \alpha_1(x - iy)$.

The following theorem can be proved similarly to Theorem 2.3.

Theorem 5.1. *Every monogenic function $\Phi : G_\zeta \to \mathbb{B}$ can be expressed in the form*

$$\Phi(\zeta) = \frac{1}{2\pi i} \int_{\Gamma_\zeta} (A\Phi)(t)(t-\zeta)^{-1}\, dt + \Phi_0(\zeta) \qquad \forall \zeta \in G_\zeta, \tag{5.8}$$

where Γ_ζ is an arbitrary closed rectifiable curve in D that embraces the point $f(\zeta)$, and $\Phi_0 : G_\zeta \to \mathcal{I}$ is a monogenic function taking values in the radical $\mathcal{I}$.

Note that the complex number $\xi = f(\zeta)$ is the spectrum of $\zeta \in \mathbb{B}$, and the integral in the equality (5.8) is the principal extension (see [8, p. 165]) of analytic function $F(\xi) = (A\Phi)(\xi)$ of the complex variable ξ into the domain G_ζ.

It follows from Theorem 5.1 that the algebra of monogenic in G_ζ functions is decomposed into the direct sum of the algebra of principal extensions of analytic functions of a complex variable and the algebra of monogenic in G_ζ functions taking values in the radical $\mathcal{I}$.

In the following theorem we describe all monogenic functions given in the domain G_ζ and taking values in the radical $\mathcal{I}$.

Theorem 5.2. *Every monogenic function $\Phi_0 : G_\zeta \to \mathcal{I}$ can be expressed in the form*

$$\Phi_0(\zeta) = F_0(\xi)\rho \qquad \forall \zeta \in G_\zeta\,, \tag{5.9}$$

where $F_0 : D \to \mathbb{C}$ is an analytic function and $\xi = f(\zeta)$.

Proof. Substituting the function (5.9) in the equality (5.7) in place of Φ and taking into account the equalities $\rho e_1 = \alpha_1 \rho$, $\rho e_2 = \pm\alpha_1 i\rho$, we get

$$\alpha_1 \frac{\partial F_0(\xi)}{\partial y}\rho = \pm\alpha_1 \frac{\partial F_0(\xi)}{\partial x}\, i\,\rho \quad \forall \xi \in D. \tag{5.10}$$

From the equality (5.10), taking into account the uniqueness of decomposition of elements of $\mathbb{B}$ with respect to the basis $\{1, \rho\}$, we obtain the equality

$$\frac{\partial F_0(\xi)}{\partial y} = \pm i \frac{\partial F_0(\xi)}{\partial x} \quad \forall \xi \in D.$$

Thus, the function F_0 is either holomorphic in D in the case where $\xi = \alpha_1(x+iy)$ or antiholomorphic in D in the case where $\xi = \alpha_1(x-iy)$, i.e., F_0 is analytic in the domain D. The theorem is proved. □

It follows from the equalities (5.8), (5.9) that any monogenic function $\Phi : G_\zeta \to \mathbb{B}$ can be constructed by means of two complex analytic in D functions F, F_0 in the form:

$$\Phi(\zeta) = \frac{1}{2\pi i} \int_{\Gamma_\zeta} F(t)(t-\zeta)^{-1} dt + F_0(f(\zeta))\rho \qquad \forall \zeta \in G_\zeta. \tag{5.11}$$

Moreover, by using the expression

$$(t-\zeta)^{-1} = \frac{1}{t-\xi} - \frac{1}{2\alpha_1}\frac{2\alpha_2\xi \pm iy}{(t-\xi)^2}\,\rho,$$
$$\forall\,\zeta = xe_1 + ye_2 \in G_\zeta \qquad \forall\,t \in \mathbb{C} : t \neq \xi = \alpha_1(x \pm iy),$$

the principal extension of an analytic in D function F into G_ζ can explicitly be constructed in the form

$$\frac{1}{2\pi i}\int\limits_{\Gamma_\zeta} F(t)(t-\zeta)^{-1}dt = F(\xi) - \frac{F'(\xi)}{\alpha_1}\left(\alpha_2\xi \pm \frac{iy}{2}\right)\rho, \qquad (5.12)$$
$$\xi = f(\zeta) \in D, \quad \forall\,\zeta = xe_1 + ye_2 \in G_\zeta.$$

Note that in a particular case, in the paper [49] principal extensions of analytic functions of a complex variable was explicitly constructed into the biharmonic plane generated by the biharmonic basis (5.4).

The following theorem can be proved similarly to Theorem 2.7.

Theorem 5.3. *Every monogenic function $\Phi : G_\zeta \to \mathbb{B}$ has derivatives of all orders in the domain G_ζ.*

5.4. Isomorphism of algebras of monogenic functions

There is an isomorphism between algebras of monogenic functions at transition from a biharmonic basis to another one. By $\mathcal{M}(\mu_{e_1,e_2}, G_\zeta)$ we denote the algebra of monogenic functions in a domain $G_\zeta \subset \mu_{e_1,e_2}$.

Theorem 5.4. *Let $\{e_1, e_2\}$ be the biharmonic basis composed of the elements* (5.4) *and $\{\tilde e_1, \tilde e_2\}$ be an arbitrary biharmonic basis composed of elements of the form* (5.3). *Let G_ζ be a domain in the biharmonic plane μ_{e_1,e_2} and $\widetilde G_{\tilde\zeta} := \{\widetilde\zeta = x\tilde e_1 \pm y\tilde e_2 : \zeta = xe_1 + ye_2 \in G_\zeta\}$ be the congruent domain in the biharmonic plane $\mu_{\tilde e_1,\tilde e_2}$. Then the algebras $\mathcal{M}(\mu_{e_1,e_2}, G_\zeta)$, $\mathcal{M}(\mu_{\tilde e_1,\tilde e_2}, \widetilde G_{\tilde\zeta})$ are isomorphic, and the correspondence $\mathcal{M}(\mu_{e_1,e_2}, G_\zeta) \ni \Phi \longleftrightarrow \widetilde\Phi \in \mathcal{M}(\mu_{\tilde e_1,\tilde e_2}, \widetilde G_{\tilde\zeta})$ are established by the equality*

$$\widetilde\Phi(\widetilde\zeta) = \Phi(\zeta) + \Phi'(\zeta)(xr_1 + yr_2)\rho,$$

where $r_1 := \alpha_2/\alpha_1$, $r_2 := i(\alpha_1^2 + 2\alpha_1\alpha_2 - 1)/(2\alpha_1^2)$ and α_1, α_2 are the same complex numbers which are situated in the equalities of the form (5.3) *for elements of the basis $\{\tilde e_1, \tilde e_2\}$.*

5.5. A representation of a biharmonic function in the form of the first component of a monogenic function

In what follows, the basic elements e_1, e_2 are defined by the equalities (5.4) and $\zeta = x\,e_1 + y\,e_2$, $z = x + iy$ and $x, y \in \mathbb{R}$.

$U : G \to \mathbb{R}$ is called a *biharmonic* function in a domain $G \subset \mathbb{R}^2$ if it satisfies equation (5.5) in G.

We shall prove that every biharmonic function $U_1(x, y)$ in a bounded simply connected domain $G \subset \mathbb{R}^2$ is the first component of some monogenic function

$$\Phi(\zeta) = U_1(x, y)\, e_1 + U_2(x, y)\, ie_1 + U_3(x, y)\, e_2 + U_4(x, y)\, ie_2, \tag{5.13}$$

in the corresponding domain $G_\zeta := \{\zeta = xe_1 + ye_2 : (x, y) \in G\}$ of the biharmonic plane μ_{e_1,e_2}, where $U_k : G \to \mathbb{R}$ for $k = \overline{1, 4}$.

At first, consider some auxiliary statements.

Lemma 5.5. *Every monogenic function* (5.13) *with* $U_1 \equiv 0$ *is of the form*

$$\begin{aligned} \Phi(\zeta) = i(-ax^2 + kx - ay^2 - by + n) + e_2(2ay^2 + 2by + c) \\ + ie_2(-2axy - bx + ky + m) \qquad \forall\, \zeta \in \mu_{e_1,e_2}, \end{aligned} \tag{5.14}$$

where a, b, c, k, m, n *are arbitrary real constants.*

To prove Lemma 5.5, taking into account the identity $U_1 \equiv 0$, one should integrate the Cauchy–Riemann condition (5.6) rewritten in expanded form:

$$\begin{aligned} 0 &= \frac{\partial U_3(x, y)}{\partial x}, \\ \frac{\partial U_2(x, y)}{\partial y} &= \frac{\partial U_4(x, y)}{\partial x}, \\ \frac{\partial U_3(x, y)}{\partial y} &= -2\frac{\partial U_4(x, y)}{\partial x}, \\ \frac{\partial U_4(x, y)}{\partial y} &= \frac{\partial U_2(x, y)}{\partial x} + 2\frac{\partial U_3(x, y)}{\partial x}. \end{aligned}$$

Lemma 5.6. *If F is a holomorphic function in a bounded simply connected domain $D \subset \mathbb{C}$, then the functions*

$$\begin{aligned} \Phi_1(\zeta) &= u(x, y) + iv(x, y) - e_2 v(x, y) + ie_2 u(x, y), \\ \Phi_2(\zeta) &= yu(x, y) + iyv(x, y) + e_2\left(\mathcal{U}(x, y) - yv(x, y)\right) \\ &\quad + ie_2\left(\mathcal{V}(x, y) + yu(x, y)\right), \\ \Phi_3(\zeta) &= xu(x, y) + ixv(x, y) + e_2\left(\mathcal{V}(x, y) - xv(x, y)\right) \\ &\quad + ie_2\left(xu(x, y) - \mathcal{U}(x, y)\right) \quad \forall\, \zeta = xe_1 + ye_2 \in G_\zeta \end{aligned}$$

are monogenic in the domain $G_\zeta \equiv \{\zeta = xe_1 + ye_2 : x + iy \in D\}$ of a biharmonic plane μ_{e_1,e_2}, where

$$\begin{aligned} u(x, y) &:= \Re F(\xi), \quad v(x, y) := \Im F(\xi), \\ \mathcal{U}(x, y) &:= \Re \mathcal{F}(\xi), \quad \mathcal{V}(x, y) := \Im \mathcal{F}(\xi) \quad \forall\, \xi \in D \end{aligned}$$

and $\mathcal{F}$ is a primitive function for the function F.

To prove Lemma 5.6, it is easy to show that the functions Φ_1, Φ_2, Φ_3 satisfy the conditions of the form (5.6).

It is well known that every biharmonic function $U_1(x,y)$ in the domain G is expressed by the Goursat formula

$$U_1(x,y) = \Re\left(\varphi(\xi) + \bar{\xi}\psi(\xi)\right), \qquad \xi = x + iy, \tag{5.15}$$

where φ, ψ are holomorphic functions in the domain $D \equiv \{x + iy : (x,y) \in G\}$, $\bar{\xi} := x - iy$.

Theorem 5.7. *Every biharmonic function $U_1(x,y)$ in a bounded simply connected domain $G \subset \mathbb{R}^2$ is the first component in the decomposition* (5.13) *of the function*

$$\Phi(\zeta) = \varphi(\xi) + \bar{\xi}\psi(\xi) + ie_2\left(\varphi(\xi) + \bar{\xi}\psi(\xi) - 2\mathcal{F}(\xi)\right), \tag{5.16}$$

monogenic in the corresponding domain G_ζ of the biharmonic plane μ_{e_1,e_2}, where φ, ψ are the same functions as in the equality (5.15) *and $\mathcal{F}$ is a primitive function for the function ψ. Moreover, all monogenic in G_ζ functions for which the first component in the decomposition* (5.13) *is the given function U_1 are expressed as the sum of the functions* (5.14) *and* (5.16).

Proof. Introducing the functions

$$u_1(x,y) := \Re\,\varphi(z), \quad u_2(x,y) := \Re\,\psi(z), \quad v_2(x,y) := \Im\,\psi(z),$$

we rewrite the equality (5.15) in the form

$$U_1(x,y) = u_1(x,y) + xu_2(x,y) + yv_2(x,y). \tag{5.17}$$

Now, it follows from equality (5.17) and Lemma 5.6 that the function (5.16) is monogenic in the domain G_ζ and its first component in the decomposition (5.13) is the given function U_1. Finally, it evidently follows from Lemma 5.5 that all monogenic in G_ζ functions for which the first component in the decomposition (5.13) is the given function U_1 are expressed as the sum of functions (5.14) and (5.16). The theorem is proved. □

5.6. Integral theorems

In contrast to the papers [12, 13], where integral theorems are established for functions differentiable in the sense of Lorch in domains of a commutative associative Banach algebra, we establish similar results for monogenic functions $\Phi : G_\zeta \to \mathbb{B}$ given only in a domain G_ζ of the biharmonic plane μ_{e_1,e_2} instead of a domain of the whole algebra $\mathbb{B}$. Moreover, note that the integral Cauchy formula established in the papers [12, 13] is not applicable to a monogenic function $\Phi : G_\zeta \to \mathbb{B}$ because it deals with an integration along a curve on which the function Φ is not given, generally speaking.

For the Euclidian norm $\|a\| := \sqrt{|\xi_1|^2 + |\xi_2|^2}$, where $a = \xi_1 e_1 + \xi_2 e_2$ and $\xi_1, \xi_2 \in \mathbb{C}$, in the algebra $\mathbb{B}$ the inequality

$$\|ab\| \le \sqrt{10}\,\|a\|\,\|b\| \quad \forall\, a, b \in \mathbb{B} \tag{5.18}$$

is fulfilled.

In the same way as in the complex plane, a rectifiable curve and an integral along a rectifiable curve are defined in the biharmonic plane μ_{e_1,e_2}.

The Cauchy integral theorem and integral formula for monogenic functions of the variable $\zeta \in \mu_{e_1,e_2}$ are proved by a classic scheme (see, for example, [26]) by using the inequality (5.18). For a proof of the Cauchy integral formula we use as well the equality (3.8) which is also fulfilled in the biharmonic plane μ_{e_1,e_2}.

Thus, the following statement is true:

Theorem 5.8. *Suppose that the boundary ∂G_ζ of domain G_ζ is a closed Jordan rectifiable curve, and a function $\Phi\colon \overline{G_\zeta} \longrightarrow \mathbb{B}$ is continuous in the closure $\overline{G_\zeta}$ of the domain G_ζ and is monogenic in G_ζ. Then the following equalities are fulfilled:*

$$\int\limits_{\partial G_\zeta} \Phi(\tau)\, d\tau = 0 \quad \textit{(the Cauchy theorem)}, \tag{5.19}$$

$$\Phi(\zeta) = \frac{1}{2\pi i} \int\limits_{\partial G_\zeta} \Phi(\tau)(\tau-\zeta)^{-1}\, d\tau \quad \forall \zeta \in G_\zeta \quad \textit{(the Cauchy formula)}. \tag{5.20}$$

For functions of the biharmonic variable ζ, the following Morera theorem can be established in the usual way (see, for example, [26]) by using Theorem 5.3 and the inequality (5.18).

Theorem 5.9. *If a function $\Phi\colon G_\zeta \longrightarrow \mathbb{B}$ is continuous in a domain G_ζ and satisfies the equality* (3.6) *for every triangle $\triangle_\zeta$ such that $\overline{\triangle_\zeta} \subset G_\zeta$, then the function Φ is monogenic in the domain G_ζ.*

5.7. The Taylor expansion

Consider a problem on an expansion of a monogenic in D_ζ function Φ in the Taylor power series. Applying to the function (5.20) a method similar to a method for expanding holomorphic functions, which is based on an expansion of the Cauchy kernel in a power series (see, for example, [26, p. 107]), we obtain immediately the following expansion of the function Φ in the power series:

$$\Phi(\zeta) = \sum_{n=0}^{\infty} b_n (\zeta - \zeta_0)^n, \tag{5.21}$$

where

$$b_n = \frac{\Phi^{(n)}(\zeta_0)}{n!} = \frac{1}{2\pi i} \int\limits_{\Gamma} \Phi(\tau)\Big((\tau-\zeta_0)^{-1}\Big)^{n+1} d\tau\,, \quad n = 0, 1, \ldots,$$

and Γ is an arbitrary closed Jordan rectifiable curve in G_ζ that embraces the point ζ_0. But in such a way it can only be proved that the series (5.21) is convergent in a disk $K_r(\zeta_0) := \{\zeta \in \mu_{e_1,e_2} : \|\zeta - \zeta_0\| < r\}$ with a radius r which is less than the distance between ζ_0 and the boundary of domain G_ζ. It is connected with that fact that the constant $\sqrt{10}$ can not be replaced by 1 in the inequality (5.18).

Nevertheless, taking into account the equality (5.12) in the plane μ_{e_1,e_2} generated by the basis (5.4) and using the equality (5.11) which is transformed now into the form

$$\Phi(\zeta) = F(\xi)e_1 - \left(\frac{iy}{2}F'(\xi) - F_0(\xi)\right)\rho \qquad \forall \zeta \in G_\zeta, \tag{5.22}$$

we can prove the convergence of the series (5.21) in the disk $K_R(\zeta_0)$ with the radius $R := \min_{\tau \in \partial D_\zeta} \|\tau - \zeta_0\|$.

Theorem 5.10. *If a function $\Phi\colon G_\zeta \longrightarrow \mathbb{B}$ is monogenic in a domain G_ζ and $\zeta_0 = x_0e_1 + y_0e_2$ is an arbitrary point in G_ζ, then Φ is expressed in the disk $K_R(\zeta_0)$ as the sum of the convergent power series* (5.21). *In this case*

$$b_n = \left(c_n + \left(c_n^{(0)} - (n+1)\frac{iy_0}{2}c_{n+1}\right)\rho\right), \tag{5.23}$$

where c_n and $c_n^{(0)}$ are coefficients of the Taylor series

$$F(\xi) = \sum_{n=0}^{\infty} c_n(\xi - \xi_0)^n, \qquad F_0(\xi) = \sum_{n=0}^{\infty} c_n^{(0)}(\xi - \xi_0)^n, \qquad \xi_0 = x_0 + iy_0, \tag{5.24}$$

for the functions F and F_0 included in the equality (5.22).

Proof. Inasmuch as in the equality (5.22) the functions F and F_0 are holomorphic in the domain $D := \{\xi = x + iy : xe_1 + ye_2 \in G_\zeta\}$, the series (5.24) are absolutely convergent in the disk $\{\xi \in \mathbb{C} : |\xi - \xi_0| < R\}$. Then we rewrite the equality (5.22) in the form

$$\Phi(\zeta) = c_0 + \sum_{n=1}^{\infty} c_n\left((\xi - \xi_0)^n - \frac{i(y - y_0)}{2}n(\xi - \xi_0)^{n-1}\rho\right)$$
$$- \frac{iy_0}{2}\sum_{n=0}^{\infty}(n+1)c_{n+1}(\xi - \xi_0)^n\rho + \sum_{n=0}^{\infty} c_n^{(0)}(\xi - \xi_0)^n\rho.$$

Now, using the relations

$$(\zeta - \zeta_0)^n = (\xi - \xi_0)^n - n\frac{i(y - y_0)}{2}(\xi - \xi_0)^{n-1}\rho, \quad (\zeta - \zeta_0)^n\rho = (\xi - \xi_0)^n\rho \tag{5.25}$$

for all $\zeta \in \mu_{e_1,e_2}$ and $n = 0, 1, \ldots$, we obtain the expression (5.21), where coefficients are defined by the equality (5.23) and the series (5.21) is absolutely convergent in the disk $K_R(\zeta_0)$. The theorem is proved. □

Now, in the same way as for holomorphic functions of a complex variable (see., for example, [26, p. 118]), we obtain the following uniqueness theorem for monogenic functions of a biharmonic variable.

Theorem 5.11. *If two monogenic in G_ζ functions coincide on a set which have at least one limit point belonging to the domain G_ζ, then they are identically equal in the whole domain G_ζ.*

5.8. Different equivalent definitions of monogenic functions

Thus, we obtain the following theorem which gives different equivalent definitions of monogenic functions in the biharmonic plane:

Theorem 5.12. *A function $\Phi\colon G_\zeta \longrightarrow \mathbb{B}$ is monogenic in the domain G_ζ if and only if one of the following conditions is satisfied:*

(I) *the components U_k, $k=\overline{1,4}$, of the expansion* (5.13) *of the function Φ are differentiable in the domain G and the condition* (5.6) *is satisfied in the domain G_ζ;*
(II) *the function Φ is expressed in the form* (5.22), *where the pair of holomorphic in D functions F and F_0 is unique;*
(III) *the function Φ is continuous in G_ζ and satisfies the equality* (3.6) *for every triangle $\triangle_\zeta$ such that $\overline{\triangle_\zeta} \subset G_\zeta$;*
(IV) *for every $\zeta_0 \in G_\zeta$ there exists a neighborhood, in which the function Φ is expressed as the sum of the power series* (5.21).

5.9. The Laurent expansion

Consider Laurent series in the biharmonic plane. Set $K_{r,R}(\zeta_0) := \{\zeta \in \mu_{e_1,e_2} : 0 \le r < \|\zeta-\zeta_0\| < R \le \infty\}$.

Theorem 5.13. *Every monogenic function $\Phi\colon K_{r,R}(\zeta_0) \longrightarrow \mathbb{B}$ is expressed in the ring $K_{r,R}(\zeta_0)$ as the sum of the convergent series*

$$\Phi(\zeta) = \sum_{n=-\infty}^{\infty} b_n(\zeta-\zeta_0)^n, \tag{5.26}$$

where $(\zeta-\zeta_0)^n := ((\zeta-\zeta_0)^{-1})^{-n}$ for $n=-1,-2,\ldots$,

$$b_n = \frac{1}{2\pi i}\int_\Gamma \Phi(\tau)(\tau-\zeta_0)^{-n-1}\,d\tau, \qquad n=0,\pm1,\pm2,\ldots, \tag{5.27}$$

and Γ is an arbitrary closed Jordan rectifiable curve in $K_{r,R}(\zeta_0)$ that embraces the point ζ_0.

Proof. Inasmuch as in the equality (5.22) the functions F and F_0 are holomorphic in the ring $\{\xi \in \mathbb{C} : r < |\xi-\xi_0| < R\}$ with its center in the point $\xi_0 = x_0 + iy_0$, they are expanded into Laurent series

$$F(\xi) = \sum_{n=-\infty}^{\infty} c_n(\xi-\xi_0)^n,$$

$$F_0(\xi) = \sum_{n=-\infty}^{\infty} c_n^{(0)}(\xi-\xi_0)^n,$$

which are absolutely convergent in the mentioned ring. Then we rewrite the equality (5.22) in the form

$$\Phi(\zeta) = c_0 + \sum_{n=1}^{\infty} c_n \left((\xi - \xi_0)^n - \frac{i(y - y_0)}{2} n(\xi - \xi_0)^{n-1}\rho \right)$$
$$- \frac{iy_0}{2} \sum_{n=0}^{\infty} (n+1)c_{n+1}(\xi - \xi_0)^n \rho$$
$$+ \sum_{n=0}^{\infty} c_n^{(0)} (\xi - \xi_0)^n \rho + c_{-1} \left(\frac{1}{\xi - \xi_0} + \frac{i(y - y_0)}{2(\xi - \xi_0)^2}\rho \right)$$
$$+ \frac{c_{-1}^{(0)}}{\xi - \xi_0}\rho + \sum_{n=-\infty}^{-2} c_n \left((\xi - \xi_0)^n - \frac{i(y - y_0)}{2} n(\xi - \xi_0)^{n-1}\rho \right)$$
$$- \frac{iy_0}{2} \sum_{n=-\infty}^{-2} (n+1)c_{n+1}(\xi - \xi_0)^n \rho + \sum_{n=-\infty}^{-2} c_n^{(0)} (\xi - \xi_0)^n \rho.$$

Further, using the equalities (5.25) for all $\zeta \in K_{r,R}(\zeta_0)$ and $n = 0, \pm 1, \pm 2, \ldots$, we obtain the expansion of the function Φ in the series (5.26), where coefficients are defined by the equalities (5.23), and, moreover, the series (5.26) is absolutely convergent in the ring $K_{r,R}(\zeta_0)$. Multiplying by $(\zeta - \zeta_0)^{-n-1}$ both parts of the equality (5.26) and integrating then along the curve Γ, we obtain the formulas (5.27) for coefficients of the series (5.26). The theorem is proved. □

5.10. The classification of isolated singular points of monogenic functions in the biharmonic plane

It is an evident fact that every convergent in $K_{r,R}(\zeta_0)$ series of the form (5.26) with coefficients from $\mathbb{B}$ is the Laurent series of its sum. Terms of the series (5.26) with nonnegative powers form its *regular part*, and terms with negative powers form the *principal part* of the series (5.26).

Let us compactify the algebra $\mathbb{B}$ by means of addition of an infinite point. Let us agree that every sequence $w_n := \xi_{1,n}e_1 + \xi_{2,n}e_2$ with $\xi_{1,n}, \xi_{2,n} \in \mathbb{C}$ converges to the infinite point in the case, where at least one of the sequences $\xi_{1,n}$, $\xi_{2,n}$ converges to infinity in the extended complex plane.

Now, for a removable singular point and a pole and an essential singular point of a function Φ which is monogenic in a pierced neighborhood $K_{0,r}(\zeta_0)$ of a point $\zeta_0 \in \mu_{e_1,e_2}$, one can give the same definitions as for appropriate notions in the complex plane (see, for example, [26]).

Moreover, an isolated singular point of monogenic function $\Phi(\zeta)$ of the biharmonic variable ζ has a relation with the form of Laurent expansion of this function. More precisely, the following statement is true:

Theorem 5.14. *If in a pierced neighbourhood $K_{0,r}(\zeta_0)$ of an isolated singular point $\zeta_0 \in \mu_{e_1,e_2}$ of a monogenic function $\Phi\colon K_{0,R}(\zeta_0) \longrightarrow \mathbb{B}$, the principal part of the Laurent series* (5.26)*:*

a) *equals to zero, then* ζ_0 *is a removable singular point;*
b) *contains only a finite number of nonzero terms, then* ζ_0 *is a pole;*
c) *contains an infinite number of nonzero terms, then* ζ_0 *is either a pole or an essential singular point.*

Indeed, it is evident that in the case a) the point ξ_0 is a removable singular point for the functions F and F_0 from the equality (5.22). It is also evident that in the case b) the point ξ_0 is a pole at least for one of the functions F, F_0, and is not an essential singular point for these functions. Therefore, the point ζ_0 is a removable singular point of the function Φ in the case a) and is a pole of this function in the case b). In the case c) the point ζ_0 can be either a pole of the function Φ (for example, in the case where the point ξ_0 is a pole of the function F and is an essential singular point of the function F_0) or an essential singular point of this function (for example, in the case where $F \equiv 0$ and the point ξ_0 is an essential singular point of the function F_0).

References

[1] F. Klein, *Vorlesungen über die Entwicklung der Mathematik im* 19. *Jahrhundert*, vol. 1, Berlin: Verlag von Julius Springer, 1926. 179

[2] P.W. Ketchum, *Analytic functions of hypercomplex variables*, Trans. Amer. Math. Soc., **30** (1928), 641–667. 179, 180

[3] I.P. Mel'nichenko, *Algebras of functionally invariant solutions of the three-dimensional Laplace equation*, Ukr. Math. J., **55** (2003), no. 9, 1551–1557. 179, 180, 181, 182

[4] I.P. Mel'nichenko and S.A. Plaksa, *Commutative algebras and spatial potential fields*, Kiev: Inst. Math. NAS Ukraine, 2008. [in Russian] 179, 180, 181, 182, 183, 184, 186, 196, 203, 205, 206, 208, 210

[5] C. Segre, *The real representations of complex elements and extensions to bicomplex systems*, Math. Ann., **40** (1892), 413–467. 180

[6] K.S. Kunz, *Application of an algebraic technique to the solution of Laplace's equation in three dimensions*, SIAM J. Appl. Math., **21** (1971), no. 3, 425–441. 180

[7] I.P. Mel'nichenko, *The representation of harmonic mappings by monogenic functions*, Ukr. Math. J., **27** (1975), no. 5, 499–505. 180, 181

[8] E. Hille and R.S. Phillips, *Functional Analysis and Semi-Groups*, American Mathematical Society, Providence, R.I., 1957. 183, 186, 205, 212

[9] S.A. Plaksa and V.S. Shpakivskyi, *Constructive description of monogenic functions in a harmonic algebra of the third rank*, Ukr. Math. J., **62** (2010), no. 8, 1078–1091. 183

[10] Ju.Ju. Trohimchuk, *Continuous mappings and conditions of monogenity*, Israel Program for Scientific Translations, Jerusalem; Daniel Davey & Co., Inc., New York, 1964. 185

[11] G.P. Tolstov, *On the curvilinear and iterated integral*, Trudy Mat. Inst. Steklov., Acad. Sci. USSR, **35** (1950), 3–101 [in Russian]. 187

[12] E.R. Lorch, *The theory of analytic function in normed abelin vector rings*, Trans. Amer. Math. Soc., **54** (1943), 414–425. 191, 215

[13] E.K. Blum, *A theory of analytic functions in Banach algebras*, Trans. Amer. Math. Soc., **78** (1955), 343–370. 191, 194, 215

[14] A. Sudbery, *Quaternionic analysis*, Math. Proc. Camb. Phil. Soc., **85** (1979), 199–225. 191, 192

[15] F. Colombo, I. Sabadini and D. Struppa, *Slice monogenic functions*, arXiv:0708.3595v2 [math.CV] 25 Jan 2008. 191, 192

[16] F. Brackx and R. Delanghe, *Duality in Hypercomplex Functions Theory*, J. Funct. Anal., **37** (1980), no. 2, 164–181. 191, 192

[17] S. Bernstein, *Factorization of the nonlinear Schrödinger equation and applications*, Complex Variables and Elliptic Equations, **51** (2006), no. 5–6, 429–452. 191, 192

[18] V.V. Kravchenko and M.V. Shapiro, *Integral representations for spatial models of mathematical physics*, Pitman Research Notes in Mathematics, Addison Wesley Longman Inc., 1996. 191, 192

[19] S.A. Plaksa and V.S. Shpakivskyi, *Integral theorems in a commutative three-dimensional harmonic algebra*, Progress in Analysis and its Applications: Proc. of the 7th Intern. ISAAC Congress, 13–18 July, 2009, World Scientific, 2010, pp. 232–239. 191

[20] V.S. Shpakivskyi and S.A. Plaksa, *Integral theorems and a Cauchy formula in a commutative three-dimensional harmonic algebra*, Bulletin Soc. Sci. et Lettr. Lódź, **60** (2010), 47–54. 191

[21] W. Sprößig, *Eigenvalue problems in the framework of Clifford analysis*, Advances in Applied Clifford Algebras, **11** (2001), 301–316. 192

[22] J. Ryan, *Dirac operators, conformal transformations and aspects of classical harmonic analysis*, J. of Lie Theory, **8** (1998), 67–82. 192

[23] W. Sprössig, *Quaternionic analysis and Maxwell's equations*, CUBO A Math. J., **7** (2005), no. 2, 57–67. 192

[24] B. Schneider and E. Karapinar, *A note on biquaternionic MIT bag model*, Int. J. Contemp. Math. Sci., **1** (2006), no. 10, 449–461. 192

[25] A.S. Meylekhzon, *On monogenity of quaternions*, Doklady Acad. Nauk SSSR, **59** (1948), no. 3, 431–434. 192

[26] B.V. Shabat, *Introduction to the complex analysis, Part 1*, Science, Moscow, 1976 [in Russian]. 196, 216, 217, 219

[27] S.A. Plaksa, *Harmonic commutative Banach algebras and spatial potential fields*, Complex Analysis and Potential Theory: Proc. of Conference Satellite to ICM-2006, Gebze Institute of Technology, Turkey, September 8–14, 2006, World Scientific, 2007, pp. 166–173. 196

[28] S.A. Plaksa, *An infinite-dimensional commutative Banach algebra and spatial potential fields*, Further progress in analysis: Proc. of 6th International ISAAC Congress, Ankara, August 13–18, 2007, World Scientific, 2009, pp. 268–277. 196

[29] I.N. Vekua, *Generalized analytic functions*, London, Pergamon Press, 1962. 202

[30] I.P. Mel'nichenko and S.A. Plaksa, *Potential fields with axial symmetry and algebras of monogenic functions of vector variable, III*, Ukr. Math. J., **49** (1997), no. 2, 253–268. 203, 205, 206

[31] S.A. Plaksa, *On integral representations of an axisymmetric potential and the Stokes flow function in domains of the meridian plane, I*, Ukr. Math. J., **53** (2001), no. 5, 726–743. 203, 208, 209

[32] S.A. Plaksa, *Dirichlet problem for an axisymmetric potential in a simply connected domain of the meridian plane*, Ukr. Math. J., **53** (2001), no. 12, 1976–1997. 203, 210

[33] S.A. Plaksa, *On an outer Dirichlet problem solving for the axial-symmetric potential*, Ukr. Math. J., **54** (2002), no. 12, 1634–1641. 203, 210

[34] S.A. Plaksa, *Dirichlet problem for the Stokes flow function in a simply connected domain of the meridian plane*, Ukr. Math. J., **55** (2003), no. 2, 197–231. 203, 210

[35] S. Plaksa, *Singular and Fredholm integral equations for Dirichlet boundary problems for axial-symmetric potential fields*, Factorization, Singular Operators and Related Problems: Proceedings of the Conference in Honour of Professor Georgii Litvinchuk, Funchal, January 28–February 1, 2002, Kluwer Academic Publishers, 2003, pp. 219–235. 203, 209, 210

[36] I.P. Mel'nichenko and S.A. Plaksa, *Outer boundary problems for the Stokes flow function and steady streamline along axial-symmetric bodies*, Complex Analysis and Potential Theory, Kiev, Institute of Mathematics of the National Academy of Sciences of Ukraine, 2003, pp. 82–91. 203, 210

[37] I.P. Mel'nichenko, *On a method of description of potential fields with axial symmetry*, Contemporary Questions of Real and Complex Analysis, Kiev, Institute of Mathematics of Ukrainian Academy of Sciences, 1984, pp. 98–102 [in Russian]. 203

[38] S. Plaksa, *Algebras of hypercomplex monogenic functions and axial-symmetrical potential fields*, Proc. of Second ISAAC Congress, Fukuoka, August 16–21, 1999, Kluwer Academic Publishers, **1** (2000), 613–622. 203

[39] I.M. Gel'fand, D.A. Raikov and G.E. Shilov, *Commutative Normed Rings*, Moscow, Fizmatgiz, 1960 [in Russian]. 204

[40] I.P. Mel'nichenko and S.A. Plaksa, *Commutative algebra of hypercomplex analytic functions and solutions of elliptic equations degenerating on an axis*, Transactions of the Institute of Mathematics of the National Academy of Sciences of Ukraine, Vol. 1 (2004), no. 3, 144–150. 206

[41] S. Plaksa, *Commutative algebras of hypercomplex monogenic functions and solutions of elliptic type equations degenerating on an axis*, More progress in analysis: Proc. of 5th International ISAAC Congress, Catania, July 25–30, 2005, World Scientific (2009), 977–986. 206

[42] Grishchuk S.V. and S.A. Plaksa, *Expressions of solutions of the Euler–Poisson–Darboux equation via components of hypercomplex analytic functions*, Dop. NAN Ukr., no. 8 (2006), 18–24 [in Ukrainian]. 207

[43] I.I. Privalov, *Boundary Properties of Analytic Functions*, Moscow, Gostekhizdat, 1950 [in Russian]. 209

[44] A.G. Mackie, *Contour integral solutions of a class of differential equations*, J. Ration. Mech. Anal., **4** (1955), no. 5, 733–750. 209

[45] P. Henrici, *On the domain of regularity of generalized axially symmetric potentials*, Proc. Amer. Math. Soc., **8** (1957), no. 1, 29–31. 209

[46] Yu.P. Krivenkov, *Representation of solutions of the Euler–Poisson–Darboux equation via analytic functions*, Dokl. Akad. Nauk SSSR, **116** (1957), no. 4, 545–548. 209

[47] G.N. Polozhii, *Theory and Application of p-Analytic and (p, q)-Analytic Functions*, Kiev, Naukova Dumka, 1973 [in Russian]. 209

[48] Grishchuk S.V. and S.A. Plaksa, *Integral representations of generalized axially symmetric potentials in a simply connected domain*, Ukr. Math. J., **61** (2009), no. 2, 195–213. 209

[49] V.F. Kovalev and I.P. Mel'nichenko, *Biharmonic functions on biharmonic plane*, Dop. AN Ukr. Ser. A., no. 8 (1981), 25–27 [in Russian]. 210, 211, 213

[50] I.P. Mel'nichenko, *Biharmonic bases in algebras of the second rank*, Ukr. Math. J., **38** (1986), no. 2, 224–226. 210

[51] Grishchuk S.V. and S.A. Plaksa, *Monogenic functions in a biharmonic algebra*, Ukr. Math. J., **61** (2009), no. 12, 1865–1876. 211

Index

S.A. Plaksa
Institute of Mathematics of the
National Academy of Sciences of Ukraine
3, Tereshchenkivska Street
Kiev, Ukraine
e-mail: plaksa@imath.kiev.ua

Advances in Applied Analysis
Trends in Mathematics, 225–253

2D Free Boundary Value Problems

Sergei V. Rogosin

Abstract. Two-dimensional free boundary value problems are considered. Different models and their connections are discussed. Main attention is paid to the celebrated Hele-Shaw model. Complex-analytic methods are applied to its study.

Mathematics Subject Classification (2010). 76D27, 46E15, 35A10, 47G10, 30E25.

Keywords. 2D free boundary value problems; Hele-Shaw cell; Hele-Shaw flow; Polubarinova-Galin equation; abstract Cauchy-Kovalevsky problem; scale of Banach spaces; singular integral operators.

1. Introduction

[1] In 1898 the British engineer Henry Selby Hele-Shaw published a paper [11] in which he described an experiment on moving a viscous fluid in a so-called cell, a narrow channel between two closely situated glass plates. By coloring the fluid he could see the behavior of the flow as it meets different obstacles in the cell. Therefore, he could confirm Stokes's prediction (see, e.g., [24]) that for low Reynolds numbers the flow velocity field $\mathbf{u}$ becomes irrotational. In this case, pressure plays the role of potential of the field, satisfying the following conditions:

$$\mathbf{u} = -\nabla p, \quad \triangle p = 0, \quad \text{in} \quad \Omega_t, \tag{1.1}$$

$$\frac{\partial p}{\partial n} = 0, \quad \text{on} \quad L, \tag{1.2}$$

where Ω_t is a domain occupied by the fluid at the time instant t, and L is the boundary of obstacles. The role of the Hele-Shaw cell as a tool for description of two-dimensional flows is especially important in the case of flows in porous media, which are sufficiently slow to satisfy the Darcy Law.

Much later (see, e.g., [20], [28]) it was understood that many filtration problems lead to the models in which the fluid domain has to be separated from the

[1] This section is partially based on the survey articles [24] and [47].

"dry" media. P.Ya. Polubarinova-Kochina came to the conclusion that in this case the pressure on the boundary has to be constant on the free boundary and thus the normal velocity of the flow $\mathbf{u}_n$ has to be proportional to the normal "geometric" velocity v_n of the free boundary. Therefore, we arrive at the problem

$$\mathbf{u} = -\nabla(p + \rho g z), \quad \triangle p = 0, \quad \text{in} \quad \Omega_t, \tag{1.3}$$

$$\frac{\partial(p + \rho g z)}{\partial n} = v_n, \quad \text{on} \quad \Gamma, \tag{1.4}$$

where Γ_t is a free boundary, and z is the vertical coordinate. For this problem were found (see [12]) several exact solutions (following pioneering work by P.Ya. Polubarinova-Kochina [28]), which are in a good agreement with experimental data. Thus, the nonlinear problem (1.3)–(1.4) is useful for description of flows under different circumstances.

In fact, this problem has been used and generalized for modeling more compound situations that can be also considered as processes with free boundary. Essential development of the theory is due to appearance of several Stefan-type models. The classical Stefan model (see [18]) describes the so-called theory of albedo, characterizing the reflective abilities of body surfaces. The prototype for this model is the problem on freezing/cooling of a material whose initial stage has a temperature of a phase transition, on condition that the heat is transferred only via conduction and there exists a fixed value of the latent melting heat. Then the value of p in (1.3)–(1.4) (in the case $g = 0$) can be interpreted as a temperature (or concentration) in the new phase, corresponding to zero melting temperature and to dimension-free latent melting heat, equal to 1. This model deals with different technological processes, namely, steel production, semi-conductors construction, foot freezing, laser welding etc. From the other side, if p is understood as electric potential, then we arrive at the model of electro-chemical machining (see, e.g., [17]). Whereas, if p is a concentration of a certain biological agent, then taking into account diffusion in the Stefan model, one can model a tumor's necrosis (see, e.g., [27]). Another interpretation of the Hele-Shaw model (1.3)–(1.4) (in the case $g = 0$) arises, if we introduce $\omega(x, y)$ as an instance of time at which the free boundary Γ reaches the point (x, y) in the Hele-Shaw cell, then the function

$$u(x, y, t) = \int_{\omega}^{t} p(x, y, \tau) d\tau, \tag{1.5}$$

satisfies $\triangle u = 1$ and thus describes transversal displacement of the membrane under a homogeneous load. Further, it follows from the relation (1.4) that $u = \frac{\partial u}{\partial n} = 0$. Therefore, (in the framework of the mechanics of contact interactions) u can be interpreted as the displacement of the pressurised membrane jammed to the rigid smooth plane. If we again suppose existence of diffusion, then a corresponding problem arises in the theory of the choice of optimal time for option realization, where the value p is connected with the price of an option and the role of spatial variable is played by the share price (see, e.g., [49]).

We can also note a series of intensively studied mathematical problems which are not really free boundary value problems. The problem (1.3)–(1.4) plays the role of a limiting singular case for these problems. Among them we single out the models described by

- Allen-Khan equation

$$\tau\frac{\partial u}{\partial t} = \varepsilon^2\triangle u + u - u^3, \quad \text{as} \quad \varepsilon, \tau \to 0,$$

- Khan-Chilliard equation

$$\tau\frac{\partial u}{\partial t} = -\triangle\left(\varepsilon^2\triangle u + u - u^3\right), \quad \text{as} \quad \varepsilon, \tau \to 0,$$

- equation of the phase field

$$\delta\frac{\partial u}{\partial t} = \varepsilon^2\triangle u + u - u^3 + \alpha T, \quad \tau\frac{\partial T}{\partial t} + \frac{\partial u}{\partial t} = \triangle T, \quad \text{as} \quad \varepsilon, \tau \to 0,$$

- filtration equation

$$\rho\frac{\partial u}{\partial t} = \nabla \cdot \left(u^m \nabla u\right), \quad \text{as} \quad m \to \infty.$$

The celebrated P.Ja. Polubarinova-Kochina's exact solution (see, e.g., [28], [10], [46]) to the Hele-Shaw problem shows that there is an essential difference between two cases of problem (1.3)–(1.4), namely, a well-posed case (when the fluid domain is extended) and an ill posed case (when the area of the fluid domain is decreasing). Namely, in the well-posed case known exact solutions (see [12]) correspond to smoothing of the free boundary in time. Vice versa, in the ill-posed case some of the exact solution corresponds to cusp formation. Such an instability illustrates an irreversibility in time the considered model. Occurrence of such an effect is demonstrated in particular by so-called "fingering" ([41], see also [40]), known also as Saffman-Taylor instability. In fact, Saffman-Taylor instability generates a series of delicate regularization analyses of the Hele-Shaw model with the aim of understanding dendrite growth and the development of two-phase domains in the case of solidification of alloys. Morphology of domains in such processes has a great level of unpredictability similar to the turbulence effect. Therefore, the Hele-Shaw model is still very important for modelling and has lead to many new considerations and results.

This paper is devoted mainly to the classical Hele-Shaw model and its modern counterpart. The article is organized as follows. Section 2 contains auxiliary results devoted to modeling fluid dynamics on the base of the Reynolds Transport Theorem. In Section 3, starting from the Navier-Stokes equation we describe a number of assumptions and simplifications, and arrive finally at the Hele-Shaw model. It is known that this model exists in two forms. Dynamics of the free boundary is studied in the framework of boundary value problems for evolution-type differential equations for real-valued functions (so-called "real Hele-Shaw model"). Another variant of the model is connected with possible parametrization of the phase domain (and thus of the free boundary) by using conformal mapping theory. The latter is derived in Section 4 in the form of a Polubarinova-Galin equation. In [30] (see also [31], [32], [33]) it was proposed to reformulate the complex Hele-Shaw model in the form of an abstract Cauchy-Kovalevsky problem and to study of the latter in the corresponding scale of Banach spaces. In order to apply this method we present in Section 5 an auxiliary result, the Ovsjannikov-Nirenberg-Nishida theorem on solvability of the abstract Cauchy-Kovalevsky problem in a proper scale of Banach spaces.

The known exact solutions show the time irreversibility of the Hele-Shaw model. Thus, the injection problem can have infinite life-time (see, e.g., [9]). Contrariwise, the suction problem has a finite-time blow-up and cusp formation at the free boundary before the moving boundary reaches the sink (see, e.g., [12]). Consequently, the above model is globally ill-posed in this respect. That is why several attempts have been made to regularise the classical Hele-Shaw model with suction. In this paper we use the so-called *kinetic undercooling regularization.* A corresponding model is presented in Section 6. This model is analysed and rewritten in the form of Problem $(Q)_\alpha$ in Section 7. It is formulated as two coupled problems – an abstract Cauchy-Kovalevsky problem and a Riemann-Hilbert-Poincaré problem.

Preliminary analysis of these two problems is performed in Section 8. In order to apply the Ovsjannikov-Nirenberg-Nishida theorem, we construct a special Banach space with the scale property. Solvability of the (time-independent) Riemann-Hilbert-Poincaré problem and analytic continuation of its solution is investigated too. The main theorem on local-in-time existence and uniqueness of the complex Hele-Shaw moving boundary value problem with kinetic undercooling regularization is presented in the last Section 9.

2. Flow of a viscous fluid

Let us introduce some basic notions of two-dimensional flows of viscous fluid, following mainly to monograph [10].

Fluid (liquid) is a substance which changes its form under the influence of an external disturbing force F. If F is acting on the domain of area A, then the ratio of the tangential component of F to A determines shear stress inside the fluid. Hence the fluid begins to flow. This stress determines the velocity of the deformation of fluid particles. If the density of the fluid changes are negligible under high external load, then the fluid is called *incompressible* (it is also said that this is a case of *incompressible flow*). If fluid particles are moving along straight lines without essential changes of their velocity, then we have *laminar flow.* In a simple case, when the deformation depends linearly on the stress, then we deal with a *Newtonian fluid.* It satisfies Newtons Law of Viscosity if the pressure in the flow in the direction x is proportional to the velocity V in the orthogonal direction y, i.e.,

$$\sigma = \frac{dF}{dA} = \mu \frac{\partial V}{\partial y}.$$

μ is called the *viscosity coefficient.*

Different approaches to determination of equations of fluid motion are based on the *Reynolds Transport Theorem.* Let the fluid occupy at the time instant t a control volume $\mathcal{V}(t)$ bounded by a control surface $S(t)$. Let $N(t)$ be a certain extensive property of the system, say mass, or moment of force, or energy. Let the spatial variables be denoted by $\mathbf{x} = (x_1, x_2, x_3)$, and let $\eta(\mathbf{x}, t)$ be an intensive property equal to the ratio of the extensive property to the unit of mass, i.e.,

$$\eta = \frac{dN}{dm}, \quad N(t) = \int\limits_{\mathcal{V}(t)} \eta\rho dv, \quad dv = dx_1 dx_2 dx_3.$$

The Reynolds Transport Theorem says that velocity of the change of the characteristic $N(t)$ in the system during a certain period of time is equal to the velocity of the change of $N(t)$ inside the control volume plus the velocity of the change of $N(t)$ on the control surface during the same period:

$$\left(\frac{dN}{dt}\right)_{\text{sys}} = \int\limits_{\mathcal{V}(t)} \frac{\partial}{\partial t}(\eta\rho) dv + \int\limits_{S(t)} \eta\rho \mathbf{V} \cdot \mathbf{n} dS. \tag{2.1}$$

By the Gauss-Ostrogradsky relation, (2.1) can be rewritten as

$$\left(\frac{dN}{dt}\right)_{\text{sys}} = \int\limits_{\mathcal{V}(t)} \left[\frac{\partial}{\partial t}(\eta\rho) + \nabla \cdot (\eta\rho\mathbf{V})\right] dv,$$

or in terms of the so-called *convective derivative* (*Euler derivative*)

$$\frac{D}{Dt} = \frac{\partial}{\partial t} + \mathbf{V} \cdot \nabla$$

the (global) Reynolds Transport Theorem is represented as

$$\left(\frac{dN}{dt}\right)_{\text{sys}} = \int\limits_{\mathcal{V}(t)} \left[\frac{D(\eta\rho)}{Dt} + \eta\rho\left(\nabla \cdot \mathbf{V}\right)\right] dv. \tag{2.2}$$

If one takes mass as an extensive property ($N \equiv m, \eta \equiv 1$), then relation (2.1) has the form

$$\left(\frac{dm}{dt}\right)_{\text{sys}} = \int\limits_{\mathcal{V}(t)} \frac{\partial \rho}{\partial t} dv + \int\limits_{S(t)} \rho\mathbf{V} \cdot \mathbf{n} dS.$$

By the mass conservation law $\left(\frac{dm}{dt}\right)_{\text{sys}} = 0$. Hence

$$\int\limits_{\mathcal{V}(t)} \left[\frac{\partial \rho}{\partial t} + \nabla \cdot (\rho\mathbf{V})\right] dv = 0.$$

Since this equality is valid for any control volume we have

$$\frac{\partial \rho}{\partial t} + \nabla \cdot (\rho\mathbf{V}) = 0. \tag{2.3}$$

In the case of an incompressible fluid relation, (2.3) yields a so-called *continuity equation*

$$\nabla \cdot \mathbf{V} = 0. \tag{2.4}$$

Let us consider in the following only the incompressible fluids. Then the linear moment of the element of mass is a vector determined by the relation $d\mathbf{P} = \mathbf{V}dm$. For the whole control volume we have

$$\mathbf{P} = \int\limits_{\mathcal{V}(t)} \rho \mathbf{V} dv.$$

By the Reynolds Transport Theorem we obtain

$$\left(\frac{d\mathbf{P}}{dt}\right)_{\text{sys}} = \int\limits_{\mathcal{V}(t)} \rho \frac{D\mathbf{V}}{Dt} dv = \int\limits_{\mathcal{V}(t)} \frac{D\mathbf{V}}{Dt} dm.$$

The Second Newton Law says that the velocity of change of $\mathbf{P}$ is equal to the force applied to the fluid in the control volume $\mathcal{V}(t)$:

$$d\mathbf{F} = \frac{D\mathbf{V}}{Dt} dm = \left(\frac{\partial \mathbf{V}}{\partial t} + (\mathbf{V} \cdot \nabla)\, \mathbf{V}\right) dm, \tag{2.5}$$

where $\mathbf{F}$ is a vector of forces.

Let us suppose for the moment that there are no shear stresses in the system. Let the surface forces F_S be due to the pressure of the surface, and internal forces F_b be due to gravitation in the direction x_3. Hence

$$d\mathbf{F} = dF_S + dF_b = -(\nabla p) dv - g(\nabla x_3)(\rho dv). \tag{2.6}$$

Substituting (2.6) into (2.5) we have

$$-\frac{1}{\rho}\nabla p - g\nabla x_3 = \frac{\partial \mathbf{V}}{\partial t} + (\mathbf{V} \cdot \nabla)\, \mathbf{V},$$

or

$$-\nabla p - \rho g \nabla x_3 = \rho \frac{D\mathbf{V}}{Dt}. \tag{2.7}$$

The latter equation is called an *Euler equation*. In terms of the control volume it has a form

$$\left(\frac{d}{dt}\right)_{\text{sys}} \int\limits_{\mathcal{V}(t)} \rho \mathbf{V} dv = -\int\limits_{\mathcal{V}(t)} (\nabla p + \rho g \nabla x_3)\, dv,$$

or

$$\left(\frac{d}{dt}\right)_{\text{sys}} \int\limits_{\mathcal{V}(t)} \rho \mathbf{V} dv = \int\limits_{S(t)} \sigma \cdot \mathbf{n} dS - \int\limits_{\mathcal{V}(t)} \rho g \nabla x_3 dv, \tag{2.8}$$

where $\sigma = (\sigma_{ij})_{i,j=1}^{3}$ is the *stress tensor*.

The Stokes Viscosity Law for incompressible fluids states

$$\sigma_{ii} = -p + 2\mu \frac{\partial V_i}{\partial x_i}, \quad \sigma_{ij} = \mu \left(\frac{\partial V_i}{\partial x_j} + \frac{\partial V_j}{\partial x_i}\right), \; i \neq j.$$

Substituting these relations into formula (2.8) and applying the Gauss-Ostrogradsky theorem we arrive at the (local) relation for the fluid motion of an incompressible Newtonian fluid with constant viscosity in the following form:

$$\frac{D\mathbf{V}}{Dt} = F_b + \frac{1}{\rho}\left(-\nabla p + \mu \Delta \mathbf{V}\right). \tag{2.9}$$

Equations (2.4), (2.9) are called the Navier-Stokes equations for an incompressible Newtonian fluid.

3. Hele-Shaw model

Hele-Shaw flow is the flow of a viscous fluid between two closely related parallel plates. Let us suppose that the narrow channel between the plates has a constant height equal to h and two other sizes of the channel are much greater than the distance between the plates. Let the flow be generated by an injection/suction through the point source/sink at one of the plates (x_1^0, x_2^0). Suppose that the velocity of the fluid is maximal at a source/sink and vanishes at the boundary of the Hele-Shaw cell.

To derive the corresponding governing equations one can start with the Navier-Stokes equations neglecting gravity (2.4), (2.9) (see, e.g., [10]),

$$\frac{\partial \mathbf{V}}{\partial t} + (\mathbf{V}\cdot\nabla)\mathbf{V} = \frac{1}{\rho}\left(-\nabla p + \mu \triangle \mathbf{V}\right), \quad \nabla\cdot\mathbf{V} = 0. \tag{3.1}$$

Assume that the injection/suction is sufficiently slow such that the fluid motion is uniform and plane parallel. It means

$$\frac{\partial \mathbf{V}}{\partial t} = 0, \quad V_3 = 0.$$

Under these conditions the equations (3.1) become

$$\begin{aligned}\left(V_1\frac{\partial}{\partial x_1} + V_2\frac{\partial}{\partial x_2}\right)V_1 &= -\frac{1}{\rho}\frac{\partial p}{\partial x_1} + \frac{\mu}{\rho}\triangle V_1,\\ \left(V_1\frac{\partial}{\partial x_1} + V_2\frac{\partial}{\partial x_2}\right)V_2 &= -\frac{1}{\rho}\frac{\partial p}{\partial x_2} + \frac{\mu}{\rho}\triangle V_2,\\ 0 &= -\frac{1}{\rho}\frac{\partial p}{\partial x_3},\end{aligned}$$

with boundary conditions

$$V_1\big|_{x_3=0,h} = V_2\big|_{x_3=0,h} = 0.$$

If h is sufficiently small and the flow is slow, then we can assume that the derivatives of V_1 and V_2 with respect to x_1 and x_2 are negligible compared to the derivatives with respect to x_3. Thus

$$\frac{\partial V_1}{\partial x_j} = \frac{\partial V_2}{\partial x_j} = \frac{\partial^2 V_1}{\partial x_j^2} = \frac{\partial^2 V_2}{\partial x_j}, \quad j = 1, 2,$$

which gives

$$\frac{\partial p}{\partial x_1} = \mu \frac{\partial^2 V_1}{\partial x_3^2}, \qquad \frac{\partial p}{\partial x_2} = \mu \frac{\partial^2 V_2}{\partial x_3^2}, \qquad 0 = \frac{\partial p}{\partial x_3}.$$

The last equation in this system shows that p does not depend on x_3, whence V_1, V_2 are polynomials of degree at most 2 as functions of x_3. The boundary conditions then imply

$$V_1 = \frac{1}{2}\frac{\partial p}{\partial x_1}\left(\frac{x_3^2}{\mu} - \frac{hx_3}{\mu}\right), \quad V_2 = \frac{1}{2}\frac{\partial p}{\partial x_2}\left(\frac{x_3^2}{\mu} - \frac{hx_3}{\mu}\right).$$

The integral means $\widetilde{V}_1$ and $\widetilde{V}_2$ of V_1 and V_2 across the gap are

$$\widetilde{V}_1 = \frac{1}{h}\int_0^h V_1 dx_3 = -\frac{h^2}{12\mu}\frac{\partial p}{\partial x_1}, \qquad \widetilde{V}_2 = \frac{1}{h}\int_0^h V_2 dx_3 = -\frac{h^2}{12\mu}\frac{\partial p}{\partial x_2},$$

so the integral mean $\widetilde{\mathbf{V}}$ of $\mathbf{V}$ satisfies

$$\widetilde{\mathbf{V}} = -\frac{h^2}{12\mu}\nabla p. \tag{3.2}$$

Equation (3.2) is called the *Hele-Shaw equation*. It is of the same form as Darcy's law, which governs flows in porous media. In the sequel we write just $\mathbf{V}$ instead of $\widetilde{\mathbf{V}}$.

The Stokes-Leibenzon model ([16]) suggests that a point sink/source (x_1^0, x_2^0) is of constant strength. The rate of area (or mass) change is given as

$$\int_{\partial U_\varepsilon} \rho \mathbf{V}\cdot \mathbf{n} ds = \text{const}, \quad \text{thus} \quad \iint_{U_\varepsilon} \frac{h^2\rho}{12\mu}\triangle p dx_1 dx_2 = \text{const}.$$

where $U_\varepsilon = \{(x,y) : (x_1 - x_1^0)^2 + (x_2 - x_2^0)^2 < \varepsilon^2\}$. Equality (3.2) together with the Green theorem imply

$$\iint_{U_\varepsilon} \frac{h^2\rho}{12\mu}\triangle p dx_1 dx_2 = \text{const}.$$

So, $\triangle p = Q\delta_{(x_1^0, x_2^0)}$ for some constant Q, where $\delta_{(x_1^0, x_2^0)}$ is the Dirac distribution, and the potential function p has a logarithmic singularity at (x_1^0, x_2^0). On the free boundary the balance of forces gives

$$p = \text{exterior air pressure} + \text{surface tension}.$$

Suppose the air pressure to be constant, and surface tension to be proportional to the curvature of the free boundary. If the distance h between the plates is small enough, and the surface tension is almost constant, then the pressure on the free boundary can be assumed to be constant (vanishing in a proper scale).

4. The Polubarinova-Galin equation and complex Hele-Shaw model

The idea of P.Ja. Polubarinova-Kochina and L.A. Galin is in parametrization of the free boundary of the flow in the Hele-Shaw cell by the time-dependent family of conformal mappings of the canonical domain on the complex plane (say, of the unit disc) onto the domain on the phase plane (occupied by the fluid). Such a model (called a *complex Hele-Shaw model*) was completely described in [48] and rediscovered in [36].

Denote by $\Omega(t)$ the bounded simply connected domain in the phase z-plane occupied by the fluid at instant t, and consider suction/injection through a single sink/source at the origin (of a constant strength Q, which is positive in the case of suction and negative in the case of injection). The dimensionless pressure p is scaled so that 0 corresponds to the atmospheric pressure. Put $\Gamma(t) = \partial\Omega(t)$ $(\Omega(0) =: \Omega_0, \Gamma(0) =: \Gamma_0)$. The potential function p is harmonic in $\Omega(t) \setminus \{0\}$ and

$$\triangle p = Q\delta_0(z), \quad z = x + iy \in \Omega(t). \tag{4.1}$$

The zero surface tension *dynamic boundary condition* is given by

$$p(z,t) = 0, \quad \forall z \in \Gamma(t). \tag{4.2}$$

The resulting motion of the free boundary $\Gamma(t)$ is given by the fluid velocity $\mathbf{V}$ on $\Gamma(t)$. The normal velocity in the outward direction is

$$v_n = \mathbf{V}\big|_{\Gamma(t)} \cdot \mathbf{n}(t), \tag{4.3}$$

where $\mathbf{n}(t)$ is the unit outer normal vector to $\Gamma(t)$. Rewriting this law of motion in terms of the potential function and using (3.2) after suitable rescaling, we get the *kinematic boundary condition*

$$\frac{\partial p}{\partial \mathbf{n}} = -v_n. \tag{4.4}$$

Let us introduce the complex potential $W(z,t), \operatorname{Re} W = p$. For each fixed t the function $W(z,t)$ is a multi-valued analytic function on $\Omega(t)$, whose real part satisfies the Dirichlet problem (4.1)–(4.2). It follows from the Cauchy-Riemann equations that

$$\frac{\partial W}{\partial z} = \frac{\partial p}{\partial x_1} - i\frac{\partial p}{\partial x_2}.$$

Since Green's function solves (4.1), (4.2), we have the representation

$$W(z,t) = \frac{Q}{2\pi}\log z + w_0(z,t), \tag{4.5}$$

where $w_0(z,t)$ is an analytic regular function in $\Omega(t)$.

In order to derive an equation of the free boundary $\Gamma(t) = \partial\Omega(t)$, we introduce some notation. Let $f(z,t) : G_1 \to \Omega(t)$ be a unique conformal mapping of the unit disc $G_1 = \{z \in \mathbb{C} : |z| < 1\}$ on the plane of parameter z, satisfying conditions $f(0,t) = 0$, $f_z(0,t) > 0$. The function $f(z,0) = f_0(z)$ parameterizes the initial boundary $\Gamma_0 = \{f_0(e^{i\theta}),\ \theta \in [0,2\pi)\}$, and the function $f(z,t)$ parameterizes the boundary $\Gamma(t) = \Gamma_t = \{f(e^{i\theta},t),\ \theta \in [0,2\pi)\}$. The velocity v_n in the direction of the outward normal to $\Gamma(t)$ is determined by (4.4). The vector of the outward normal to $\Gamma(t)$ is defined by the relation

$$\mathbf{n} = z\frac{f_z}{|f_z|}, \quad z \in \partial G_1.$$

Hence the normal velocity can be written as

$$v_n = \mathbf{V}\cdot\mathbf{n} = -\mathrm{Re}\left(\frac{\partial W}{\partial \zeta}\, z\, \frac{f_z}{|f_z|}\right).$$

Since the Green's function is invariant under conformal mapping we have

$$(W \circ f)(z,t) = \frac{Q}{2\pi}\log z, \qquad \text{and thus} \qquad \frac{\partial W}{\partial \zeta}\frac{\partial f}{\partial z}(z,t) = \frac{Q}{2\pi z}.$$

From the other side, the normal velocity on the free boundary can be represented in the form

$$v_n = \mathrm{Re}\left[\frac{\partial f}{\partial t}\, z\, \overline{\frac{f_z}{|f_z|}}\right].$$

Finally we have got the following *Polubarinova-Galin equation*:

$$\mathrm{Re}\left[\frac{\partial f}{\partial t}(z,t)\, z\, \overline{\frac{\partial f}{\partial z}(z,t)}\right] = -\frac{Q}{2\pi}, \quad z = e^{i\phi}. \tag{4.6}$$

A complex mathematical model for Hele-Shaw flows with a free boundary produced by injection/suction into/from a narrow channel supposing constant atmospheric pressure on the moving boundary. This model can be represented in terms of the following form, where the Riemann mapping functions $f = f(z,t)$, from the unit disc G_1 onto the region occupied by fluid at the time t, and the rate of injection/suction is $Q(t)$:

Let the functions $f_0 = f_0(z)$ and $Q = Q(t)$ be given. Let f_0 be holomorphic and univalent in a neighbourhood of the unit disk G_1, $f_0(0) = 0$, and Q be continuous in a right-sided neighbourhood of $t = 0$. Find a function $f = f(z,t)$, holomorphic and univalent as a function of z in a neighbourhood of G_1, continuously differentiable in t such that with respect to t in a right-sided neighbourhood of $t = 0$, satisfies

$$\mathrm{Re}\left(\frac{1}{z}\frac{\partial f}{\partial t}(z,t)\overline{\frac{\partial f}{\partial z}(z,t)}\right) = Q(t), \qquad z \in \partial G_1, \quad t \in [0,T); \tag{4.7}$$

$$f(z,0) = f_0(z), \qquad z \in G_1; \tag{4.8}$$

$$f(0,t) = 0, \qquad t \in [0,T). \tag{4.9}$$

Under the additional assumption $f'_z(0,t) > 0$ one gets with the results of [48] the local existence and uniqueness of solutions in the case $Q(t) \equiv 1$, where the solution depends analytically on t.

In [8] a more elementary proof of local existence and uniqueness of solutions for the above model is given in the case that f_0 is a polynomial or a rational function. In particular, it was shown (see, e.g., [10]) that the solution remains polynomial (rational) with respect to z if the initial function f_0 is.

5. Nirenberg-Nishida theorem

Let us formulate an auxiliary result which will be used in our further analysis. This theorem is called the *Nirenberg-Nishida theorem* (see [22]–[23]), but the basic result of such a type was obtained earlier by L.V. Ovsjannikov [26], and thus it is more correct to call it the *Ovsjannikov theorem* (or at least *Ovsjannikov-Nirenberg-Nishida theorem*). This result is applied for the study of a so-called abstract Cauchy-Kovalevsky problem.

The abstract Cauchy-Kovalevsky problem means the following. We consider the Cauchy problem

$$d_t w = F(t, w), \quad w(0) = 0, \tag{5.1}$$

for a differential equation with (an abstract) nonlinear operator $\mathcal{F} : t \mapsto F(t, w(t))$. If the operator $\mathcal{F}$ is acting in a locally compact space then the classical Picard-Lindelöf approach is already inapplicable. However if this operator is acting in a certain scale of Banach spaces, then one can use a form of the Cauchy-Kovalevsky theorem (*e.g.*, that of Nirenberg-Nishida type). By the scale of Banach spaces we understand here a family of Banach spaces $\{B_s, \|\cdot\|_s\}_{0<s\leq 1}$ such that for each $0 < s' \leq s \leq 1$ the norm of the canonical imbedding operator $\mathcal{I}_{s\to s'}$ is not greater than 1. In [35] the corresponding scale is constructed to provide a "conical evolution" of the solution to (5.1) (see, *e.g.*, [45]). It is convenient to present here the Nirenberg-Nishida theorem which is applied below.

Theorem 5.1. *Let us consider an abstract Cauchy-Kovalevsky problem*

$$d_t w = F(t, w), \quad w(0) = 0, \tag{5.2}$$

in a scale of Banach spaces $\{B_s, \|\cdot\|_s\}_{0<s\leq 1}$. *Let for certain absolute (i.e., independent of* s, s', t*) constants* C, K, R, T *the following conditions be satisfied:*

- *for each fixed* $s, 0 < s \leq 1$, *the mapping* $F(t, w)$ *of* $[0, T] \times \{w \in B_s : \|w\|_s < R\}$ *to* $B_{s'}$ *is continuous with respect to* t*;*
- *for all* $0 < s' < 1$ *the continuous function* $F(t, 0)$ *satisfies the inequality*

$$\|F(t, 0)\|_{s'} \leq \frac{K}{1 - s'};$$

- *for all* $0 < s' < s \leq 1$, $t \in [0, T]$, $w_1, w_2 \in \{\|w\|_s < R\}$ *the inequality*

$$\|F(t, w_1) - F(t, w_2)\|_{s'} \leq \frac{C}{s - s'} \|w_1 - w_2\|_s$$

is valid.

Then the problem (5.2) *has a unique solution*

$$w \in \mathcal{C}^1\left([0, a_0(1 - s)), B_s\right)_{0<s<1}, \quad \|w(t)\|_s < R,$$

where a_0 *is a suitable positive constant.*

6. Kinetic undercooling regularization

The first exact solutions for Hele-Shaw flows driven by a single sink at the origin were constructed in [28], [4]. They have finite-time blow-up and cusp formation at the free boundary before the moving boundary reaches the sink (see, e.g., [12]). Consequently, the above model is globally ill-posed in this respect. The physical meaning of this situation is that the velocity of points on the moving boundary tends to infinity at the cusp.

There are different approaches to the regularization of the suction problem by incorporating extra terms in the free boundary conditions to penalize large curvatures or large normal derivatives (see [44] and [13]). The most common corrections are to include a Gibbs-Thomson term proportional to the curvature or a kinetic undercooling term proportional to the normal velocity at the moving boundary.

This paper is concerned with kinetic undercooling regularization. For Hele-Shaw flows this regularization first appeared in the doctoral thesis [39] (see also the review in [42]). A local linear stability analysis shows that this regularization successfully penalises the short wavelength growth which is usually associated with blow-up [14]. The following problem (P_α) was derived for a complex model which describes Hele-Shaw flows with kinetic undercooling regularization (see, e.g., [35] and [34]):

Problem (P_α): *Let the two functions $f_0 = f_0(z)$ and $Q = Q(t)$ be given. Suppose that $f_0 = f_0(z)$ is a given holomorphic and univalent in a neighbourhood of $G_1, f_0(0) = 0$, $Q(t)$ is a given continuous function in a right-sided neighbourhood of $t = 0$.*

The problem is to find two functions $f = f(z,t)$, $w_{\text{reg}} = w_{\text{reg}}(z,t)$, f is holomorphic and univalent as a function of z in a neighbourhood of G_1, continuously differentiable in t in a right-sided neighbourhood of $t = 0$, w_{reg} is holomorphic in z in a neighbourhood of G_1, continuous in t in a right-sided neighbourhood of $t = 0$, such that for all $z \in \partial G_1$, $t \in [0,T)$,

$$\operatorname{Re}\left(\frac{1}{z}\frac{\partial f}{\partial t}(z,t)\overline{\frac{\partial f}{\partial z}(z,t)}\right) = Q(t) + \operatorname{Re}\left(z w_{\text{reg}}(z,t)\right), \tag{6.1}$$

$$f(z,0) = f_0(z), \tag{6.2}$$

$$f(0,t) = 0, \tag{6.3}$$

$$\operatorname{Im}\left(z w_{\text{reg}}(z,t)\right) = \alpha\partial_\theta\left(\left|\frac{\partial f}{\partial z}\right|^{-1}\left(Q(t) + \operatorname{Re}\left(z w_{\text{reg}}(z,t)\right)\right)\right), \tag{6.4}$$

where $\alpha > 0$, $z = re^{i\theta}$, and $Q(t) < 0$ in the case of suction.

If $\alpha = 0$, then $\operatorname{Im}\left(z w_{\text{reg}}(z,t)\right) = 0, z \in \partial G_1$. Consequently, $\operatorname{Re}\left(z w_{\text{reg}}(z,t)\right) = \text{const}, z \in \partial G_1$, that is $w_{\text{reg}}(z,t) \equiv 0, z \in \overline{G_1}$. This means our starting model (4.7)–(4.9) coincides with (P_0).

Remark 6.1. An approach called kinetic undercooling regularization takes into account the kinematic boundary condition $V_n = U_n$ and the dynamic boundary condition $p \sim V_n$ at the moving boundary $\Gamma(t)$ in the physical plane. Here p is the pressure, V_n is the normal component of velocity of the fluid and U_n the normal component of the velocity of $\Gamma(t)$. Then using $f = f(z,t)$ and the complex potential $\chi = \chi(f(z,t),t)$ we have that both conditions can be written in the mathematical plane as

$$\operatorname{Re}\left(\frac{1}{z}\frac{\partial f}{\partial t}\overline{\frac{\partial f}{\partial z}}\right) = \operatorname{Re}\left(z\frac{\partial \chi}{\partial z}\right),$$

$$\operatorname{Re}\chi = -\alpha\left|\frac{\partial f}{\partial z}\right|^{-1}\operatorname{Re}\left(\frac{1}{z}\frac{\partial f}{\partial t}\overline{\frac{\partial f}{\partial z}}\right),\ \alpha > 0.$$

The ansatz for $\chi = Q(t)\log z + \chi_{\mathrm{reg}}$ and for the conjugate $\overline{w} := \partial_z\chi = Q(t)/z + w_{\mathrm{reg}}$ of the complex velocity w gives immediately the kinematic boundary condition (6.1). After differentiation of the second condition with respect to θ we obtain some calculations that lead to the dynamic boundary condition (6.4).

7. On structure of the problem (P_α)

Let us discuss the structure of (6.1)–(6.4), and derive an equivalent problem to (P_α) to which we will apply complex-analytic methods. Under an additional assumption

$$\operatorname{Im}\left(\frac{1}{z}\frac{\partial f}{\partial t}(z,t)\left(\frac{\partial f}{\partial z}\right)^{-1}(z,t)\right)(0,t) = 0, \tag{7.1}$$

one can rewrite (6.1) by using Schwarz's integral formula as follows:

$$\frac{\partial f}{\partial t}(z,t) - z\frac{\partial f}{\partial z}(z,t)\frac{1}{2\pi i}\int_{|\zeta|=1}\left|\frac{\partial f}{\partial \zeta}\right|^{-2}(Q(t) + \operatorname{Re}(\zeta w_{\mathrm{reg}}))\frac{\zeta + z}{\zeta - z}\frac{d\zeta}{\zeta} = 0.$$

Differentiating and setting

$$\omega(z,t) := z w_{\mathrm{reg}}(z,t), \qquad \phi(z,t) := \left(\frac{\partial f}{\partial z}\right)^{-1}(z,t)$$

gives the following equivalent form of equation (6.1) (cf. [30] for the model (4.7)–(4.9), (7.1)):

$$\frac{\partial \phi}{\partial t}(z,t) - z\frac{\partial \phi}{\partial z}(z,t)\mathbf{T}_t(\phi,\omega) + \phi(z,t)\frac{\partial}{\partial z}(z\mathbf{T}_t(\phi,\omega)) = 0,$$

$$\phi_0(z) := \phi(z,0) = \left(\frac{\partial f_0}{\partial z}\right)^{-1},$$

where

$$\phi_0(z) := \phi(z,0) = \left(\frac{\partial f_0}{\partial z}\right)^{-1}, \mathbf{T}_t(\phi,\omega) := \frac{1}{2\pi i}\int_{|\zeta|=1}|\phi|^2(Q(t) + \operatorname{Re}\omega)\frac{\zeta + z}{\zeta - z}\frac{d\zeta}{\zeta}. \tag{7.2}$$

Equation (6.4) is a Riemann-Hilbert-Poincaré problem for $\omega = \omega(z,t)$:

$$\operatorname{Im}(\omega(z,t)) = \alpha\partial_\theta\big(|\phi(z,t)|\big(Q(t) + \operatorname{Re}(\omega(z,t))\big)\big), \quad \omega(0,t) = 0.$$

In the following we will study the next problem which is equivalent to problem (P_α) (cf. [30] for the model (4.7)–(4.9), (7.1)):

Problem (Q_α): *Let ϕ_0 be given holomorphic and non-vanishing in a neighbourhood of G_1, and Q be given continuous in a right-sided neighbourhood of $t = 0$.*

The problem is to find two functions, namely $\phi = \phi(z,t)$ being holomorphic and non-vanishing in a neighbourhood of G_1, continuously differentiable with respect to t in a neighbourhood of $t = 0$, and $\omega = \omega(z,t)$ being holomorphic in z in a neighbourhood of G_1, continuous with respect to t in a right-sided neighbourhood of $t = 0$, both functions satisfy

- *the abstract Cauchy-Kovalevsky problem (or Cauchy problem for the abstract evolution equation)*

$$\frac{\partial \phi}{\partial t}(z,t) - z\frac{\partial \phi}{\partial z}(z,t)\mathbf{T}_t(\phi,\omega) + \phi(z,t)\frac{\partial}{\partial z}\left(z\mathbf{T}_t(\phi,\omega)\right) = 0, \tag{7.3}$$

$$\phi(z,0) = \left(\frac{\partial f_0}{\partial z}\right)^{-1}, \quad \text{in a cylinder } \{(z,t) \in G_1 \times [0,T)\}; \tag{7.4}$$

- *the Riemann-Hilbert-Poincaré problem with respect to ω ($\omega(0,t) = 0$)*

$$\operatorname{Im}\left(\omega(z,t)\right) = \alpha\partial_\theta\Big(|\phi(z,t)|\big(Q(t) + \operatorname{Re}\left(\omega(z,t)\right)\big)\Big) \text{ on } \partial G_1 \times [0,T) \tag{7.5}$$

under the additional constraint

$$\omega(0,t) = 0 \tag{7.6}$$

where $\mathbf{T}_t(\phi,\omega)$ is defined by (7.2), $\alpha > 0$, and $z = re^{i\theta}$.

8. Mathematical treatment of problem (Q_α)

In order to study the problem (Q_α) we introduce a function space for $\phi = \phi(z,t)$ and $\omega = \omega(z,t)$. To define this space we need constants. Let us fix constants r_0, r_1, $1 < r_0 < r_1$, a positive constant b, and a parameter $s \in (0,1)$. By $\mathcal{H}(G_{(s)})$ we denote the space of functions which are holomorphic in $G_{(s)}$, where $G_{(s)} := \{z \in \mathbb{C} : |z| < r_0 + s(r_1 - r_0)\}$. Then we define the space

$$\mathbf{B} := \Bigg\{ g = g(z,t) \in \bigcup_{0<s<1} \mathcal{C}\left([0, b(1-s)), \mathcal{H}(G_{(s)}) \cap \mathcal{C}^{1,\lambda}(\overline{G}_{(s)})\right) :$$

$$\|g\|_{\mathbf{B}} = \max\Bigg\{ \sup_{s\in(0,1),h<b(1-s)} \max_{t\in[0,h]} \|g(\cdot,t)\|_{\mathcal{C}^\lambda(\overline{G}_{(s)})};$$

$$\sup_{s\in(0,1),t<b(1-s)} \left\|\frac{\partial g}{\partial z}(\cdot,t)\right\|_{\mathcal{C}^\lambda(\overline{G}_{(s)})} \left(1 - \frac{t}{b(1-s)}\right)^{\frac{1}{2}} \Bigg\} < \infty \Bigg\}.$$

Lemma 8.1. *The function space $\mathbf{B}$ is a Banach space and an algebra with*

$$\|g \cdot h\|_{\mathbf{B}} \leq 2\|g\|_{\mathbf{B}}\|h\|_{\mathbf{B}} \quad \text{for all } g, h \in \mathbf{B}.$$

9. The abstract Cauchy-Kovalevsky problem

In order to study the evolution problem (7.3), (7.4) we want to understand how the operators ∂_z, $\mathbf{T}_t(\phi,\omega)$ act on $\mathbf{B}$, $\mathbf{B}\times\mathbf{B}$, respectively.

It is clear that ∂_z is an unbounded operator on $\mathbf{B}$. If we use the operator $(\mathbf{J}\psi)(\cdot,t) := \int\limits_0^t \psi(\cdot,\tau)d\tau$, then it turns out that $\mathbf{A} := \mathbf{J}\circ\partial_z$ is a continuous operator on $\mathbf{B}$ (see, e.g., [45]).

Lemma 9.1. *Let $\phi \in \mathbf{B}$. Then the operator*

$$\mathbf{A} : \phi \mapsto \mathbf{J}\circ\partial_z\phi(\cdot) = \int\limits_0^t \partial_z\phi(\cdot,\tau)d\tau, \qquad t\in[0,b(1-s)), \quad s\in(0,1),$$

is a continuous operator mapping $\mathbf{B}$ into itself, and satisfying the estimate

$$\|\mathbf{A}\phi\|_{\mathbf{B}} \le Cb\|\phi\|_{\mathbf{B}}, \ \ C = C(r_0,r_1) \tag{9.1}$$

where $C = C(r_0,r_1)$.[2]

Proof. The main ideas of this proof are taken from [45]. The idea is to consider the differential operator ∂_z in the scale of Banach spaces $\{\mathcal{H}(G_{(s)})\cap\mathcal{C}^\lambda(\overline{G}_{(s)})\}_{0<s<1}$. The operator of integration with respect to time acts as a regularizing one, this means it compensates the unboundedness of ∂_z.[3] □

To study the nonlinear operator $\mathbf{T}_t(\phi,\omega)$ for given $\phi,\omega\in\mathbf{B}$ we introduce another function space (Banach space)

$$\mathbf{B}_a := \Bigg\{ g = g(z,t) \in \bigcup_{0<s<1} \mathcal{C}\left([0,b(1-s)),\mathcal{H}(A_{(s)})\cap\mathcal{C}^{1,\lambda}(\overline{A}_{(s)})\right) :$$

$$\|g\|_{\mathbf{B}_a} = \max\Bigg\{ \sup_{s\in(0,1),h<b(1-s)} \max_{t\in[0,h]} \|g(\cdot,t)\|_{\mathcal{C}^\lambda(\overline{A}_{(s)})};$$

$$\sup_{s\in(0,1),t<b(1-s)} \left\|\frac{\partial g}{\partial z}(\cdot,t)\right\|_{\mathcal{C}^\lambda(\overline{A}_{(s)})} \left(1-\frac{t}{b(1-s)}\right)^{\frac{1}{2}} \Bigg\} < \infty \Bigg\},$$

where

$$A_{(s)} = \left\{ z\in\mathbb{C} : \frac{1}{r_0+(r_1-r_0)s} < |z| < r_0+(r_1-r_0)s \right\}, \ 0<s<1.$$

The following lemma is evident.

Lemma 9.2. *If ϕ belongs to $\mathbf{B}$, then the function $\tilde{\phi} = \tilde{\phi}(z,t) := \overline{\phi(\frac{1}{\bar z},t)}$ as well as the product $\phi\tilde{\phi}$ belongs to $\mathbf{B}_a$. Besides,*

$$\|\tilde{\phi}\|_{\mathbf{B}_a} \le C\|\phi\|_{\mathbf{B}}.$$

---

[2]Up to the end of the paper we use C as a universal constant.

[3]The goal of this paper is to prove the local existence in time. So, we have no need to minimize the constant Cb in (9.1).

Now we have all tools for the consideration of $\mathbf{T}_t(\phi,\omega)$ on $\mathbf{B}$.

Lemma 9.3. *The nonlinear operator* $\mathbf{T}_t(\phi,\omega)$ *is a continuous operator mapping* $\mathbf{B}\times\mathbf{B}$ *into* $\mathbf{B}$. *Moreover there exists a constant* $C = C(\lambda, Q, r_0, r_1)$ *such that*

$$\|\mathbf{T}_t(\phi,\omega)\|_{\mathbf{B}} \le C\|\phi\|^2_{\mathbf{B}}(1+\|\omega\|_{\mathbf{B}}).$$

Proof. Let the given functions ϕ and ω belong to $\mathbf{B}$. Then

$$\begin{aligned}\mathbf{T}_t(\phi,\omega)(z,t) &:= \frac{1}{2\pi i}\int_{|\zeta|=1} |\phi|^2\,(Q(t)+\mathrm{Re}\omega)\,\frac{\zeta+z}{\zeta-z}\frac{d\zeta}{\zeta}\\ &= \frac{1}{2\pi i}\int_{|\zeta|=1} \phi(\zeta,t)\overline{\phi(\frac{1}{\overline{\zeta}},t)}\left(Q(t)+\frac{\omega(\zeta,t)+\overline{\omega(\frac{1}{\overline{\zeta}},t)}}{2}\right)\frac{\zeta+z}{\zeta-z}\frac{d\zeta}{\zeta}.\end{aligned}$$

Due to Lemma 9.2 the functions $\tilde{\phi} = \tilde{\phi}(z,t) := \overline{\phi(\frac{1}{\overline{z}},t)}$ and $\tilde{\omega} = \tilde{\omega}(z,t) := \overline{\omega(\frac{1}{\overline{z}},t)}$ belong to $\mathbf{B}_a$ (but not to $\mathbf{B}$). This is the motivation for us to consider the Banach space $\mathbf{B}_a$. For our purpose it is necessary to use the domains $A_{(s)}, s\in(0,1)$, because the product $\phi\tilde{\phi}$ is defined only on these sets. Due to the definition of the tilded functions $\tilde{\phi},\tilde{\omega}$ and their properties in $A_{(s)}$, $0<s<1$, we have

$$\begin{aligned}\mathbf{T}_t(\phi,\omega)(z,t) &= \frac{1}{2\pi i}\int_{|\zeta|=1} \phi(\zeta,t)\tilde{\phi}(\zeta,t)\left(Q(t)+\frac{\omega(\zeta,t)+\tilde{\omega}(\zeta,t)}{2}\right)\frac{\zeta+z}{\zeta-z}\frac{d\zeta}{\zeta}\\ &= \frac{1}{2\pi i}\int_{\partial G_{(s)}} \phi(\zeta,t)\tilde{\phi}(\zeta,t)\left(Q(t)+\frac{\omega(\zeta,t)+\tilde{\omega}(\zeta,t)}{2}\right)\frac{\zeta+z}{\zeta-z}\frac{d\zeta}{\zeta}\end{aligned}$$

for all $s\in(0,1), t\in[0,b(1-s))$ and all $z\in G_{(s)}$. The last formula gives us immediately $\mathbf{T}_t(\phi,\omega)\in\mathcal{C}\left([0,b(1-s)),\mathcal{H}(G_{(s)})\cap\mathcal{C}^{1,\lambda}(\overline{G}_{(s)})\right)$ for all $s\in(0,1)$. Using Hölder estimates for the Schwarz's integral we have with a constant $C = C(\lambda,Q,r_0,r_1)$ the following estimate

$$\|\mathbf{T}_t(\phi,\omega)(\cdot,t)\|_{\mathcal{C}^\lambda(\overline{G}_{(s)})} \le C\|\phi\|_{\mathbf{B}}\|\tilde{\phi}\|_{\mathbf{B}_a}\left(1+\frac{1}{2}(\|\omega\|_{\mathbf{B}}+\|\tilde{\omega}\|_{\mathbf{B}_a})\right).$$

It follows from this inequality together with Lemma 9.2 that

$$\|\mathbf{T}_t(\phi,\omega)(\cdot,t)\|_{\mathcal{C}^\lambda(\overline{G}_{(s)})} \le C\|\phi\|^2_{\mathbf{B}}(1+\|\omega\|_{\mathbf{B}}),$$

$$\sup_{s\in(0,1),h<b(1-s)}\ \max_{t\in[0,h]}\|\mathbf{T}_t(\phi,\omega)(\cdot,t)\|_{\mathcal{C}^\lambda(\overline{G}_{(s)})} \le C\|\phi\|^2_{\mathbf{B}}(1+\|\omega\|_{\mathbf{B}}). \tag{9.2}$$

For the derivative we have

$$\partial_z\mathbf{T}_t(\phi,\omega)(z,t) = I_1+I_2,$$

where

$$I_1 := \frac{1}{2\pi i}\int_{\partial G_{(s)}} \phi(\zeta,t)\tilde{\phi}(\zeta,t)\left(Q(t)+\frac{\omega(\zeta,t)+\tilde{\omega}(\zeta,t)}{2}\right)\frac{\zeta+z}{(\zeta-z)^2}\frac{d\zeta}{\zeta}$$

and

$$I_2 := \frac{1}{2\pi i}\int_{\partial G_{(s)}} \phi(\zeta,t)\tilde{\phi}(\zeta,t)\left(Q(t)+\frac{\omega(\zeta,t)+\tilde{\omega}(\zeta,t)}{2}\right)\frac{1}{(\zeta-z)}\frac{d\zeta}{\zeta}.$$

The second integral I_2 can be estimated in the same way as $\mathbf{T}_t(\phi,\omega)(\cdot,t)$. Thus

$$\|I_2\|_{C^\lambda(\overline{G}_{(s)})} \le C\|\phi\|^2_{\mathbf{B}}(1+\|\omega\|_{\mathbf{B}}). \tag{9.3}$$

After using the holomorphy of $\tilde{\phi},\tilde{\omega}$ with respect to ζ in $A_{(s)}$ we have for all $z\in\overline{G}_{(s)}$, $s\in(0,1)$ and $t\in[0,h]$,

$$\begin{aligned} I_1 &= -\frac{1}{2\pi i}\int_{\partial G_{(s)}} \phi(\zeta,t)\tilde{\phi}(\zeta,t)\left(Q(t)+\frac{\omega(\zeta,t)+\tilde{\omega}(\zeta,t)}{2}\right)\frac{\zeta+z}{\zeta}\partial_\zeta\left(\frac{1}{\zeta-z}\right)d\zeta \\ &= \frac{1}{2\pi i}\int_{\partial G_{(s)}} \partial_\zeta\left(\phi(\zeta,t)\tilde{\phi}(\zeta,t)(Q(t)+\frac{\omega(\zeta,t)+\tilde{\omega}(\zeta,t)}{2})\frac{\zeta+z}{\zeta}\right)\frac{1}{\zeta-z}d\zeta. \end{aligned}$$

Consequently, the integral I_1 is equal to the sum of Schwarz's integral with the density

$$\partial_\zeta\left(\phi(\zeta,t)\tilde{\phi}(\zeta,t)\left(Q(t)+\frac{\omega(\zeta,t)+\tilde{\omega}(\zeta,t)}{2}\right)\right)$$

and Cauchy's integral with the density

$$-\phi(\zeta,t)\tilde{\phi}(\zeta,t)\Big(Q(t)+\frac{\omega(\zeta,t)+\tilde{\omega}(\zeta,t)}{2}\Big)\frac{z}{\zeta^2}.$$

For both types of integrals, estimates in Hölder spaces are well known. Hence,

$$\begin{aligned} \|I_1\|_{C^\lambda(\overline{G}_{(s)})} \le C\Big(&\|\frac{\partial\phi}{\partial z}\|_{C^\lambda(\overline{G}_{(s)})}\|\tilde{\phi}\|_{\mathbf{B}_a}+\|\frac{\partial\tilde{\phi}}{\partial z}\|_{C^\lambda(\overline{A}_{(s)})}\|\phi\|_{\mathbf{B}}\Big) \\ &\times\Big(|Q(t)|+\|\omega\|_{\mathbf{B}}+\|\tilde{\omega}\|_{\mathbf{B}_a}\Big) \\ &+C\|\phi\|_{\mathbf{B}}\|\tilde{\phi}\|_{\mathbf{B}_a}\left(\|\frac{\partial\omega}{\partial z}\|_{C^\lambda(\overline{G}_{(s)})}+\|\frac{\partial\tilde{\omega}}{\partial z}\|_{C^\lambda(\overline{A_s})}\right) \\ &+C\|\phi\|_{\mathbf{B}}\|\tilde{\phi}\|_{\mathbf{B}_a}\Big(|Q(t)|+\|\omega\|_{\mathbf{B}}+\|\tilde{\omega}\|_{\mathbf{B}_a}\Big), \end{aligned}$$

$$\|I_1\|_{C^\lambda(\overline{G}_{(s)})}\left(1-\frac{t}{b(1-s)}\right)^{\frac{1}{2}} \le C\|\phi\|_{\mathbf{B}}\|\tilde{\phi}\|_{\mathbf{B}_a}\Big(|Q(t)|+\|\omega\|_{\mathbf{B}}+\|\tilde{\omega}\|_{\mathbf{B}_a}\Big), \tag{9.4}$$

respectively. Applying Lemma 9.2 and summarizing (9.3) and (9.4) gives

$$\sup_{s\in(0,1),t\in[0,b(1-s))}\|\partial_z\mathbf{T}_t(\phi,\omega)(\cdot,t)\|_{C^\lambda(\overline{G}_{(s)})}\left(1-\frac{t}{b(1-s)}\right)^{\frac{1}{2}} \le C\|\phi\|^2_{\mathbf{B}}(1+\|\omega\|_{\mathbf{B}}). \tag{9.5}$$

The inequalities (9.2) and (9.5) yield

$$\|\mathbf{T}_t(\phi,\omega)\|_{\mathbf{B}} \le C\|\phi\|^2_{\mathbf{B}}(1+\|\omega\|_{\mathbf{B}}),$$

that is, $\mathbf{T}_t(\phi,\omega)$ belongs to $\mathbf{B}$. Following the same approach implies moreover

$$\|\mathbf{T}_t(\phi_1,\omega)-\mathbf{T}_t(\phi_2,\omega)\|_{\mathbf{B}} \le C\max\{\|\phi_1\|_{\mathbf{B}},\|\phi_2\|_{\mathbf{B}}\}(1+\|\omega\|_{\mathbf{B}})\|\phi_1-\phi_2\|_{\mathbf{B}}, \tag{9.6}$$

$$\|\mathbf{T}_t(\phi,\omega_1)-\mathbf{T}_t(\phi,\omega_2)\|_{\mathbf{B}} \le C\|\phi\|^2_{\mathbf{B}}\|\omega_1-\omega_2\|_{\mathbf{B}}. \tag{9.7}$$

Thus $\mathbf{T}_t$ depends continuously on ϕ and ω. All the statements of our lemma are proved. □

Corollary 9.4. *For each $\omega \in \mathbf{B}$ there exists a constant $b = b(\omega)$ such that the abstract Cauchy-Kovalevsky problem* (7.3)–(7.4) *has a uniquely determined solution $\phi \in \mathbf{B}$.*

Proof. The problem (7.3)–(7.4) can be rewritten in the following operator form:

$$\mathbf{I}(\phi) = \mathbf{I}(\phi_0) + \mathbf{A} \circ \mathbf{M}\Big(\phi, \mathbf{M}(z, \mathbf{T}_t(\phi,\omega))\Big) - 2\mathbf{J} \circ \mathbf{M}\Big(\phi, \partial_z \circ \mathbf{M}(z, \mathbf{T}_t(\phi,\omega))\Big), \quad (9.8)$$

where $\mathbf{M}(\phi,\omega) := \phi \cdot \omega$ is the operator of multiplication, $\mathbf{J}\psi := \int\limits_0^t \psi(\cdot,\tau)d\tau$ is the operator of integration, $\mathbf{I}$ is the identity operator, and $\circ$ denotes the superposition operator.

Lemmas 8.1, 9.1, 9.3 imply immediately

$$\|\mathbf{A} \circ \mathbf{M}\Big(\phi, \mathbf{M}(z, \mathbf{T}_t(\phi,\omega))\Big)\|_{\mathbf{B}} \leq Cb\|\phi\|_{\mathbf{B}}^3(1 + \|\omega\|_{\mathbf{B}}), \quad (9.9)$$

$$\|2\mathbf{J} \circ \mathbf{M}\Big(\phi, \partial_z \circ \mathbf{M}(z, \mathbf{T}_t(\phi,\omega))\Big)\|_{\mathbf{B}} \leq Cb\|\phi\|_{\mathbf{B}}^3(1 + \|\omega\|_{\mathbf{B}}), \quad (9.10)$$

respectively, where in the last two formulas the constants C are independent of ϕ and ω.

The assumptions on ϕ and ω from problem (Q_α) guarantee the existence of r_0, r_1 such that $\phi_0 \in \mathbf{B}$. Let us fix a constant $R > 0$. Then a suitable small b makes sure together with (9.8), (9.9), and (9.10) that the right-hand side of (9.8) maps $\{\phi \in \mathbf{B} : \|\phi - \phi_0\|_{\mathbf{B}} \leq R\}$ into itself. By (9.1) this mapping is a contractive one if we eventually choose a smaller b. Banach's fixed point theorem leads to a uniquely determined solution $\phi \in \mathbf{B}$ of (7.3)–(7.4). This completes the proof. □

9.1. Riemann-Hilbert-Poincaré problem

Here we discuss the holomorphy of the second component $w_{\text{reg}} = w_{\text{reg}}(z,t)$ of solutions of problem (P_α) (or equivalently $\omega = \omega(z,t)$ of problem (Q_α)). As we have already shown, the solution $\phi = \phi(z,t)$ of (7.3)–(7.4) admits an analytic continuation into a bigger domain if we suppose a corresponding property for ω. Without loss of generality we can suppose that this domain is a disk $G_d = \{z : |z| < d\}, d > 1$, and it is a common domain of holomorphy for all functions $\phi = \phi(\cdot,t)$, $t \in [0, T_d)$. In problem (7.5)–(7.6) the variable t is only a parameter. For this reason let us omit it till the end of this section (and use for sake of brevity the notations $\phi(z), \omega(z), Q$ etc.)

Let ϕ be a given holomorphic and non-vanishing function in G_d. Our goal is to prove that ω connected with ϕ by

$$\operatorname{Im}(\omega(z)) = \alpha\partial_\theta\Big(|\phi(z)|\big(Q + \operatorname{Re}(\omega(z))\big)\Big) \quad on \quad \partial G_1, \quad (9.11)$$

$$\omega(0) = 0, \quad (9.12)$$

is also holomorphic in G_d.

Therefore we need the following auxiliary result:

Lemma 9.5. *Let the non-vanishing function* ϕ *belong to* $\mathcal{H}(G_1) \cap \mathcal{C}^{1,\lambda}(\overline{G}_1)$. *Then the boundary value problem* (9.11)–(9.12) *possesses a uniquely determined solution* $\omega(z) \in \mathcal{H}(G_1) \cap \mathcal{C}^{1,\lambda}(\overline{G}_1)$.

Proof. Using $\text{Im}(\omega(0)) = 0$ we can rewrite the boundary condition (9.11) in the equivalent form

$$-\mathbf{H}\,(\text{Re}\,\omega)\,(z) = \alpha Q \partial_\theta |\phi(z)| + \alpha \partial_\theta \Big(|\phi(z)| \text{Re}\,(\omega(z)) \Big), \; z = re^{i\theta} \in \partial G_1, \quad (9.13)$$

where $\mathbf{H}$ is the Hilbert transform for the unit circle,

$$\mathbf{H}g(e^{i\eta}) := \frac{1}{2\pi} \int_0^{2\pi} g(e^{i\sigma}) \cot \frac{\eta - \sigma}{2} d\sigma, \qquad \eta \in [0, 2\pi).$$

Using $\text{Re}\,(\omega(0)) = 0$ and applying $\mathbf{H}$ once more we get

$$\text{Re}\,(\omega(z)) = \alpha Q \partial_\theta \mathbf{H}\,(|\phi(z)|) + \alpha \partial_\theta \Big(\mathbf{H}\,(|\phi(z)| \text{Re}\,(\omega(z))) \Big). \quad (9.14)$$

Let us introduce a new function $U = U(z, \bar{z})$, which is harmonic in the unit disk, and which satisfies the boundary relation

$$U(z, \bar{z}) = |\phi(z)| \text{Re}\,(\omega(z)), \qquad z \in \partial G_1. \quad (9.15)$$

Using the Cauchy-Riemann equations on the unit circle we get finally, that (9.11)–(9.12) can be rewritten as the third kind of boundary problem for U,

$$\partial_r U + \frac{1}{\alpha |\phi(z)|} U = Q\mathbf{H}\,(\partial_\theta |\phi|)\,(z), \qquad z \in \partial G_1.$$

The coefficient $(\alpha|\phi(z)|)^{-1}$, and the right-hand side of this boundary condition belong to $\mathcal{C}^\lambda\,(\partial G_1)$. It is known that this problem is uniquely solvable (see, e.g., [15]) in the space $\mathcal{C}^{1,\lambda}\,(\overline{G}_1)$. By Poisson's formula and (9.15) we determine the unique harmonic function $\text{Re}\,(\omega)$. The formula (9.14) implies $\int\limits_0^{2\pi} \text{Re}\,(\omega(e^{i\theta}))d\theta = 0$, i.e., $\text{Re}\,(\omega(0)) = 0$. The corresponding unique solution ω of problem (9.11)–(9.12) is obtained by Schwarz's formula under the additional constraint $\text{Im}(\omega(0)) = 0$. Using the assumption for ϕ and the regularity of U gives immediately the statement of the lemma. □

Let the function ϕ be holomorphic and non-vanishing in the disk G_d, $d > 1$.[4] One can introduce the following holomorphic function:

$$h^+(z) := (\phi(z))^{\frac{1}{2}}, \qquad z \in G_d,$$

[4]We will use further the same notations for functions given in the unit disk and for their analytic continuations into bigger domains.

choosing any branch of the root function. We can define another holomorphic function by

$$h^-(z) := \overline{h^+\left(\frac{1}{\bar z}\right)} \quad \text{for} \quad \left\{z : |z| > \frac{1}{d}\right\}.$$

Both of these functions $h^+(z)$, and $h^-(z)$ as well as their product $m(z) := h^+(z)h^-(z)$ are holomorphic in the annulus $A = \left\{z : \frac{1}{d} < |z| < d\right\}$. Besides we have on the unit circle

$$|\phi(z)| = h^+(z)h^-(z), \qquad z \in \partial G_1.$$

In the same manner we can introduce

$$\omega^-(z) := \overline{\omega^+\left(\frac{1}{\bar z}\right)} \quad \text{for } \{z : |z| > 1\},$$

where $\omega^+(z) := \omega(z)$ in the unit disk G_1. With these notations the boundary condition (9.11) can be rewritten on ∂G_1 in the equivalent form

$$\omega^+(z) - \omega^-(z) = 2i\alpha Q\partial_\theta m(z) + i\alpha\partial_\theta\Big(m(z)\left(\omega^+(z) + \omega^-(z)\right)\Big),$$

or

$$\begin{aligned} &\alpha z m(z) d_z\omega^+(z) + (1 + \alpha z d_z m(z))\omega^+(z) + \alpha Q z d_z m(z) \\ &\quad = -\alpha z m(z) d_z\omega^-(z) + (1 - \alpha z d_z m(z))\omega^-(z) - \alpha Q z d_z m(z). \end{aligned} \tag{9.16}$$

Let us note that all the terms at the left-hand side of (9.16) are in fact holomorphic on the internal annulus $A^i := \left\{z : \frac{1}{d} < |z| < 1\right\}$ and Hölder-continuous up to ∂G_1, but those at the right-hand side of (9.16) are holomorphic on the external annulus $A^e := \{z : 1 < |z| < d\}$ and Hölder-continuous up to ∂G_1. Therefore, due to the theorem on analytic continuation the left- (right-) hand side of (9.16) is the restriction of a holomorphic function (say $F = F(z)$) defined on the annulus A to A^i (A^e). This means that (9.16) is equivalent to the system of differential equations

$$\begin{aligned} \alpha z m(z) d_z\omega^+ + (1 + \alpha z d_z m(z))\omega^+ + \alpha Q z d_z m(z) &= F(z),\ z \in A^i, \\ -\alpha z m(z) d_z\omega^- + (1 - \alpha z d_z m(z))\omega^- - \alpha Q z d_z m(z) &= F(z),\ z \in A^e. \end{aligned} \tag{9.17}$$

By Lemma 9.5 the function $\omega^+(z)$ is uniquely determined in $\overline{G}_1$. We have, due to Schwarz's Reflection Principle, that

$$F(z) = \overline{F\left(\frac{1}{\bar z}\right)}, \quad \text{for} \quad z \in A^e; \tag{9.18}$$

taking into consideration that the values of $F(z)$ on ∂G_1 are real, this means (9.18) holds on ∂G_1. Consequently, we can use the differential equation

$$\alpha z m(z) d_z\omega^+ + (1 + \alpha z d_z m(z))\omega^+ + \alpha Q z d_z m(z) = F(z), \quad z \in A, \tag{9.19}$$

for the continuation of ω^+ from $A^i \cup \partial G_1$ into A^e. Now we are in position to prove the next statement.

Lemma 9.6. *Let the function $\omega^i = \omega^i(z)$ be a single-valued holomorphic function in an open annulus A^i whose boundary consists of two disjoint circles Γ_i, Γ_0 ($\Gamma_i \subset int\,\Gamma_0$). Let ω^i satisfy on A^i the first-order differential equation*

$$d_z\omega + a(z)\omega = b(z), \tag{9.20}$$

where $a, b \in \mathcal{H}(A^i)$.

Let us additionally suppose that a, b admit analytic continuations into an annulus A^e whose boundary consists of two disjoint circles Γ_0, Γ_e ($\Gamma_0 \subset int\,\Gamma_e$). Then ω^i admits an analytic continuation ω into A^e, too.

Proof. Let us cover $\overline{A}^e$ by a finite number $\{V_1, \ldots, V_n\}$ of open sets in such a way that

$$V_j \cap A^e \neq \emptyset; \bigcup_{j=1}^{n} V_j \supset \overline{A}^e; V_j \cap V_{j+1} \cap A^i \neq \emptyset, j = 1, \ldots, n; V_{n+1} = V_1.$$

Then we choose a point $z_1 \in V_n \cap V_1 \cap A^i$ and solve the Cauchy problem

$$\omega(z_1) = \omega^i(z_1)$$

for the differential equation (9.20) on $V_1 \cap A^e$. It has a unique holomorphic solution $\omega_1(z)$, which in fact coincides with $\omega^i(z)$ on $V_1 \cap V_n \cap A^i$. Hence $\omega_1(z)$ is an analytic continuation of $\omega^i(z)$ into $V_1 \cap A^e$. In the same manner we get a continuation into all domains $V_j \cap A^e, j = 1, \ldots, n$, choosing corresponding points $z_j \in V_{j-1} \cap V_j \cap A^i$. At last $\omega_n(z)$ coincides with $\omega_1(z)$ on $V_1 \cap V_n \cap \Omega_e$ because both functions are analytic continuations of the single-valued holomorphic function $\omega^i(z)$ from the domains $V_n \cap A^i$ and $V_1 \cap A^i$, respectively. Moreover, $V_n \cap A^i$ and $V_1 \cap A^i$ have a common non-empty domain. □

Corollary 9.7. *If in* (7.5) *the function $\phi = \phi(z,t)$ admits an analytic continuation to a bigger disk G_d, $d > 1$, for certain $t \in [0, T)$, then the corresponding solution $\omega = \omega(z,t)$ of* (7.5)–(7.6) *admits an analytic continuation into G_d for the same t, too.*

10. Existence result

We suppose that $\phi_0 = \phi_0(z)$ is non-vanishing in a neighbourhood of the unit disk $\overline{G}_1$. Thus there exists $r_1 > 1$ such that

$$\rho_0 := \inf_{z \in G_{r_1}} |\phi_0(z)| = \inf_{z \in G_{r_1}} \left| \left(\frac{\partial f_0}{\partial z}(z) \right)^{-1} \right| > 0. \tag{10.1}$$

Using Corollary 9.7 the solution $\omega_0 = \omega_0(z)$ of the problem (9.11)–(9.12) with $\phi := \phi_0$ is holomorphic in $G_{(1)} = G_{r_1}$, too. The next lemma describes properties of continuations of solutions for (9.11)–(9.12). Let us introduce the function space

$$\mathbf{B}_1 = \{\phi \in \mathcal{H}\left(G_{(1)}\right) : \|\phi\|_1 := \sup_{s \in (0,1)} \|\phi\|_{C^{1,\lambda}(\overline{G}_{(s)})} < \infty\}.$$

Moreover, let $B_1(\phi_0, \rho) \subset \mathbf{B}_1$ be the ball around ϕ_0 with radius ρ.

Lemma 10.1. *If $\phi \in B_1(\phi_0, \rho), \rho < \rho_0$, then the continuations $\omega = \omega(z)$ of solutions for* (9.11)–(9.12) *belong to $\mathbf{B}_1$ and satisfy*

$$\|\omega\|_1 \leq C, \quad \textit{where} \quad C = C(\lambda, Q, r_0, r_1, \rho, \rho_0).$$

Proof. By the aid of Lemma 9.5 and following its proof we obtain with the same notations

$$\|U(z)\|_{C^{1,\lambda}(\overline{G}_1)} \leq C\|\phi(z)\|_{C^{1,\lambda}(\overline{G}_1)}.$$

Using (10.1) and the assumption $\rho < \rho_0$ we have the estimate

$$\||\phi(z)|^{-1}\|_{C^{1,\lambda}(\overline{G}_1)} \leq C.^5$$

Taking into account (9.9), (9.10) and the properties of the Hilbert transform we get immediately

$$\|\omega\|_{C^{1,\lambda}(\overline{G}_1)} \leq C. \tag{10.2}$$

From Corollary 9.7 we get the existence of ω in $G_{(1)}$. The continuation of ω from $\overline{G}_1$ to $G_{(1)}$ is defined in the annulus $A = \{z \in \mathbb{C} : \frac{1}{r_1} < |z| < r_1\}$ as the solution of the differential equation

$$d_z\omega^+ + \frac{1 + \alpha z d_z m(z)}{\alpha z m(z)}\omega^+ = \frac{F(z)}{\alpha z m(z)} - \frac{Q d_z m(z)}{m(z)}. \tag{10.3}$$

We remember that $m(z) = (\phi(z))^{\frac{1}{2}}(\overline{\phi(\frac{1}{\overline{z}})})^{\frac{1}{2}}, z \in A$, and $F(z)$ is defined in $A^i = \{z \in \mathbb{C} : \frac{1}{r_1} < |z| < 1\}$ by (9.17) and in $A^e = \{z \in \mathbb{C} : 1 < |z| < r_1\}$ by (9.18).

By the aid of points $z_1, \dots, z_n$, n sufficiently large, we choose an equidistant partition of ∂G_1 and overlapping sectors $S_1, \dots, S_n$ defined as follows:

$$S_k := \left\{z \in \mathbb{C} : \frac{1}{r_1} < |z| < r_1, -\frac{\pi}{n} - \delta < \arg z - \delta_k < \frac{\pi}{n} + \delta\right\},$$

where $\delta_k = \arg z_k$, δ is a sufficiently small positive number. It is enough to get an estimate for $\|\omega\|_{C^{1,\lambda}(\overline{G}_{(s)} \cap S_k)}$ uniformly for all $s \in (0,1)$, and $\phi \in B_1(\phi_0, \rho)$. All these estimates together yield an estimate for $\|\omega\|_{C^{1,\lambda}(\overline{G}_{(s)})}$.

The solution of (10.3) can be represented in S_k in the form

$$\omega(z) := J_1(z) + J_2(z) := \omega(z_k)\exp\left\{-\int_{z_k}^{z} \frac{1 + \alpha\zeta d_\zeta m(\zeta)}{\alpha\zeta m(\zeta)} d\zeta\right\}$$
$$+ \int_{z_k}^{z}\left(\frac{F(\zeta)}{\alpha\zeta m(\zeta)} - \frac{Q d_\zeta m(\zeta)}{m(\zeta)}\right)\exp\left\{\int_{z}^{\zeta} \frac{1 + \alpha\xi d_\xi m(\xi)}{\alpha\xi m(\xi)} d\xi\right\} d\zeta.$$

For $J_1(z)$ we have

$$J_1(z) = \omega(z_k)\frac{m(z_k)}{m(z)}\exp\left\{-\int_{z_k}^{z} \frac{d\zeta}{\alpha\zeta m(\zeta)}\right\}.$$

[5] In this section the constant C is independent of $\phi \in B_1(\phi_0, \rho)$ and of $s \in (0,1)$.

The function m^{-1} belongs to $\mathcal{C}^{1,\lambda}(G_{(1)} \cap S_k)$, where $\|m^{-1}\|_{\mathcal{C}^{1,\lambda}(\overline{G}_{(s)} \cap S_k)} \leq C$. Here we use $|m(z)| \geq \rho_0 - \rho > 0$ for all $\phi \in B_1(\phi_0, \rho)$. Consequently,

$$\|J_1(z)\|_{\mathcal{C}^{1,\lambda}(\overline{G}_{(s)} \cap S_k)} \leq C. \tag{10.4}$$

The discussion of $J_2(z)$ brings no new difficulties besides the consideration of the integral $\int\limits_{z_k}^{z} \frac{F(\zeta)}{\alpha \zeta m(\zeta)} d\zeta$. From (9.17) and (10.2) we get

$$\|F\|_{\mathcal{C}^{\lambda}(G_1 \cap S_k)} \leq C \quad \text{for all} \quad \phi \in B_1(\phi_0, \rho).$$

Hence, by (9.18) we have

$$\|F\|_{\mathcal{C}^{\lambda}(\overline{G}_{(s)} \cap S_k)} \leq C.$$

But this gives immediately that $J_2(z) \in \mathcal{C}^{1,\lambda}(G_{(1)} \cap S_k)$ with

$$\|J_2(z)\|_{\mathcal{C}^{1,\lambda}(\overline{G}_{(s)} \cap S_k)} \leq C. \tag{10.5}$$

The inequalities (10.4), and (10.5) imply $\|\omega\|_{\mathcal{C}^{1,\lambda}(G_{(1)} \cap S_k)} \leq C$, that is $\omega \in \mathbf{B}_1$ with $\|\omega\|_1 \leq C$ for all $\phi \in B_1(\phi_0, \rho)$. This completes the proof. □

Corollary 10.2. *If $\phi \in B_1(\phi_0, \rho)$, then $\omega \in B_1(\omega_0, \eta)$ with a constant η depending on ρ. If ρ tends to 0, then η tends to 0, too.*

Proof. Let ω, ω_0 be the solutions of (9.11)–(9.12) for ϕ, ϕ_0, respectively. Then $v := \omega - \omega_0$ solves the following Riemann-Hilbert-Poincaré Problem on ∂G_1:

$$\begin{aligned} \operatorname{Im}(v(z)) &= \alpha \partial_\theta \Big(|\phi(z)| \operatorname{Re}(v(z)) \Big) + \alpha \partial_\theta \Big((|\phi(z)| - |\phi_0(z)|) \big(Q + \operatorname{Re}(\omega_0(z)) \big) \Big), \\ v(0) &= 0. \end{aligned}$$

The statement follows from the fact that $v \equiv 0$ is a solution of

$$\operatorname{Im}(v(z)) = \alpha \partial_\theta \Big(|\phi(z)| \operatorname{Re}(v(z)) \Big), \; v(0) = 0 \text{ on } \partial G_1,$$

and $\alpha \partial_\theta \big((|\phi(z)| - |\phi_0(z)|)(Q + \operatorname{Re}(\omega_0(z))) \big)$ can be considered as a small perturbation. To complete the proof we follow the same approach as in the proof of Lemma 10.1. □

Remark 10.3. The statements of Lemma 10.1 and Corollary 10.2 remain true if we replace $\mathbf{B}_1$ by $\mathbf{B}_{s_0}$, $s_0 \in (0,1)$, where

$$\mathbf{B}_{s_0} := \{ \phi \in \mathcal{H}(G_{(s_0)}) : \|\phi\|_{s_0} := \sup_{s \in (0, s_0)} \|\phi\|_{\mathcal{C}^{1,\lambda}(\overline{G}_{(s)})} < \infty \}.$$

10.1. Main result

The results of the previous sections serve as preparations to prove our main result.

Theorem 10.4. *There exists an (in general) small interval of time* $[0, b)$ *such that the problem* (Q_α) *has a uniquely determined solution* (ϕ, ω)*. The first component* $\phi = \phi(z, t)$ *has no zeros on* $\overline{G}_{r_0} \times [0, b)$ *and belongs to the space*

$$\mathcal{C}^1\left([0, b), \mathcal{H}(G_{r_0}) \cap \mathcal{C}^{1,\lambda}(\overline{G}_{r_0})\right).$$

The second component $\omega = \omega(z, t)$ *belongs to the space*

$$\mathcal{C}\left([0, b), \mathcal{H}(G_{r_0}) \cap \mathcal{C}^{1,\lambda}(\overline{G}_{r_0})\right).$$

The constant r_0 *is taken as in the definition of* $\mathbf{B}$.

Corollary 10.5. *There exists an (in general) small interval of time* $[0, b)$ *such that the problem* (P_α) *has a uniquely determined solution* (f, w_{reg})*. The first component* $f = f(z, t)$ *is univalent with respect to* z *on* G_{r_0} *for* $t \in [0, b)$ *and belongs to* $\mathcal{C}^1\left([0, b), \mathcal{H}(G_{r_0}) \cap \mathcal{C}^{2,\lambda}(\overline{G}_{r_0})\right)$*. The second component* $w_{\text{reg}} = w_{\text{reg}}(z, t)$ *belongs to* $\mathcal{C}\left([0, b), \mathcal{H}(G_{r_0}) \cap \mathcal{C}^{1,\lambda}(\overline{G}_{r_0})\right)$*. The constant* r_0 *is taken as in the definition of* $\mathbf{B}$.

This corollary is a direct consequence of the theorem. Univalence of the function f follows from the properties of initial function f_0, the choice of $\rho < \rho_0$ (see (10.1)) and the chosen norm of $\mathbf{B}$ which is stronger than the sup-norm.

Proof of Theorem 10.4. Let us consider the problem (Q_α), that is (7.3)–(7.6). Let ω_0 be the solution of (7.5)–(7.6) for $\phi = \phi_0(z)$. One can find constants r_0, r_1 such that ϕ_0 and ω_0 belong to $\mathbf{B}$. We will prove the existence of solutions $\phi = \phi(z, t)$ and $\omega = \omega(z, t)$ belonging to $M_R(\phi_0) := \{\phi : \|\phi - \phi_0\|_{\mathbf{B}} \leq R\}$ and $M_K(\omega_0) := \{\omega : \|\omega - \omega_0\|_{\mathbf{B}} \leq K\}$, respectively, where the constants R and K will be chosen later.

Step 1: The abstract Cauchy-Kovalevsky problem

Using Corollary 1, to each $\omega \in M_K(\omega_0)$ corresponds a constant $b = b(\omega)$ such that (7.3)–(7.4) has a uniquely determined solution $\phi \in \mathbf{B}$. We can choose the constant b in such a way that $\phi \in M_R(\phi_0)$ uniformly for all $\omega \in M_K(\omega_0)$. Let us define the operator

$$\mathbf{P}_1 : \omega \in M_K(\omega_0) \mapsto \phi = \phi(\omega) \in M_R(\phi_0),$$

which maps $\omega \in M_K(\omega_0)$ to the uniquely determined solution $\phi = \phi(\omega)$ of (7.3)–(7.4). Taking into consideration (9.6), (9.7) the operator $\mathbf{P}_1$ depends continuously on ω. The inequalities (9.6), (9.7) and (9.9), (9.10) yield the existence of a constant $C_1 = C_1(R, K)$ independent of b such that

$$\|\mathbf{P}_1(\omega_1) - \mathbf{P}_1(\omega_2)\|_{\mathbf{B}} \leq C_1 b \|\omega_1 - \omega_2\|_{\mathbf{B}}. \tag{10.6}$$

Consequently, this inequality remains valid if we choose a smaller b.

Discussion of the choice of b: Let us choose $R > 0$. From (9.9), (9.10) we obtain

$$\|\phi - \phi_0\|_{\mathbf{B}} \leq Cb\big(\|\phi_0\|_{\mathbf{B}} + R\big)^3(1+K).$$

Hence,

$$b \leq \frac{R}{C\big(\|\phi_0\|_{\mathbf{B}} + R\big)^3(1+K)} \tag{10.7}$$

guarantees that $\phi = \phi(\omega) \in M_R(\phi_0)$ for all $\omega \in M_K(\omega_0)$. Using (10.1) a sufficiently small choice of R gives additionally that $\phi \in M_R(\phi_0)$ has no zeros in $\bigcup\limits_{0<s<1} G_{(s)} \times [0, b(1-s))$.

Step 2: The Riemann-Hilbert-Poincaré problem

Due to Corollary 2 we can define a mapping

$$\mathbf{P}_2 : \phi \in M_R(\phi_0) \mapsto \tilde{\omega},$$

where $\tilde{\omega}$ is the uniquely determined solution of (7.5)–(7.6). Due to the results from Section 3.3 this mapping takes values in $\mathbf{B}$. Repeating the proof of Lemma 8, taking into account Corollary 3, leads to the next statement:

Lemma 10.6. *If*

$$\max\left\{\sup_{s\in(0,1),h<b(1-s)} \max_{t\in[0,h]} \|\phi(\cdot,t)\|_{C^\lambda(\overline{G}_{(s)})};\right.$$
$$\left.\sup_{s\in(0,1),h<b(1-s)} \left\|\frac{\partial\phi}{\partial z}(\cdot,t)\right\|_{C^\lambda(\overline{G}_{(s)})} C_{t,s}\right\} < \infty,$$

then the same is true for $\tilde{\omega} = \tilde{\omega}(z,t)$.

Proof. It is clear because the derivatives $\partial_z J_1(z)$, $\partial_z J_2(z)$ and the left-hand side of the differential equation (9.19) depend linearly on the derivatives $\partial_z\phi$ and $\partial_z\tilde{\phi}$. □

Using $C_{t,s} = \big(1-\frac{t}{b(1-s)}\big)^{\frac{1}{2}}$ gives that $\mathbf{P}_2$ maps $\phi \in M_R(\phi_0)$ to $\tilde{\omega} = \mathbf{P}_2(\phi) \in \mathbf{B}$. By Corollary 3 the function $\tilde{\omega}$ belongs to $M_K(\omega_0)$ for all $\phi \in M_R(\phi_0)$. The proof of this corollary implies the existence of a constant $C_2 = C_2(R,K)$ independent of b such that

$$\|\mathbf{P}_2(\phi_1) - \mathbf{P}_2(\phi_2)\|_{\mathbf{B}} \leq C_2 b\|\phi_1 - \phi_2\|_{\mathbf{B}}. \tag{10.8}$$

The inequalities (10.6), (10.8) give for the mapping

$$\mathbf{P}_2 \circ \mathbf{P}_1 : \omega \in M_K(\omega_0) \mapsto \tilde{\omega} \in M_{\tilde{K}}(\omega_0)$$

the estimate

$$\|\tilde{\omega}_1 - \tilde{\omega}_2\|_{\mathbf{B}} = \|\mathbf{P}_2 \circ \mathbf{P}_1(\phi_1) - \mathbf{P}_2 \circ \mathbf{P}_1(\phi_2)\|_{\mathbf{B}} \leq C_2C_1 b\|\omega_1 - \omega_2\|_{\mathbf{B}}.$$

Discussion of the choice of K, R *and* b: If $\tilde{K} > K$, then we choose $K := \tilde{K}$, otherwise it is unchanged. The constant R is determined by Corollary 3 in such a way, that $\phi \in M_R(\phi_0)$ is mapped by $\mathbf{P}_2$ into $M_K(\omega_0)$. Then (10.7) and $b <$

$(C_1C_2)^{-1}$ (the constants C_1, C_2 are independent of b) ensure that $\mathbf{P}_2 \circ \mathbf{P}_1$ is a contractive mapping on $M_K(\omega_0)$. Consequently, there exists a uniquely determined fixed point $\omega_{fix} \in M_K(\omega_0)$. This fixed point and $\phi := \mathbf{P}_1(\omega_{fix})$ form the unique solution $(\phi, \omega) := (\mathbf{P}_1(\omega_{fix}), \omega_{fix})$ of (Q_α). All statements of the theorem are proved. □

References

[1] M.V. Dubatovskaya, E.V. Pesetskaya, S.V. Rogosin, On Hele-Shaw flow in a bounded domain, *Proceedings of N.I.Lobachevsky Mathematical Center. Kazan*, **14**, 2002, 113–129.

[2] J. Duchon, R. Robert, Estimation d'opérateur intégraux du type de Cauchy dans les eschelles d'Ovsjannikov et application, *Ann. Inst. Fourier, Grenoble*, **36**, No. 1, 1986, 83–95.

[3] F.D. Gakhov, *Boundary Value Problems*, Nauka, Moscow, (3rd ed.) 1977 (in Russian).

[4] L.A. Galin, Unsteady filtration with the free boundary, *Dokl. Akad. Nauk SSSR*, **47**, No. 4, 1945, 246–249 (in Russian). 236

[5] R.P. Gilbert, Zhenyuan Xu, Several Hele-Shaw flows associated with injection molding, in *Complex Analysis and its Applications*, (C.C. Yang, G.C. Wen, K.Y. Li and Y.M. Chiang (eds.)), Longman, Harlow, 1994, pp. 26–37.

[6] I. Gohberg, N. Krupnik, *Introduction to the Theory of One-Dimensional Singular Integral Equations*, vol. I, II, Birkhäuser Verlag, Basel-Boston-Berlin, 1992.

[7] G.M. Goluzin, *Geometric Theory of Functions of Complex Variables*, Amer. Math. Soc., Providence, RI, 1969.

[8] B. Gustafsson, On a differential equation arising in a Hele-Shaw flow moving boundary problem, *Arkiv för Matematik*, **22**, 1984, 251–268. 234

[9] B. Gustafsson, D. Prokhorov, A. Vasil'ev, Infinite lifetime for the starlike dynamics in Hele-Shaw cells, *Proc. Amer. Math. Soc.*, **132**, No. 9, 2004, 2661–2669. 228

[10] B. Gustafsson, A. Vasil'ev, *Conformal and Potential Analysis in Hele-Shaw Cells*, Birkhäuser Verlag, Basel-Stockholm, 2006. 227, 228, 231, 234

[11] H.S. Hele-Shaw, The flow of water, *Nature.* **58** (1489), 1898, 33–36. 225

[12] Yu.E. Hohlov, S.D. Howison, On the classification of solutions to the zero-surface-tension model for the Hele-Shaw free boundary flows, *Quart. J. Appl. Math.*, **51**, No. 4, 1993, 777–789. 226, 227, 228, 236

[13] Yu.E. Hohlov, S.D. Howison, C. Huntingford, J.R. Ockendon, A.A. Lacey, A model for non-smooth free boundaries in Hele-Shaw flows, *Quart. J. Mech. Math.*, **47**, No. 1, 1994, 107–128. 236

[14] S.D. Howison, Complex variable methods in Hele-Shaw moving boundary problems. *Euro. J. Appl. Math.*, **3**, 1992, 209–224. 236

[15] O.A. Ladyzhenskaya, N.N. Uraletseva, *Linear and Quasilinear Elliptic Equations*, Nauka, Moscow, 1964 (in Russian; Engl. transl. by Academic Press, New York, 1968). 243

[16] L.S. Leibenzon, *Oil producing mechanics*, in two volumes, Nefteizdat, Moscow, 1934 (in Russian). 232

[17] J.A. McGeough, H. Rasmussen, On the derivation of the quasi-steady model in electro-chamical machining, *J. Inst. Math. and Appl.*., **13**, No. 1, 1974, 13–21. 226

[18] A.M. Meirmanov, *Stefan Problem*, Nauka, Moscow, 1986 (in Russian). 226

[19] A.M. Meirmanov, B. Zaltzman, Global in time solution to the Hele-Shaw problem with a change of topology. *Euro. J. Appl. Math.*, **13**, No. 4, 2002, 431–447.

[20] M. Muskat, *The flow of the homogeneous fluids through porous media.* McGraw-Hill, N.Y., 1937. 225

[21] N.I. Muskhelishvili, *Singular Integral Equations*, P. Noordhoff N.V., Groningen, 1953.

[22] L. Nirenberg, An abstract form of the nonlinear Cauchy-Kowalevski theorem, *J. Differential Geom.*, **6**, 1972, 561–576. 235

[23] T. Nishida, A note on a theorem of Nirenberg, *J. Differential Geom.*, **12**, 1977, 629–633. 235

[24] J.R. Ockendon, S.D. Howison, P.Ja. Kochina and Hele-Shaw in modern mathematics, natural sciences and technology, *Prikl. Mat. i Mech.*, **66**, vyp. 3, 2002, 515–524 (in Russian). 225

[25] H. Oertel (ed.), *Prandtl's essentials of fluid mechanics*, Springer Verlag, New York, 2004.

[26] L.V. Ovsjannikov, A singular operator in a scale of Banach spaces, *Dokl. AN SSSR*, **163**, No. 4, 1965, 819–822 (in Russian). 235

[27] C.P.Please, G. Pettet, D.L.S. McElwain, A new approach to modelling the formation of necrotic regions in tumours, *Appl. Math. Letter*, **11**, No. 3, 1998, 89–94. 226

[28] P.Ya. Polubarinova-Kochina, On the motion of the oil contour, *Dokl. Akad. Nauk SSSR*, **47**, No. 4, 1945, 254–257 (in Russian). 225, 226, 227, 236

[29] M. Reissig, About a nonstationary mixed problem for holomorphic functions arising by the study of a potential flow past a circular cylinder with permeable surface, *Math. Nachr.*, **164**, 1993, 283–297.

[30] M. Reissig, L. von Wolfersdorf, A simplified proof for a moving boundary problem for Hele-Shaw flows in the plane, *Arkiv för Math.*, **31**, No. 1, 1993, 101–110. 228, 237, 238

[31] M. Reissig, The existence and uniqueness of analytic solutions for moving boundary value problem for Hele-Shaw flows in the plane, *Nonlinear Anal. Theory, Methods & Appl.*, **23**, No. 5, 1994, 565–576. 228

[32] M. Reissig, A generalized theorem of Peano in scale of Banach spaces with completely continuous imbedding, *Funkc. Ekvacioj*, **37**, 1994, 521–530. 228

[33] M. Reissig, Leray-Volevich conditions for systems of abstract evolution equations of Nirenberg/ Nishida type, *Tsukuba J. Math.*, **18**, No. 1, 1994, 193–202. 228

[34] M. Reissig, F. Hübner, Analytical and numerical treatment of Hele-Shaw models with and without regularization, in *Generalized Analytic Functions* (H. Florian et al., eds.), Kluwer AP, Amsterdam, 1998, pp. 271-187. 236

[35] M. Reissig and S.V. Rogosin, with an appendix of F. Huebner, Analytical and numerical treatment of a complex model for Hele-Shaw moving boundary value problems

with kinetic undercooling regularization, *Euro J. Appl. Math.*, **10**, 1999, 561–579. 235, 236

[36] S D. Richardson, Hele-Shaw flows with a free boundary produced by injection of fluid into a narrow channel. *J. Fluid Mech.*, **56**, 1972, 609–618. 233

[37] S.V. Rogosin, On a scale of Banach spaces. *Proceedings of Institute of Mathematics. Minsk*, **12**, No. 1, 2004, 126–133.

[38] S.V. Rogosin, Hele-Shaw Moving Boundary Value Problem in a Bounded Domain. Local in Time Solvability, *Complex Variables*, **50**, No. 7–11, 2005, 745-764.

[39] L.A. Romero, *The Fingering Problem in a Hele-Shaw Cell.* PhD Thesis, California Institute of Technology, 1981. 236

[40] P.G. Saffman, Viscous fingering in Hele-Shaw cells. *J. Fluid Mech.*, **173**, 1986, 73–94. 227

[41] P.G. Saffman, G.I. Taylor, The penetration of a fluid into porous medium or Hele-Shaw cell containing a more viscous liquid, *Proc. Roy. Soc. London. Ser. A*, **245**, No. 1242, 1958, 312-329. 227

[42] P.G. Saffman, Viscous fingering in Hele-Shaw cells, *J. Fluid Mech.*, **173**, 1986, 73–94. 236

[43] V.A. Solonnikov, Boundary and initial-boundary value problems for the Navier-Stokes equations in domains with non-compact boundaries / In: *Mathematical Topics in Fluid Mechanics* (J.-F. Rodrigues, A. Sequeira (eds.)). Longman, Harlow, 1993, pp. 117–162.

[44] S. Tanveer, Evolution of Hele-Shaw interface for small surface tension, *Phil. Trans. R. Soc. Lond.*, **A 343**, 1993, 155–204. 236

[45] W. Tutschke, *Solution of Initial Value Problems in Classes of Generalized Analytic Functions*, Teubner, Leipzig, 1989. 235, 239

[46] A.N. Varchenko, P.I. Etingof, *Why the Boundary of a Round Drop Becomes an Inverse Image of an Ellipse*, Nauka, Moscow, 1995 (in Russian). 227

[47] A. Vasil'ev, From the Hele-Shaw Experiment to Integrable Systems: A Historical Overview. *Coml. Anal. Oper. Theory*, **3**, 2009, 551–585. 225

[48] Yu.P. Vinogradov, P.P. Kufarev, On a filtration problem. *Prikl. Mat. Mech.* **12**, 1948, 181–198 (in Russian) (English translation: University of Delaware, Applied Mathematics Institute, Technical Report 182A, 1984). 233, 234

[49] P. Willmott, S.D. Howison, J. Dewynne, *The Mathematics of Financial Derivatives: A Student Introduction*, Cambridge University Press, Cambridge, 1996. 226

Index

Sergei V. Rogosin
Department of Economics
Belarusian State University
Nezavisimosti ave, 4
220030 Minsk, Belarus

Fax: +375 17 220 22 84

e-mail: rogosinsv@gmail.com
rogosin@bsu.by

Progress in Mathematics (PM)

Progress in Mathematics is a series of books intended for professional mathematicians and scientists, encompassing all areas of pure mathematics. This distinguished series, which began in 1979, includes research level monographs, polished notes arising from seminars or lecture series, graduate level textbooks, and proceedings of focused and refereed conferences. It is designed as a vehicle for reporting ongoing research as well as expositions of particular subject areas.

PM 301: Ruzhansky, M.; Sugimoto, M.; Wirth, J. (Eds.)
Evolution Equations of Hyperbolic and Schrödinger Type. Asymptotics, Estimates and Nonlinearities (2012).
ISBN 978-3-0348-0350-2

PM 300: Bump, D.; Friedberg, S.; Goldfeld, D. (Eds.)
Multiple Dirichlet Series, L-functions and Automorphic Forms (2012).
ISBN 978-0-8176-8333-7

PM 299: Müller-Hoissen, F.; Pallo, J. M.; Stasheff, J. (Eds.)
Associahedra, Tamari Lattices and Related Structures. Tamari Memorial Festschrift (2012).
ISBN 978-3-0348-0404-2

PM 298: Getz, J.; Goresky, M.
Hilbert Modular Forms with Coefficients in Intersection Homology and Quadratic Base Change (2012).
ISBN 978-3-0348-0350-2

PM 297: Dai, X.; Rong, X. (Eds.)
Metric and Differential Geometry. The Jeff Cheeger Anniversary Volume (2012).
ISBN 978-3-0348-0256-7

PM 296: Itenberg, I.; Jöricke, B.; Passare, M. (Eds.)
Perspectives in Analysis, Geometry, and Topology. On the Occasion of the 60th Birthday of Oleg Viro (2012).
ISBN 978-0-8176-8276-7

PM 295: Joseph, A.; Melnikov, A.; Penkov, I. (Eds.)
Highlights in Lie Algebraic Methods (2012).
ISBN 978-0-8176-8273-6

PM 294: Barreira, L.
Thermodynamic Formalism and Applications to Dimension Theory (2011).
ISBN 978-3-0348-0205-5

PM 293: Mazzucchelli, M.
Critical Point Theory for Lagrangian Systems (2011).
ISBN 978-3-0348-0162-1

PM 292: van den Ban, E. P.; Kolk, J.A.C. (Eds.)
Geometric Aspects of Analysis and Mechanics. In Honor of the 65th Birthday of Hans Duistermaat (2011).
ISBN 978-0-8176-8243-9

PM 291: Greene, R.E.; Kim, K.-T.; Krantz, S.G.
The Geometry of Complex Domains (2011).
ISBN 978-0-8176-4139-9

PM 290: Mantegazza, C.
Lecture Notes on Mean Curvature Flow (2011).
ISBN 978-3-0348-0144-7

PM 289: Colombo, F.; Sabadini, I.; Struppa, D. C.
Noncommutative Functional Calculus (2011).
ISBN 978-3-0348-0109-6

PM 288: Neeb, K.-H.; Pianzola, A. (Eds.)
Developments and Trends in Infinite-Dimensional Lie Theory (2011).
ISBN 978-0-8176-4740-7

PM 287: Cattaneo, A.S.; Giaquinto, A.; Xu, P. (Eds.)
Higher Structures in Geometry and Physics (2011).
ISBN 978-0-8176-4734-6

PM 286: Abbes, A.
Éléments de Géométrie Rigide.
Volume 1: Construction et Étude Géométrique des Espaces Rigides (2011).
ISBN 978-3-0348-0011-2

PM 285: Soifer, A.
Ramsey Theory. Yesterday, Today, and Tomorrow (2010).
ISBN 978-0-8176-8091-6

MIX
Papier aus verantwortungsvollen Quellen
Paper from responsible sources
FSC® C105338

If you have any concerns about our products,
you can contact us on
ProductSafety@springernature.com

In case Publisher is established outside the EU,
the EU authorized representative is:
Springer Nature Customer Service Center GmbH
Europaplatz 3, 69115 Heidelberg, Germany

Printed by Libri Plureos GmbH
in Hamburg, Germany